THE
GLOBAL WARMING
DESK REFERENCE

THE
GLOBAL WARMING
DESK REFERENCE

Bruce E. Johansen

GREENWOOD PRESS
Westport, Connecticut • London

Library of Congress Cataloging-in-Publication Data

Johansen, Bruce E. (Bruce Elliott), 1950–
    The global warming desk reference / by Bruce E. Johansen.
      p.   cm.
    Includes bibliographical references and index.
    ISBN 0–313–31679–1 (alk. paper)
      1. Global warming. I. Title.
  QC981.8.G56J64   2002
  363.738'74—dc21   2001016108

British Library Cataloguing in Publication Data is available.

Library of Congress Catalog Card Number: 2001016108
ISBN: 0–313–31679–1

First published in 2002

Greenwood Press, 88 Post Road West, Westport, CT 06881
An imprint of Greenwood Publishing Group, Inc.
www.greenwood.com

Printed in the United States of America

The paper used in this book complies with the
Permanent Paper Standard issued by the National
Information Standards Organization (Z39.48–1984).

10 9 8 7 6 5 4 3 2 1

Contents

Preface: Diary of a Warm Winter

Nebraska winters have been notable for their severity. During my 18 years in Omaha, winter has usually paid its first visit by the first week of November, with heavy snow an occasional possibility through the third week of April. As I set to work researching and writing *The Global Warming Desk Reference* (and not, of course, because I had any role in determining the weather), I was met with a November notable for spring-like warmth.

On November 8, 1999, the high temperature reached 82 degrees F., the highest November reading in the 127 years records have been kept in Omaha. On November 11, our two housecats began to shed, off cycle. A few days later, the record monthly high of 82 was eclipsed by a new monthly high temperature of 84. It was a Saturday and the Nebraska Cornhuskers were playing at home as television sports commentators in Lincoln amused themselves by recalling Husker games that had been played in ice, snow, and howling cold winds.

Most people in Omaha remembered that November as a very nice time, not the harbinger of a wrenchingly hot future portended by rising levels of greenhouse gases in the atmosphere. We had a sparkling, mild fall with cool nights, trees rich in broadleaf color, and salmon-colored sunsets. The squirrels became as fat as Pillsbury doughboys, some of them nearly tottering over, as they stored up fat for the cold weather, ice, and snow that didn't arrive on time. The city crows also became fat and sleek. "Nice day," people remarked all around me, but all I could think of was the angle of the temperature and carbon dioxide curves. That November turned out to be the warmest on record in Omaha and in many other Midwestern locations. Temperatures in Omaha averaged 46.8 degrees F., 8.2 degrees above average (Rosman 1999, 11).

I watched the Huskers play football, remarking that their Memorial Stadium is a shrine to greenhouse gas generation. There was, among the 77,000 people in the stadium, nothing that produced oxygen. Not even the stadium turf (plastic "Astroturf") produced oxygen. At least the Huskers play in the open air, however. Domed stadiums, with their gigantic heating and cooling requirements, may someday earn a plaque in the Greenhouse Gas Museum, which will be heated, cooled, and lit with solar power.

One afternoon that November, as I was riding my bicycle one mile from campus to Crossroads, an Omaha shopping mall, I realized that I was riding without my jacket, which is very unusual for November in Omaha. As I worked up a sweat, I found myself wishing I was wearing shorts. A gnat hit my glasses, also unusual for November.

At the mall, I was struck by a similarity to the football stadium: nothing in the mall produces oxygen. At Crossroads, I witnessed all manner of ways in which the resources of the Earth have been turned into merchandise and greenhouse gases. In the mall, even the plastic plants were manufactured from petroleum by-products. I wondered how many tons of greenhouse gases this palace of human consumption was adding to the world's heat budget each day. At about the same time, the local newspaper featured a picture of the Shanghai skyline—a row of sparkling new skyscrapers worthy of Donald Trump's New York City—and I wondered whether Shanghai had shopping malls yet. China must be heating and air conditioning its new skyline with electricity generated from coal, I thought. A little later, I read that China is now home to nine of the ten most polluted urban areas in the world, as a billion Chinese undergo the same sort of coal-fired industrual revolution that filled London's skies with smoke two centuries ago.

As I rode my bicycle on a sidewalk along Dodge Street, Omaha's busiest roadway (other than local interstate highways), I watched a stream of exhaust pipes pass me. One way to really get noticed in Omaha, I have discovered, is to ride a bicycle along Dodge Street (or anywhere else). The automobile—those little factories of greenhouse gases—has remade Omaha in its own image. I had just finished reading urgent appeals from various international climate experts that the world's generation of greenhouse gases should be reduced by at least half in 30 years. I wondered whether any of these experts had tried to take a bus in Omaha or ride a bicycle on Dodge Street. Navigating the stream of tailpipes on Dodge Street, I thought: the Cold War has ended. No real economic or social alternatives remain to what we loosely call "the American way," an intoxicating mixture of hedonistic affluence and individual gratification. The "American way" also is the world's most potent creator of atmospheric carbon dioxide (and other greenhouse gases) in the history of the planet. The spread of acquisitive capitalism compounds the fecundity of our species as more people around the world come to enjoy the fruits of human ingenuity, which build and fuel machines turning carbon-based fuels into comfort, convenience, waste heat, and greenhouse gases.

On December 15, 1999, a month later than usual, the first snow of the millennial winter frosted Omaha, covering the dead browns of drought and dampening the dust. In Omaha, in mid-December, we had our first decisively cold days of the year, and our first decidedly below-average temperatures since early

October. Three days later, more snow fell, and the National Weather Service office warned of a major cold wave crossing the Canadian border with Montana, heading south and east. I recalled the words of my master's degree advisor, Donald Gillmor, who had been raised on the Canadian prairies and taught in Minneapolis. Fresh from mild, soggy Seattle, I walked into his office one mid-November morning in 1974 and exclaimed, "It's cold!" "Sure, it's cold," Gillmor replied, "but it's not painful—yet." The prospect of face-biting cold in Omaha was a balm after months of reading greenhouse science and speculation and watching local temperatures topple record highs. It was not yet so warm that we couldn't get uncomfortably cold in a traditional Nebraska way—at least for a few days. Even after two weeks of customary Nebraska cold and snow, however, temperatures for the month still averaged 6.8 degrees F. above normal on Omaha's official thermometer.

I picked up the *New York Times* near the end of December and was greeted by a front page headline: "1998 and 1999 Warmest Years Ever Recorded." The article reported that early spring flowers were sprouting at the New York Botanical Garden in the Bronx four to five months early, as well as in the suburbs, adding "a premature touch of spring to December" (Stevens, Warmest Years, 1999, 1). The same article reported that the previous November had been the warmest in 105 years of record keeping across the United States.

Omaha's December cold spell abated at Christmas in spasms of sudden warmth—from a high of about 20 degrees F. one day, to about 60 the next. A large ridge in the upper atmosphere took up residence over the Pacific Ocean, pushing warm air northward into Alaska, causing the temperature to rise from minus 35 F. to plus 45 F. in Fairbanks during the week before Christmas. The ridge moved slowly east, as temperatures quickly plunged back to minus 30 degrees F. in Fairbanks. Two days later, the high hit 64 in Calgary, Alberta. On December 29, the temperature hit 61 in Omaha, breaking the old daily record of 59, under sparkling-blue skies. The same day, the high and low in Calgary were 63 and 45, almost 40 degrees F. above average. Many high-temperature records were set the same day across the prairies and plains of the United States and Canada.

The day after the high temperature reached 61 in Omaha, the city's daily newspaper published a summary of the year's local weather headlined "Year of the Drought (and the Floods): Extremes Mark Weather of '99." Even in an area where frequent extremes are not unusual, the variations during 1999 provoked comment. The article read like an anecdotal almanac of the Intergovernmental Panel on Climate Change's projections for the region 50 to 100 years in the future. Wintertime temperatures in Omaha were averaging 6 to 10 degrees F. above average, month after month.

At the same time, deluge alternated with drought. A year which had ended with the most intense drought since the Dust Bowl years also had included (on August 7), a one-day storm which dumped 10.5 inches of rain on Omaha, the most intense deluge in the city's history. Beginning in early September, however, Omaha and its hinterland endured a rainless spell of between 44 days (in Omaha) and 49 days (in Lincoln), the fourth longest on record (Range 1999, 11).

During the first week of the new century, a tornado was reported in Indiana during an extended spell of unusual warmth in central and eastern North America. The National Weather Service office in Portland, Maine, also announced that the city had experienced its longest period without snow since records began there during the winter of 1881–1882. The same week (to illustrate how varied weather can be), a cold snap of unusual severity killed at least 67 people in Bangladesh.

Back at home in Nebraska, farmers were becoming more concerned that a continuing lack of moisture could presage an intense drought during the coming growing season. Reflecting on the continuing drought and lack of snow cover, Nebraska State Climatologist Al Dutcher told the *Omaha World-Herald*, "Farmers are in a serious world of hurt right now. It's going to be a long year if it keeps up like this" (Sherry 2000, 9). The next day, Dutcher said that soil moisture levels in Nebraska and nearby areas had dropped as low as those which had helped create the Dust Bowl during the 1930s. "We're going to need 150 percent of normal precipitation this spring just to get us back up to the point where crops can grow," he said (Range 2000, 18). The severity of the drought was illustrated by the fact that 95 percent of the marshes used by migrating birds along the Platte River were dry or carrying less than five percent of their usual water. Even with near normal precipitation in February, Nebraska and Iowa were having their driest six-month period in a century.

On January 13, 2000, Boston's Logan Airport reported its first measurable snowfall of the season. In Omaha, January unfolded much like the previous November—mild, dry, and warm enough to meet many climate models' specifications for an Omaha winter 50 to 100 years from now. Dutcher's comparison of Nebraska's developing drought to the Dust Bowl years caught the eyes of many people; within a few days, a front page article in the *Omaha World-Herald* was cataloging stories describing the effects of the winter's unusual warmth. Armadillos were reported to be migrating northward from Kansas. Free tee times were scarce at Omaha golf courses in January.

Shortly after the dawn of the new year, a broad mass of relatively cold air descended on the United States and Canada and much of Europe and Asia, putting an end, for a time, to widespread unseasonable warmth. A surprise blizzard streaked up the eastern seaboard of the United States, burying some areas (notably parts of North Carolina) in up to 20 inches of snow. Raleigh-Durham, which averages eight inches of snow a year, received 18 inches in one storm. At about the same time, an Associated Press photograph in the *Omaha World-Herald* showed a group of Muslim worshippers bowing toward Mecca from the steps of a mosque in Jerusalem that had just been cleared of a very rare 15-inch snowstorm (Mideast Snow 4).

The freeze was short-lived, however. As in December, the cold air ebbed, and by February 3, 2000, the temperature in Omaha had reached the 50s again, as I rode my bike home with my jacket unbuttoned and my gloves in my pockets. The next morning, the daily newspaper reported that January temperatures in Omaha had averaged 6 degrees F. above average. The newspaper's editorial cartoonist drew a hilly local park, a favorite site for sledding, as a father explained to a child holding a skateboard that back in the old days he used to sled down the same hill most of the winter.

On Valentine's Day, February 14, 2000, eighteen people died in Georgia's worst outbreak of tornadoes (at any season) in 50 years. Four days later, Omaha experienced its first substantial snowfall of the winter, six to eight inches, every flake richly appreciated by the thirsty earth. This snow represented Omaha's heaviest precipitation since early September, five and a half months. Three days after the snow fell, it melted to puddles as temperatures rose to the middle 60s, 25 degrees above average for Omaha during the third week of February. Temperatures for the month averaged 9 degrees F. above average. February also brought Omaha its first near-average precipitation since August.

By the beginning of March, winter seemed to have flopped on its back and died. The high-altitude westerlies (the "jet stream") had receded northward into Canada, bringing much of North America an early spring. Temperatures soared to 65, then 75, before an anemic (for the season) cold air mass turned the thermometer around. In Omaha, early flowers shot out of the earth and bloomed. The grass greened with February's moisture, and many trees eagerly put out buds, as if they were accustomed to greeting spring in Omaha during the first week of March, instead of late April.

On March 6, a bright, sunny, Monday, I rode home in shorts, pushed by a southerly wind that felt like the breath of May. The golden blooms of early crocuses waved in the wind, as bulbs poked above the soil. A carpet of green was beginning to envelop Omaha with our average date of last killing frost six weeks away. The Omaha newspaper mused on March 6, "It was another typical early summer afternoon in the Midlands" (Avok 2000, 11). On March 8, a tornado touched down near the Milwaukee airport; and, at Reagan National Airport in Washington, D.C., the high temperature reached 85. In our front yard, several plants which usually do not survive the winter (including snap dragons and dianthus) sprang to life after our early March warm spell. Omaha temperatures for March averaged six degrees F. above average.

During April, Omaha broke its several month string of much-above normal monthly average temperatures. The average temperature for the month was 51.8 degrees F., only 0.7 degrees above average. Rainfall during the month was near long-term averages, but the subsoil remained very dry due to previous lack of precipitation.

In early May, warmth and humidity that would have been more typical of mid-summer enveloped Omaha. One day early in the month, the high hit 93 degrees F. Neighbor Sue Fenton observed that she has been taking peonies from her yard to a local cemetery every Memorial Day for a half-century. By the late

1990s, however, the peonies had lost their blooms by the last week of May. Meanwhile, I was editing 150,000 words worth of research into a book manuscript, wondering, weatherwise, what the future would bring after a winter that nearly wasn't, in a place where, until now, the "painful" season had always shown its face.

ACKNOWLEDGMENTS

Special thanks are due Theresa Cervantes and Christine Kasel for research assistance. Thanks also are due the staff of the University of Nebraska at Omaha Interlibrary Loan, as well as my wife, Pat Keiffer, and my editors at Greenwood, Cynthia Harris and Rebecca Ardwin. Thanks for help at various stages of this journey also are due James E. Hansen and Andrew A. Lacis of the Goddard Institute for Space Studies, New York City. Stephen H. Schneider sent advance copies of his papers. Friends Barbara A. Mann, John Kahionhes Fadden, and others helped gather information via newspapers and the Internet. This book also benefited from relief of part of my teaching load (thanks to department chair Deb Smith-Howell and Dean of Arts and Sciences John Flocken). Jose Barreiro and Leslie Logan, editors at *Native Americas*, provided me the idea for this book with an assignmnent for the *Native Americas* (1999) special issue on global warming. Thanks are due everyone for their enduring efforts on behalf of the Seventh Generation.

REFERENCES

Avok, Michael. 2000. "Where's My Sunblock? Midlanders Relish Taste of Summer in March." *Omaha World-Herald*, 6 March, 11.
"Mideast Snow Disrupts Life, Prayers." 2000. *Omaha World-Herald*, 29 January, 4.
Native Americas. 1999. Fall/Winter.
Range, Stacey. 1999. "Year of the Drought (and the Floods): Extremes Mark Weather of '99." *Omaha World-Herald*, 30 December, 11, 15.
———. 2000. "Climatologists Say Midlands in Dust Bowl-like Drought." *Omaha World-Herald*, 12 January, 18.
Rosman, Veronica. 1999. "November Among Warmest Ever." *Omaha World-Herald*, 2 December, 11.
Sherry, Mike. 2000. "Western High Presure Keeps Snow, Cold Out." *Omaha World-Herald*, 10 January, 9.
Stevens, William K. 1999. "1998 and 1999 Warmest Years Ever Recorded." *New York Times*, 19 December, 1, 38.

Introduction

Global temperatures spiked in the late 1980s and 1990s, repeatedly breaking records set only a year or two earlier. The warmest year in recorded history was 1998, breaking the record set in 1996, which exceeded 1995's new benchmark. According to NASA's Goddard Institute for Space Studies, the ten warmest years since reliable records have been kept on a global scale (roughly 1890) occurred after 1980.

Carbon dioxide and other "greenhouse gases," such as methane, nitrous oxides, and chlorofluorocarbons (CFCs), retain heat in the atmosphere. The proportion of carbon dioxide in the atmosphere has risen from 280 parts per million (p.p.m.) to roughly 365 p.p.m. as the new millennium dawned on the Christian calendar. Other greenhouse gases also have risen in similar proportion. During the 1990s, a vivid public debate grew around the world regarding how much warmer the earth has become, and how much warmer it may become. A considerable body of scientific evidence and speculation has been published in a field which, to date, has produced very few reference works for undergraduate research. *The Global Warming Desk Reference* is an attempt to begin to fill that void.

Given the advent of global warming, coming generations may be subjected to something resembling an experiment in junior high school biology: put a frog in a beaker full of water and raise the temperature slowly. The frog's nervous system will not tell it the water is too hot until it is too late. Humankind's collective nervous system seems to be serving us no better. What's more, most natural climate changes occur over long periods of time, by human standards. Global warming which is due at least partially to rising levels of atmospheric greenhouse gases is taking place on a much shorter time scale, one which human beings can recognize, in some cases, within a lifetime.

Since the beginning of the industrial age three centuries ago, humankind has been altering the composition of the atmosphere. We are carbon-creating creatures. Throughout the twentieth century, we have assembled an increasing array of machines, all of which produce carbon dioxide and other gases which have been changing the atmospheric balance in ways that retain an increasing amount of heat in the atmosphere. In 1910, the average American commanded about 1.5 horsepower worth of mechanized energy; in 1990, largely due to the general acquisition of automobiles, the average person commanded the power (and the greenhouse gas effluent) of 130 horsepower (Dornbusch 1991, 53).

George M. Woodwell, director of the Woods Hole Research Institute, presents the problem in a nutshell:

Over the past 260,000 years, the period for which a detailed record now exists from ice cores, temperature and atmospheric concentrations of carbon dioxide and methane have been closely correlated.... While cause and effect cannot be proven from such a record, the fact of the correlation is indisputable.... [During the last] 150 years, the carbon dioxide concentration has risen as a result of human activities from about 270 parts per million to about 360 p.p.m. or about 30 percent above what it was in the middle of the last century and more than 20 percent above the highest concentration in 260,000 years. ...The cause is the burning of fossil fuels and changes in land use, especially the destruction of forests. (Epstein et al. 1996)

Scholarly literature on global warming exploded during the 1980s and 1990s, emerging from several scientific fields, including climatology, oceanography, and several of the Earth sciences, as well as sociology, anthropology, and economics. The field's development is so recent, and so interdisciplinary, that a body of standard reference works for undergraduates does not yet exist. One problem which vexes anyone who attempts to write such a reference (as well as textbooks in this area) is that material is accumulating very quickly. For example, the age of the oldest measurements taken from Antarctic ice cores increased from 260,000 years (noted above by George Woodwell) to 420,000 years as this book was being written. A few months after first reports of 420,000 year-old ice cores, Paul Pearson of the University of Bristol and Martin Palmer of Imperial College, London, reported in *Nature* August 17, 2000 that they had developed proxy methods for measuring the atmosphere's carbon dioxide level to about 60 million years before the present. The upshot of Pearson and Palmer's studies is that carbon dioxide levels at the year 2000 were as high as they have been in at least the last 20 million years. According to their records, however, carbon dioxide levels reached the vicinity of 2,000 p.p.m. during "the late Palaeocene and earliest Eocene periods (from about 60 to 52 million years ago" (Pearson, and Palmer 2000, 695). Many of the chapters which follow evoke debates which will continue to evolve after this reference is published. Students are advised to keep up with the scientific literature, especially accounts published in *Nature* and *Science*, as well as in more specialized journals.

This volume is designed to provide a survey of the climate-change literature from a wide array of academic perspectives. It is meant to serve undergraduates who are interested in researching the subject of climate change due to infrared forcing ("the greenhouse effect") and its implications for the future. This book combines a survey of the science of global warming (and disputes attending it) with reporting from around the world, from sinking Pacific islands to thawing Arctic permafrost, which indicate that significant global warming already has begun.

This book begins with a survey of the problem: ever-larger numbers of human beings experiencing greater affluence, overloading the Earth's atmosphere with carbon dioxide, methane, and other "trace" gases which cause the lower atmosphere to retain more heat. The reader is taken, during this survey, on a brief tour of the world, in terms of energy consumption and greenhouse-gas emissions. The rest of Chapter 1 comprises evidence of how deeply fossil-fuel consumption (with its convenience and abundance) is tied into present day, worldwide economic activity. The automobile and jet aircraft are examined as particular sources of rising greenhouse-gas levels in the atmosphere. The outline of the problem concludes with some thoughts about political and philosophical problems which will be crucial to any effort to avert significant warming of the Earth: capitalism's penchant for growth, humankind's assumed superiority to nature, and the power of special interests in governance (and climate diplomacy).

Having sketched the problem, *The Global Warming Desk Reference* proceeds, in Chapter 2, to a summary of the scientific basis for theories that rising levels of greenhouse gases may, in coming decades, cause the Earth to warm beyond the tolerances of many plants and animals. This chapter also includes a history of the greenhouse effect as an idea, beginning within decades of the steam engine's invention and refinement early in the eighteenth century. The idea that burning fossil fuels warms the Earth has been a matter of debate for more than a century, but only during a notably hot summer of 1988 did the idea become a matter of broad societal concern. Since that time, an enormous amount of research material and popular commentary has been produced, some of which is surveyed in this chapter.

Special note should be made of the "biotic feedbacks" section of Chapter 2, because scientists in this area, most notably George M. Woodwell, have described ways in which human-induced increases of atmospheric greenhouse gases could, during the next few centuries, help to trigger reactions in the natural world which may accelerate the rate of greenhouse-gas emissions, along with rises in global temperatures. For example, most of the Earth's methane is not stored in the atmosphere, but in solid form at the bottom of the world's oceans, as well as in the permafrost of the Arctic. Scientists have expressed concern that as the world warms, some proportion of this methane will turn to gas, accelerating the rate of warming. That's the possible scenario if the permafrost remains moist as it thaws. If water levels drop, massive underground deposits of peat could catch fire underground, creating smoldering engines of carbon and methane generation, which would add to the atmosphere's overload of greenhouse gases. This chapter also describes recent developments in the science of global warming, such as ways in which warming near the surface of the Earth may be accelerating depletion of stratospheric ozone.

The third chapter of this reference considers the minority of scientists who do not believe global warming is a threat, that it will not be as bad as feared, and could be beneficial to the flora and fauna of the Earth. This chapter surveys the ideological landscape of the "skeptics" (as they are often called), along with some of the debates they have helped to engender regarding the accuracy of climate models, such as the role of "urban heat islands," and the role of sulfur aerosols in the climate.

Chapters 4 though 8 offer detailed summaries of reporting and research related to the effects of global warming on various parts of the Earth's ecosystem. This survey begins with the causes and consequences of melting ice (in the Arctic, Antarctic, and mountain glaciers), then proceeds to the ecological results of warming oceans, seas, and rivers. Contemporary reporting in these sections indicates that some consequences of warming are already becoming apparent in such diverse places as British Columbia, where salmon have been dying in record numbers of heat stress, and in the tropical oceans, where many corals have been dying because of warming waters. Chapter 6 surveys problems posed by warming for the flora and fauna of the Earth. Chapter 7 describes expected effects on human health. Chapter 8 describes global warming's already evident toll on the indigenous peoples of the Arctic.

Recent reporting of conditions around the world in these chapters may leave haunting images in a reader's mind—of hungry polar bears pacing Arctic beaches in Alaska, waiting for the ice, and its food supply; of the ocean around the Pacific island nation of Kiribati rising, inundating people's homes, on clear days, with no wind or storm. Another image is the "drunken forests" of Alaska, where enough of the permafrost has thawed to partially topple trees and power lines. In British Columbia, hundreds of Canadian salmon fishing boats circle the mouth of the Fraser River, near Vancouver, not unlike the polar bears pacing the beaches of the Arctic, as the boats' owners wait fruitlessly for the sea to bring them food.

This reference also examines the many variables of global warming, notably its role in shaping ocean circulation, which may have such generally unanticipated effects as cooling Western Europe (through diversion of the Gulf Stream) as most of the rest of the world warms. Chapter 9 contains summaries of studies that attempt to forecast the regional weather phenomena that may develop as the Earth becomes generally warmer, concentrating on different regions of the United States of America.

This book's textual matter concludes with a number of proposed solutions for global warming, a varied mixture of the prosaic and fantastic, from painting roofs white to building photo-voltaic, solar-energy colonies on the Moon. In between, many alternatives are described to fossil-fueled energy, along with steps to make existing power generation and consumption more efficient. This reference's bibliography, at more than 25,000 words, is one of the most comprehensive in the field, and an indication of the literature's scope and explosive growth in recent years.

REFERENCES

Casten, Thomas R. 1998. *Turning Off the Heat: Why America Must Double Energy Efficiency to Save Money and Reduce Global Warming*. Amherst, N.Y.: Prometheus Books.

Dornbusch, Rudiger, and James M. Poterba, eds. 1991. *Global Warming: Economic Policy Reponses*. Cambridge, Mass.: MIT Press.

Epstein, Paul, Georg Grabbher, Tom Karl, Ellen Mosley-Thompson, Kevin Trenberth, and George M. Woodwell. 1996. Current Effects: Global Climate Change. An Ozone Action Roundtable, 24 June, Washington, D.C. [http://www.ozone.org/curreff.html]

Pearson, Paul N., and Martin R. Palmer. 2000. "Atmospheric Carbon Dioxide Concentrations Over the Past 60 Million Years." *Nature* 406 (17 August): 695–699.

CHAPTER 1

A Sketch of the Problem

For the past two centuries, at an accelerating rate, the basic composition of the Earth's atmosphere has been materially altered by the fossil-fuel effluvia of machine culture. Human-induced warming of the Earth's climate is emerging as one of the major scientific, social, and economic issues of the twenty-first century, as the effects of climate change become evident in everyday life in locations as varied as small island nations of the Pacific Ocean and the shores of the Arctic Ocean.

"The risks of global warming are real, palpable, the effects are accumulating daily, and the costs of correcting the trend rise with each day's delay," warns Dr. George M. Woodwell, Director of the Woods Hole Research Center (Eco Bridge N.d.). Dean Edwin Abrahamson, a early leader in the field, comments: "Fossil fuel burning, deforestation, and the release of industrial chemicals are rapidly heating the earth to temperatures not experienced in human memory. Limiting global heating and climatic change is the central environmental challenge of our time" (Abrahamson 1989, xi).

Evidence has been accumulating that sustained, human-induced warming of the Earth's lower troposphere has been in progress since about 1980, accelerating during the 1990s. During 1997 and 1998, the global temperature set records for 15 consecutive months; July of 1998 averaged 0.6 of a degree F. higher than July of 1997, an enormous increase if maintained year to year. The year 1998 was the warmest of the millennium, topping 1997 by a quarter of a degree F. (Christianson 1999, 275).

Alarm bells have been ringing regarding global warming in the scientific community for the better part of two decades. A statement issued in Toronto during June, 1988, representing the views of more than 300 policymakers and scientists from 46 countries, the United Nations, and other international organizations warned

Humanity is conducting an unintended, uncontrolled, globally pervasive experiment whose ultimate consequences could be second only to nuclear war. The earth's atmosphere is being changed at an unprecedented rate by pollutants resulting from human activities, inefficient and wasteful fossil fuel use and the effects of rapid population growth in many regions. These changes are already having harmful consequences over many parts of the globe. (Abrahamson, Global Warming, 1989, 3)

Michael Meacher, speaking as Great Britain's environment minister, has said, "Combating climate change is the greatest challenge of human history" (P. Brown 1999, 44). If the atmosphere's carbon dioxide level doubles over pre-industrial levels, which is likely (at present rates of increase) before the year 2100, climate models indicate that temperatures may rise 1.9 to 5.2 degrees C. (3.4 to 9.4 degrees F.) within a century, producing "a climate warmer than any in human history. The consequences of this amount of warming are unknown and could include extremely unpleasant surprises" (National Academy 1991, 2).

The problem is at once very simple, and also astoundingly complex. Increasing human populations, rising affluence, and continued dependence on energy derived from fossil fuels are at the crux of the issue. The complexity of the problem is illustrated by the degree to which the daily lives of machine-age peoples depend on fossil fuels. This dependence gives rise to an array of local, regional, and national economic interests. These interests cause tensions between nations attending negotiations to reduce greenhouse-gas emissions. The cacophony of debate also illustrates the strength and diversity of established interests which are being assiduously protected. Add to the human elements of the problem the sheer randomness of climate (as well as the amount of time which passes before a given level of greenhouse gases is actually factored into climate), and the problem becomes complex and intractable enough to (thus far) seriously impede any serious, unified effort by humankind to fashion solutions.

The "greenhouse effect" is not an idea which is new to science. It has merely become more easily detectable in our time as temperatures have risen and scientists have devised more sophisticated ways to measure and forecast atmospheric processes. The atmospheric balance of "trace" gases actually started to change beyond natural bounds at the dawn of the industrial age, with the first large-scale burning of fossil fuels. It became noticeable in the 1880s, and an important force in global climate change by about 1980. After an intensifying debate, the idea that human activity is warming the earth in potentially damaging ways became generally accepted in scientific circles by 1995.

Taken to extremes, an atmosphere beset by the greenhouse effect can be very unpleasant—witness perpetually cloudy Venus, with an atmosphere in which carbon dioxide is the dominant element. The surface temperature on Venus, warmed considerably by runaway infrared forcing, is roughly 800 degrees F. The planet Mars, with nearly no infrared forcing in its atmosphere, averages minus 53 degrees C. Earth's moderate degree of infrared forcing (along with its blanket of liquid water) keeps the planet habitable.

RISING LEVELS OF GREENHOUSE GASES DUE TO FOSSIL FUEL EFFLUVIA

Since the beginning of the industrial age, roughly three centuries ago, an increasing world human population has made more widespread use of fossil fuels to aid economic development, as well as to augment human comfort, convenience, and financial profit. Combustion of oil, coal, and natural gas has been changing the atmospheric balance of carbon dioxide, methane, nitrous oxides, and other naturally occurring "trace gases," as well as chemicals created by industry, such as chlorofluorocarbons (CFCs). Detectable increases in most of these gases, all of which retain heat in the atmosphere, can be traced to the middle of the nineteenth century. The rise in greenhouse gas levels was small at the time, with effects which were very difficult to separate from the natural variability of climate.

The Earth's atmosphere is comprised of 78.1 percent nitrogen and 20.9 percent oxygen. All the other gases, including those responsible for the greenhouse effect, make up only about one percent of the atmosphere. Carbon dioxide (CO_2) is 0.035 percent; methane (CH_4) is 0.00017 percent, and ozone 0.000001–0.000004 percent. To a certain extent, the greenhouse effect is necessary to keep the Earth at a temperature that sustains life as we know it. Without "infrared forcing" (popularly, the greenhouse effect), the average temperature of the Earth would be about 33 degrees C. (60 degrees F.) colder than today's averages, too cold to sustain the Earth's existing plant and animal life.

During 1860, human-induced carbon emissions stood at about one-tenth of a gigaton (billion metric tons) per year. Between 1900 and 1940, human carbon production rose from roughly 1.0 to 1.5 gigatons. During the 1940s, usage began to increase more rapidly, passing 3.0 gigatons about 1960, 5.0 about 1970, and more than 8.0 gigatons by the late 1980s. Between 1950 and 1980, worldwide emissions of carbon dioxide increased 219 percent, or 7.3 percent a year. The rate of increase slowed during the 1980s, but the world total still grew almost 13 percent between 1980 and 1988 (Kane and South 1991, 193). Between 1850 and 2000, human combustion of fossil fuels has risen 50-fold. Contemporary observations indicate that carbon dioxide content of the atmosphere is increasing at an average of between 1.0 and 1.5 parts per million per year, with large annual variations. Some years, such as 1972 and 1988, atmospheric carbon dioxide has risen 2.5 to 3.0 percent, while during one year, 1973, it did not increase at all. Roughly two-thirds of the increase in carbon dioxide since industrialism began has taken place since about 1950.

Carbon dioxide is only the best known of several gases that contribute to global warming, but it is responsible for about half of the atmosphere's human-induced infrared forcing. Several other gases, the most notable being methane, contribute the other half. Water vapor, which also retains heat in the atmosphere, also plays a major role in the infrared forcing of the atmosphere. As the air warms, it holds more water vapor. The atmosphere holds six percent more water vapor for each degree rise in temperature.

Today, due mainly to the combustion of fossil fuels, the balance of heat-retaining gases in the atmosphere is increasing rapidly. In a century and a half of rapid worldwide industrialization, the proportion of carbon dioxide has risen from roughly 280 to about 370 p.p.m. By the year 2000, scientists had tested the composition of the atmosphere to roughly 60 million years in the past. The level of carbon dioxide today is believed, according to such measurements, to be as high as it has been in at least 20 million years.

As scientists explore the complexities of atmospheric chemistry, some long-held assumptions have been cast aside. One such assumption has been that the Earth's carbon cycle persists in steady state over time. A. Indermuhle and several co-authors published results of ice-core tests from Alaska which indicate that "the global carbon cycle has not been in steady state during the past 11,000 years. . . . Changes in terrestrial biomass and sea-surface temperature were largely responsible for the observed millennial-scale changes in atmospheric CO_2 concentrations" (Indermuhle et al. 1999, 121). Carbon dioxide levels in the atmosphere have now been measured (most often from ice cores) to more than 400,000 years in the past, varying from about 200 p.p.m. to about 285 p.p.m., roughly the level when the industrial age began. By contrast, the carbon-dioxide level by the year 2000 was about 370 p.p.m. Indermuhle and co-authors find that "The rate of change of atmospheric CO_2 concentration over the Holocene is two orders of magnitude smaller than the anthropogenic CO_2 increase since industrialization" (Indermuhle et al. 1999, 125).

The levels of carbon dioxide, methane, and other greenhouse gases in the atmosphere during the last years of the twentieth century were higher than at any time since humankind has walked the earth, a circumstance that has pro-voked one student of the greenhouse effect to remark that "We are headed for rates of temperature rise unprecedented in human history; the geological record screams a warning to us of just how unprecedented . . . [the stresses on] the natural environment will be" (Leggett 1990, 22). In 1995, The Intergovernmental Panel on Climate Change's *Second Assessment* found that the average temper-atures for the period 1901–1990 were higher than for any 90-year interval since at least 914 A.D. The panel also has stated that the combined effect of all green-house gases is likely to produce a warming greater, in terms of its speed, than any other climatic event in the last 10,000 years (Bolin et al. 1995).

The speed with which temperatures are expected to rise in the next century may be so rapid, according to Brian Huntley, that

There is little likelihood that many organisms, and trees in particular, will be able to migrate fast enough to remain in equilibrium with future climate. For many organisms the situation may be made considerably worse by patterns of human land use and their already depleted populations. (Huntley 1990, 144)

Emissions of carbon dioxide and other "greenhouse gases" are built into our everyday lives—our modes of transportation, production, and consumption—to the extent that they are enabled by the combustion of fossil fuels. Roughly 80 percent of human industrial activity on Earth is fueled by energy which produces carbon dioxide (and, oftentimes, other greenhouse gases as well) when it is burned. The same industrial processes also produce waste heat in addition to greenhouse gases. Sometimes these manufactured goods (such as automobiles) also produce waste heat and greenhouse gases as they are operated.

The fossil fuels being burned to power today's machine culture were created by natural processes during the past 600 million years. The geologic record suggests that the richest deposits came into being during times when the earth was much warmer, and its atmosphere was much richer in carbon dioxide than today. Carbon dioxide levels were as much as seven times higher than today during the epoch of the dinosaurs, 65 million to 200 million years ago (Schneider 1989, 21). During much of this period, the earth had no permanent ice at all, not even at the poles. The temperature of the earth also was markedly warmer, and sea levels were about 300 feet above those of the present day. Stephen H. Schneider estimates that if today's Antarctic ice cap were to melt, world sea levels would rise by about 230 feet (Schneider 1989, 162–163). During the glaciations that have characterized the last several million years, carbon dioxide levels have risen and fallen with the temperature. At the end of the last great ice age, more than 10,000 years ago, carbon dioxide levels were about 60 percent of today's levels. Such a level, slightly under 200 p.p.m., is close to the lower limit for successful photosynthesis in many green plants.

Ecologist Barry Commoner writes

The amounts of these fuels burned to provide society with energy represent the carbon captured by photosynthesis over millions of years. So, by burning them . . . we have returned carbon dioxide to the atmosphere thousands of times faster than the rate at which it was removed by the early tropical forests. (Commoner 1990, 6)

"We are reinjecting our fossil carbon legacy into the atmosphere at an incredibly accelerated pace," writes Schneider. According to Schneider, human activity is changing the temperature of the atmosphere at between 10 and 100 times as quickly as natural processes at the end of the last ice age (Schneider 1989, 28).

CO2 is dumped into the atmosphere at a much faster rate than it can be withdrawn or absorbed by the oceans or living things in the biosphere. CO2 buildup in the next few decades to centuries could well be one of the principal controlling factors of the near-future climate. (Schneider 1989, 21)

The oceans absorb large amounts of carbon dioxide. At the surface, the gas dissolves in water. Through complex chemical processes, some of the carbon becomes part of the tissues of marine organisms. Carbon-containing seashells eventually settle to the ocean floor, forming carbonate sediments such as limestone. The vast capacity of the oceans and other "carbon sinks" is being overwhelmed, however, by the volume of greenhouse gases being poured into the atmosphere as human ingenuity invents new ways to put fossil fuel-generated energy to work, and as populations and material living standards rise around the world.

In addition to adding carbon dioxide, during the 1990s, human activity was adding about 550 million tons of methane to the atmosphere each year. Methane's pre-industrial range in the atmosphere was 320 to 780 parts per billion; by 2000, that level had risen to about 1,700 p.p.b., a steeper rise, in proportional terms, than carbon dioxide. Carbon dioxide is 200 times more plentiful in the atmosphere than methane, but a molecule of methane can retain ten times as much heat as one of carbon dioxide.

Atmospheric methane is produced by many human activities, from transporting natural gas to raising meat animals, dumping of garbage in landfills, and growing rice. Methane contributes about half as much retained heat to the atmosphere as carbon dioxide. The rate of increase in atmospheric methane (one percent a year) was about twice as rapid as that of carbon dioxide for much of the twentieth century (Jager and Ferguson 1991, 79). In the 1990s, however, as carbon dioxide levels continued to rise, methane levels stagnated. Atmospheric scientists are not yet sure why the proportion of methane leveled out, or whether the trend will continue.

Methane also interacts with temperature increases prompted by rising levels of carbon dioxide; raise the temperature by one degree C. and methane production rises 20 to 30 percent. As the Earth warms, there is a palpable fear among students of global climate change that rapid warming will cause methane deposits to gassify from the depths of the oceans to the tundra of Siberia, Alaska, and Canada. (This possibility is explored in Chapter 2, under Biotic Feedbacks.)

Carbon monoxide shares greenhouse gas properties with carbon dioxide; its presence in the atmosphere has been rising 0.8 to 1.5 percent a year. Most of the carbon monoxide in the atmosphere is human produced, much of it by the burning of fossil fuels and deforestation. By the late 1980s, roughly one billion metric tons of human-produced carbon monoxide was being added to the global atmospheric inventory each year.

Add to this mix of trace gases tropospheric ozone, which is produced photochemically in the atmosphere from the oxidation of carbon monoxide, methane, or other hydrocarbons in the presence of nitrogen oxides, which act as catalysts. Tropospheric ozone was increasing about one percent a year by the late 1980s. Levels in the air have increased 20 percent to 50 percent during the twentieth century. This type of ozone, which is contributed to the atmosphere by several industrial processes, acts to absorb infrared radiation and thereby, like carbon dioxide and methane, contributes to infrared forcing.

A few other gases, such as nitrous oxide ("laughing gas") add to infrared forcing as well. Greenhouse gases include the chlorofluorocarbon (CFC) family, several synthetic chemicals which have been implicated in the destruction of the earth's stratospheric ozone layer. The CFCs also retain heat in the atmosphere even more efficiently than methane, contributing about 20 percent as much retained heat as atmospheric carbon dioxide. Molecule for molecule, CFCs retain thousands of times as much heat as carbon dioxide. A molecule of sulfur hexafluoride, one of the chemicals subject to controls under the Kyoto Protocol, is 23,900 times more potent over 100 years than a molecule of carbon dioxide. Most CFC use is being phased out under international protocols, but its effects "will only decrease very slowly next century," writes John T. Houghton (1997, 37). As will be discussed in Chapter 2, global warming near the Earth's surface may be contributing to destruction of stratospheric ozone.

The level of greenhouse gases in the atmosphere influences surface temperature, but not in a simple one-on-one relationship. As Michael E. Mann, Raymond S. Bradley, and Michael K. Hughes point out in *Nature*, the level of greenhouse gases in the atmosphere is one part in a cacophony of influences on a given climate at any particular time and place.

In addition to the possibility of warming due to greenhouse gases during the past century, there is evidence that both solar irradiance and explosive volcanism have played an important role in forcing climate variations over the past several centuries. (Mann et al. 1998, 779)

One may add the role of the oceans to Mann's list, including cyclical changes in sea-surface circulations and temperatures (such as El Niño and La Niña), as well as longer-term changes in deep-water movement, such as the thermohaline circulation of the North Atlantic Ocean, which bathes Europe in the warmth of the Gulf Stream. Among these influences, "greenhouse gases emerg[ed] as the dominant forcing during the twentieth century" (Mann et al. 1998, 779, 785). Mann, et al. describe a history of climate from roughly 1400 A.D. to the present, compiling data "from tree rings, ice cores, ice-melt indices, and long historical records of temperature and precipitation" (Hegerl 1998, 758). They find that greenhouse forcings have become powerful enough to make the decade of the 1990s the warmest decade since at least A.D. 1400 (Mann et al. 1998, 779).

A BRIEF SURVEY OF WORLDWIDE GREENHOUSE GAS GENERATION

During the 1990s, even as international protocols called for freezes or reductions in fossil-fuel use, the burning of carbon-based fuels continued to increase around the world. Nearly every country and region experienced substantial increases in fossil-fuel use: 20 percent in Brazil, 28 percent in India, 40 percent in Indonesia, and 27 percent in China (Ridenour 1998). The only regions to experience the type of decline that scientists say will be required to forestall substantial global warming were Russia and Ukraine, whose fossil-fuel use declined 40 to 45 percent during the 1990s because of widespread economic collapse. Several more prosperous nations in Europe also stabilized or slightly reduced their generation of greenhouse gases.

Ninety percent of the world's remaining fossil-fuel reserves are in the form of coal, the most dangerous fossil fuel from a greenhouse point of view, because most coals produce roughly 70 percent more carbon dioxide per unit of energy generated than natural gas, and about 30 percent more than oil. Coal is also the most plentiful fossil fuel, especially in places with large populations, such as China, which controls 43 percent of remaining reserves. During the 1980s, China passed the Soviet Union as the world's largest coal producer. Coal poses environmental problems other than carbon emissions. The mining of coal also produces methane; its combustion produces sulfur dioxide and nitrous oxides, as well as carbon dioxide. Transport of coal also usually requires more energy than that of any other fossil fuel.

While emissions of carbon dioxide in the United States increased from 2,858 million tons in 1960 to 4,804 million tons during 1988, China's carbon dioxide production rose from 790 million tons in 1960 to 2,236 million tons in 1988, nearly tripling. While energy consumption in the United States, Europe, and Japan rose by about 28 percent between 1970 and 1990, consumption rose by almost 10 times that amount—208 percent—during the same period in China (Gelbspan 1997b). At the same time, alarms over global warming were beginning to ring in some of China's official agencies. Oceanographers at China's State Oceanic Administration argue that a sea level rise of three feet a century could cause flooding in many of China's coastal areas, home to half of China's large cities, and 40 percent of its population.

Energy consumption in China rose from 293 million metric tons of coal equivalent (MTCE) in 1970 to 845 million MTCE in 1987. Projections at that time forecast that China's energy consumption would pass one billion MTCE late in the 1990s (Barbier et al. 1991, 137). China by 1988 was the third largest producer of carbon dioxide among the nation-states of the world. The Soviet Union produced 1,454 tons in 1960 and 3,982 million tons in 1988 (National Academy 1991, 7). After the breakup of the Soviet Union, economic activity plummeted and carbon dioxide emissions fell sharply, as indicated above.

China's methane emissions, 0.2 million tons in 1950, rose to a world-leading 5.6 million by 1988; the United States produced 2.3 million tons of methane in 1950 and 2.8 million in 1988, third in the world behind the former Soviet Union (4.1 million tons) and China. The United Kingdom cut its methane production in half during the 1950–1988 period, from 1.6 million tons (then third in the world) to 0.7 million, mainly due to decreased burning of coal for heating (Hariss et al. 1993, 343).

The level of energy use and its efficiency vary widely around the world. In 1990, the United States generated more carbon per capita (5.03 tons per person) than any other nation, with Canada (4.24), Australia (4.00), the Soviet Union (3.68) and Saudi Arabia (3.60) being next highest. The world's average was

1.08, with Zaire (0.03), Nigeria (0.09), Indonesia (0.16) and India (0.19) generating the least. By the late 1980s, the United States was producing six times as much economic activity per unit of carbon dioxide generated as China. The United States, in turn, used three times as much energy per unit of output as France and Japan (National Academy 1991, 8). With regard to energy efficiency, the world average in 1987 was 327 grams of carbon produced per dollar of gross national product (GNP). The United States produced a dollar of GNP with 276 grams of carbon emissions, while worldwide, this measure ranged from 147 in Italy and 156 in Japan to 655 in India and 2,024 in China. The figures for China and India are inflated by widespread use of dirty, low-power coal. This type of usage is accelerating and will play a major role in worldwide greenhouse-gas levels during the twenty-first century (Whalley and Wigle 1990, 238).

A baby born into the machine culture of the United States of America in the year 2000 will consume about a hundred times the natural resources (including fossil-fuel energy) of a baby born in Bangladesh. The wealthiest one-quarter of the world's population consumes 80 percent of its aggregate energy resources. The average resident of an industrialized country consumes the energy contained in roughly 32 barrels of oil per year, about seven times the average consumption by residents of Third World countries (Silver and DeFries 1990, 53). According to Anita Gordon and David Suzuki, "The average North American uses the energy equivalent of 10 tons of coal a year. Bangladeshis use less than 100 kilograms (220 pounds)" (Gordon and Suzuki 1991, 106). In other words, the average North American (in the United States and Canada) contributes as much greenhouse effluent to the atmosphere as 60 Angolans or 25 residents of India.

The type of coal-powered industrialization which began in Britain three centuries ago is now taking place in China and India. The major ecological difference is that England was inhabited by about 20 million people during the dawn of its industrial revolution. China's population at the dawn of the twenty-first century was about 1.3 billion, and India's was nearing one billion.

China surpassed the United States as the world's largest burner of coal during the late 1980s; by 1990, China was responsible for 10 percent of the world's carbon dioxide emissions. By the middle of the twenty-first century, China may become the world's largest national source of greenhouse gases. Much of this emission load will be produced by the development of energy systems in rural areas. In the late 1980s, according to Prof. Lu Ying-zhong of Beijing's Institute for Techno-economics and Energy Systems, only 10 percent of China's oil and coal was being consumed in rural areas, where 80 percent of the people live (Oppenheimer and Boyle 1990, 141). World levels of greenhouse gases may jump as rural China hooks into power grids powered by its extensive deposits of low-energy coal.

Greenhouse-gas emissions also are increasing in countries which are nearly fully industrialized, such as Japan. According to a report by Japan's Environment Agency, carbon dioxide emissions in that country increased 8.3 percent between 1990 and 1995. The agency warned that, without restrictions, emissions will rise by 20 percent between 1995 and 2010.

Energy consumption is skewed widely by region. North American energy usage per capita is roughly five to six times as high north of the Rio Grande River as in Latin America. Energy consumption per person in the United States is more than twice what it is in Europe, and roughly 15 times the amount in sub-Saharan Africa. The largest factor in this difference is the pervasiveness in any given area of private automobiles and other motorized vehicles.

Canada has one of the worst records in greenhouse gas emissions growth among industrialized countries, as sketched by Christopher Rolfe during the middle 1990s:

Natural Resources Canada estimates that Canada's emissions will grow by 13 percent by 2000. In British Columbia, between 1990 and 1994, carbon dioxide emissions increased by almost 9 percent. Both Canada and British Columbia are clearly failing the international community by failing to live up to our international commitments. We are failing to meet the needs of the global and our own ecosystems. And we are failing our own economic wellbeing by not paving the way to a smooth transition to a sustainable economy. (Rolfe 1996)

Based on projected growth in vehicle miles traveled, as well as the failure of vehicle manufacturers to improve fuel efficiency, the Canadian Ministry of Environment, Lands, and Parks projected a 65 percent increase in greenhouse-gas emissions from the light-duty vehicle fleet in British Columbia between 1990 and 2020. Carbon dioxide emissions in British Columbia from electrical power generation have grown from 1,227 kilotons in 1990 to 2,400 kilotons in 1994.

In 1950, the United States produced 40 percent of the world's industrial carbon dioxide. That proportion fell to 25 percent in 1975 and 22 percent in 1988. Emissions in the United States were growing during this period, but those of the rest of the world, on balance, were increasing even more rapidly.

Human life and labor have become more mechanized than that of even a few decades ago; agriculture, for example, has ceased, for the most part, to be the province of family farmers, as it was a century ago. It has, instead, become industrialized on a large, mass-production scale, with increasing utilization of fossil fuels and fertilizers which produce greenhouse gases. The industrialization of agriculture has been accompanied by a gradual increase in agricultural-industrial scale that has made many of the small towns of the U.S. Midwest economically obsolete. In Nebraska, for example, the cities of Omaha and Lincoln thrive, with unemployment rates of about 2 percent, among the lowest in the world, while many small farming towns in its hinterland crumble. Animal protein and cereal crops, such as wheat and corn, are now produced on a factory model, with attendant air, soil, and water pollution.

THE ABUNDANCE AND CONVENIENCE OF FOSSIL FUELS

In *The Economics of Global Warming*, William R. Cline estimated that 3,500 to 7,000 gigatons (ggt) of coal could be mined and burned "before the market mechanism chokes off further atmospheric buildup through high prices caused by scarcity of a severely depleted resource" (Cline 1992, 46). Since the total stock of carbon in the atmosphere as of 1990 was roughly 750 ggt, the market economy is prepared, if unrestrained, to increase carbon dioxide levels in the atmosphere five- to ten-fold before scarcity becomes a factor in the price of coal and other fossil fuels. There is some climatic irony in the fact that such a level of greenhouse gases in the atmosphere would take the Earth back, roughly, to the atmospheric days of the dinosaurs, a time when temperatures were much higher than today, when the planet had no permanent pack ice, and the dominant animal species, including most of the dinosaurs, were cold blooded.

Following spot oil shortages of the 1970s, the petroleum industry improved its extractive technology to the point where the supply of oil (and natural gas, often a byproduct of oil drilling) will last for centuries, even at increasing rates of consumption. Offshore oil drilling, for example, was restricted to roughly 500 feet below the water's surface during the 1960s. By the middle 1990s, the limit had reached 10,000 feet. As one commentator has written,

"Will we run out of gas?"—a question we began asking during the oil shocks of the 1970s—is now the wrong question. The earth's supply of carbon-based fuels will last a long time. But if humans burn anywhere near that much carbon, we'll burn up the planet. (Hertsgaard 1999, 111)

If the Earth's entire storehouse of fossil fuels that could be harvested at a cost that the market will bear is burned to provide energy in the fashion of the twentieth century, carbon dioxide levels in the atmosphere would reach roughly 2,800 parts per million, or ten times the pre-industrial level, according to one estimate (Baar et al. 1992, 164).

According to Paul Epstein, Harvard University epidemiologist, industrial society's dependence on fossil fuels creates many problems in addition to rising levels of greenhouse gases.

Extraction damages terrestrial and offshore ecosystems. Drilling in Ecuador, for example, harms the forest and the headwaters to the Amazon. Transport and the accompanying leaks and spills have enormous impacts on coastal ecosystems. The benzene by-products of refining and impacts on surrounding communities is not healthy. Finally, combustion has multiple hazards. Nitrates in acid precipitation harm forest and lake stocks. Sulfates and hydrocarbons contribute to lung disease, heart disease and cancer. And now the aggregate of burning fossil fuels and cutting forests (a chief carbon sink) may be destabilizing the climate system itself. Thus, from a health perspective, ending our civilization's addiction to fossil fuels is a healthy "no-regrets" policy. Because they are finite, it is also a necessary step. The only question is, will we change soon enough to reduce the enormous and mounting [climatic] costs of not changing from "business as usual?" (Epstein et al. 1996)

The sheer economic power of fossil-fuel energy generation (and the implied difficulty of ending the "addiction" alluded to by Epstein) is described by Ross Gelbspan:

The energy industries now constitute the largest single enterprise known [to human history]. With annual sales in excess of one trillion dollars and daily sales of more than two billion dollars, the oil industry alone supports the economies of the Middle East and large segments of the economies of Russia, Mexico, Venezuela, Nigeria, Indonesia, Norway, and Great Britain. Begin to enforce restriction on the consumption of oil and coal, and the effects on the global economy—unemployment, depression, social breakdown, and war—might lay waste to what we have come to call civilization. (Gelbspan 1995)

John Whalley and Randall Wigle (1990) project that countries comprising the Organization of Petroleum Exporting Countries (OPEC) would sustain an 18.7 percent drop in gross national product if the world were to reduce its fossil fuel use by 50 percent, the minimum recommended by the Intergovernmental Panel on Climate Change (IPCC) to significantly forestall global warming.

DIPLOMACY AND FOSSIL-FUELED REALITY

Ross Gelbspan surveyed international climate diplomacy during the 1990s, and finds it "at once laudably noble and regrettably futile" (Gelbspan 1998). Gelbspan believes that ecologically meaningful reductions in greenhouse-gas emissions will require the following:

The world's political leaders must forge a binding treaty to phase out virtually all of our fossil-fuel use within a decade and create a new worldwide energy economy based on energy efficiency and climate-friendly, renewable technologies. The imposition of draconian carbon taxes in the United States and elsewhere would be politically unacceptable and economically inadequate. Instead, I would envision an energy-based global public-works program, funded by the profits from fossil fuels, under a stringent regime of international governmental regulation. (Gelbspan 1998)

Judging by the rhetoric of many leaders of powerful countries, global warming is high on the worldwide political agenda. While political rhetoric favoring reduction of greenhouse-gas emissions flowed freely during the late 1990s, actual emissions, as described above, were still rising in all except a few nations of the world.

During the 1990s, various political leaders warned of global warming as the people they led continued to increase their emissions of greenhouse gases. U.S. President Bill Clinton spoke about global warming to the National Geographic

Society on October 22, 1997, a few weeks before the Kyoto Protocol was negotiated:

Today we have a clear responsibility and a golden opportunity to conquer one of the most important challenges of the twenty-first century—the challenge of climate change—with an environmentally sound and economically strong strategy, to achieve meaningful reductions in greenhouse gases in the United States and throughout the industrialized and the developing world. It is a strategy that, if properly implemented, will create a wealth of new opportunities for entrepreneurs at home, uphold our leadership abroad, and harness the power of free markets to free our planet from an unacceptable risk. . . .

America can stand up for our national interest and stand up for the common interest of the international community. America can build on prosperity today and ensure a healthy planet for our children tomorrow. In so many ways the problem of climate change reflects the new realities of the new century. Many previous threats could be met within our own borders, but global warming requires an international solution. Many previous threats came from single enemies, but global warming derives from millions of sources.

Many previous threats posed clear and present danger; global warming is far more subtle, warning us not with roaring tanks or burning rivers but with invisible gases, slow changes in our surroundings, increasingly severe climatic disruptions that, thank God, have not yet hit home for most Americans. But make no mistake. The problem is real. And if we do not change our course now, the consequences sooner or later will be destructive for America and for the world.

First, the United States proposes at Kyoto that we commit to the binding and realistic target of returning to emissions of 1990 levels between 2008 and 2012. And we should not stop there. We should commit to reduce emissions below 1990 levels in the five-year period thereafter, and we must work toward further reductions in the years ahead. The industrialized nations tried to reduce emissions to 1990 levels once before with a voluntary approach, but regrettably, most of us—including especially the United States—fell short. We must find new resolve to achieve these reductions, and to do that we simply must commit to binding limits. . . .

[B]oth industrialized and developing countries must participate in meeting the challenge of climate change. The industrialized world must lead, but developing countries also must be engaged. The United States will not assume binding obligations unless key developing nations meaningfully participate in this effort. As President Carlos Menem stated forcefully last week when I visited him in Argentina, a global problem such as climate change requires a global answer. If the entire industrialized world reduces emissions over the next several decades, but emissions from the developing world continue to grow at their current pace, concentrations of greenhouse gasses in the atmosphere will continue to climb. Developing countries have an opportunity to chart a different energy future consistent with their growth potential and their legitimate economic aspirations. What Argentina, with dramatic projected economic growth, recognizes is true for other countries as well: We can and we must work together on this problem in a way that benefits us all. Here at home, we must move forward by unleashing the full power of free markets and technological innovations to meet the challenge of climate change.

I propose a sweeping plan to provide incentives and lift road blocks to help our companies and our citizens find new and creative ways of reducing greenhouse gas emissions. First, we must enact tax cuts and make research and development investments worth up to $5 billion over the next five years—targeted incentives to encourage energy efficiency and the use of cleaner energy sources. Second, we must urge companies to take early actions to reduce emissions by ensuring that they receive appropriate credit for showing the way. Third, we must create a market system for reducing emissions wherever they can be achieved most inexpensively, here or abroad; a system that will draw on our successful experience with acid rain permit trading. Fourth, we must reinvent how the federal government, the nation's largest energy consumer, buys and uses energy. Through new technology, renewable energy resources, innovative partnerships with private firms and assessments of greenhouse gas emissions from major federal projects, the federal government will play an important role in helping our nation to meet its goal. Today, as a down payment on our million solar roof initiative, I commit the federal government to have 20,000 systems on federal buildings by 2010.

Fifth, we must unleash competition in the electricity industry, to remove outdated regulations and save Americans billions of dollars. We must do it in a way that leads to even greater progress in cleaning our air and delivers a significant down payment in reducing greenhouse gas emissions. Today, two-thirds of the energy used to provide electricity is squandered in waste heat. We can do much, much better.

Sixth, we must continue to encourage key industry sectors to prepare their own greenhouse-gas reduction plans. And we must, along with state and local government, remove the barriers to the most energy efficient usage possible. There are ways the federal government can help industry to achieve meaningful reductions voluntarily, and we will redouble our efforts to do so. (Clinton 1997)

Clinton's rhetoric, posed mainly in the future tense, does not come to grips with the fact that his 270 million constituents were increasing their fossil fuel consumption as he spoke.

Japanese Prime Minister Ryutaro Hashimoto touched on similar themes six. months before his nation hosted the 1997 Kyoto climate conference:

I would like to stress two points: our responsibility to future generations, and global human security. Bearing these points in mind, it is necessary that each of us develop a strong consciousness and shoulder our responsibilities. We must change our lifestyles. Moreover, it is necessary to develop innovative environmental technologies and to promote their transfer to developing countries in order to foster sustainable development. . . . It [a solution to global warming] will consist of two pillars: Green Technology and Green Aid. Under Green Technology, we would promote the efforts of developed countries in the development and dissemination of energy conservation technologies; the introduction of non-fossil energy sources such as photovoltaic power generation; the development of innovative energy and environmental technologies; and worldwide reforestation and preservation of forests. Under Green Aid, we would utilize . . . financial resources to cope with the issues of energy and global warming and promote cooperation with developing countries through the development of human resources. (Hashimoto 1997)

Hashimoto described a detailed program under the rubric of "Green Technology":

(a) Development and diffusion of energy-saving technologies.

   —Highly efficient energy-saving technologies in the industrial sector;

   —Energy-saving consumer appliances and energy efficiency standards.

(b) Introduction of non-fossil fuel energy.

   —Very low cost solar energy;

   —Ultra-low-emission-vehicles and those using electricity, etc.;

   —R[esearch] and D[evelopment] for super high-efficiency solar cells;

   —R&D on technologies for effective use of bio-energy.

(c) Promotion of global forestation and forest conservation.

   —R&D on plants with substantial durability for severe conditions such as the desert and combat desertification;

   —Enhanced activities of public sector and private business sector in global forestation.

(d) Development of innovative energy and environment technologies.

   —CO2 sequestration (into ocean and aquifers) technologies;

   —Chemical and biological CO2 fixation and utilization technologies. (Hashimoto 1997)

Echoing both Clinton and Hashimoto was Wim Kok, prime minister of the Netherlands, on behalf of the European Union, June 23, 1997:

A fair judgment of the present situation obliges us to be extremely attentive on environmental issues. We are in danger of passing thresholds beyond which serious damage will occur, some of it irreversible. And even if part of the damage would be reparable, it would be against an unnecessarily, or even unaffordable, high price. To safeguard future generations from this danger and burden, it is our duty to act now. . . .

The industrialized world should take the lead in reducing its emissions of greenhouse gases. The developed countries should conclude a legally binding commitment in Kyoto. The European Union has agreed to a phased reduction of the emissions of greenhouse gases of 15 percent below the 1990 level by the year 2010. Mandatory and recommended policies and measures, including harmonized ones, must ensure that this target is achieved. (Kok 1997)

Diplomats from the United States of America and approximately 160 other nations agreed in negotiations held in Kyoto, Japan, during December, 1997, to reduce greenhouse-gas emissions. The Kyoto Protocol places binding limits on each industrial country's combined emissions of the six principal categories of greenhouse gases: carbon dioxide, nitrous oxide, methane, sulfur hexafluoride, perfluorocarbons, and hydrofluorocarbons. These limits apply to the 38 so-called Annex I (industrialized) countries. Under the Kyoto Protocol, each industrial country's baseline for calculating reductions is its 1990 emissions of carbon dioxide, methane, and nitrous oxides and its choice of 1990 or 1995 levels in the other three categories of gases. The United States agreed to a target of seven percent below this baseline by the period between 2008 and 2012.

The Kyoto Climate Convention and refinements negotiated at Buenos Aires during 1998 have not done much to curtail global greenhouse-gas emissions because some industrialized countries (most notably the United States) will not accept cuts in their own emissions if Third World (a.k.a. "developing" or "industrializing") nations are not forced to reduce their fossil-fuel consumption as well. Many Third World countries reply that holding high-consumption and low-consumption countries to the same reductions amounts to energy colonialism, which locks all players in place over time. An across-the-board cut in allowable emissions would, for example, lock Bangladesh at roughly one percent of the per capita greenhouse effluent of the United States.

Anil Agarwal and Sunita Narain, in *Global Warming in an Unequal World: A Case of Environmental Colonialism*, quote the Worldwatch Institute on the nature of this problem:

There remains the extraordinarily difficult question of whether carbon emissions should be limited in developing countries, and, if so, at what level. It is a simple fact of atmospheric science that the planet will never be able to support a population of 10 billion people emitting carbon at, say, the rate of Western Europe today. This would imply carbon emissions of four times the current level, or as high as 23 billion tons per year. (Agarwal and Narain 1991, 1)

One very large problem in climate diplomacy is the reluctance of people representing the diverse interests of many of the world's nation-states to place those interests aside in pursuit of a common goal of reducing greenhouse gas emissions. National rather than global interests dominated climate diplomacy throughout the 1990s (despite the rhetoric of many national leaders, cited above), as atmospheric greenhouse gases continued to rise.

A few days after the Buenos Aires conference on climate change ended in 1998, the United States showed just how illusionary much of its work had been. The U.S. Energy Information Administration, an office of the Department of Energy, published a draft version of its 1999 annual energy outlook. The outlook projected that United States emissions of carbon dioxide from energy use will rise 33 percent by 2010, compared to 1990 levels. By 2020, the Energy Department projected that emissions would rise 47 percent compared to 1990 levels. Between 1990 (the Kyoto baseline year) and 1996, United States greenhouse gas emissions already had increased by nine percent. Given the figures provided by its own Energy Department, the United States left little at the table of climate diplomacy except lip service.

The problem which the world faces is illustrated by comparing a passage on page 110 and another on page 114 of Ross Gelbspan's *The Heat is On: The High Stakes Battle Over Earth's Threatened Climate* (1997). On page 110, Gelbspan reports that a team of scientists from the Netherlands (which is very vulnerable to rising sea levels provoked by global warming) has reported that the industrialized world must cut its emissions by more than 50 percent below 1990 levels by the year 2010, "to head off serious disruption." On page 114, Gelbspan cites a study by the environment ministry of Japan, issued in 1996, saying that "Carbon dioxide and other greenhouse gases from fifteen Asian nations will more than double over the next thirty years" (Gelbspan, *The Heat*, 1997, 110, 114). The same study indicates that Asia will create a third of the world's human-generated greenhouse gases by 2025 and more than half by the end of the twenty-first century (Gelbspan, *The Heat*, 1997, 115).

Gelbspan calls the results of climate diplomacy at the Kyoto conference "a yawning disjunction between what may be politically feasible and the natural requirements of the planet's inflamed atmosphere . . ." (Gelbspan, *A Global Warming*, 1997).

All of this, in the end, leaves the resolution of perhaps the most ominous environmental problem ever confronting humanity to the mercy of the global marketplace. . . . Meanwhile, the world's glaciers are melting, the world's oceans are warming, plants and insects are migrating northward, the zooplankton are dying in the Pacific Ocean, the Antarctic ice shelves are breaking up, and the planet continues to heat at a faster rate than at any time in the last ten thousand years. (Gelbspan, *A Global Warming*, 1997)

John Prescott, the United Kingdom's deputy prime minister, in April, 2000, challenged the United States to tackle the threat posed by global warming, warning that "no country, however big" can afford to ignore its consequences (Newman 2000, 17). While British emissions of greenhouse gases by the year 2000 had fallen between five and six percent compared to the Kyoto Protocol 1990 targets, Prescott noted that emissions in the United States rose 11 percent between 1990 and 1998. Canada's greenhouse gas emissions rose 13 percent during the 1990s, while several European countries (including Britain) have made progress toward meeting the goals of the Kyoto Protocol.

HUMANKIND'S ASSUMED SUPERIORITY TO NATURE

In assuming that humankind is separate from nature (and superior to it), the cultural architecture of the fossil fuel age has sowed some worrisome climatic seeds. The philosophic attitudes supporting this sense of superiority are much older than the industrial revolution. In Christianity's Bible (Genesis 9:1–3), God instructs Noah and his sons to

Be fruitful, and multiply, and fill the Earth. The fear of you and the dread of you shall be upon every beast of the Earth, and upon every bird of the air, and upon everything that creeps on the ground, and all the fish in the sea; into your hand, they are delivered. (McNeill 2000, 327)

Modern capitalism was born in Europe as the Protestant Reformation told the entrepreneurs of the early fossil-fueled machine age that God helps those who help themselves. Such attitudes have become entrenched in western industrial belief systems and institutions. Through "globalization," during the twentieth century, these beliefs have become a world-girding ideological currency.

In addition to a belief system that alienates machine-age humankind from nature, we live in an economic and political culture which regards our environmental heritage as an ownerless commons which anyone may defile for profit without moral sanction or fiscal penalty. Ian H. Rowlands quotes Aristotle as the first philosopher to wrestle with the concept that private individuals will defile a freely held common good without concern for humanity's shared welfare:

What is common to the greatest number gets the least amount of care. Men pay most attention to what is their own; they care less for what is common. . . . When everyone has his own sphere of interest . . . the amount of interest will increase, because each man will feel that he is applying himself to what is his own. (Rowlands 1995, 4, quoted in Hardin and Baden 1977, xi)

Anita Gordon and David Suzuki's *It's a Matter of Survival* (1991) questions several "sacred truths," assumptions of machine age culture:

The assumptions that we've made about how the natural world operates and what our relationship is to it are no longer tenable. These *sacred truths* that we've grown up with— "nature is infinite;" "growth is progress;" "science and technology will solve our problems;" "all nature is at our disposal;" "we can manage the planet"—offer no comfort as we enter the last decade of this [the twentieth] century. In fact, we're being told that to continue to subscribe to these assumptions is to ensure the destruction of civilization as we know it. (Gordon and Suzuki 1991, 1)

INCREASING HUMAN POPULATIONS AND MATERIAL AFFLUENCE

Human transformation of the global environment, including climate change, is driven to an important degree by ever-larger numbers of people. The human population of the earth reached six billion in 1999, and has been increasing by almost 100 million a year. Of all the human beings ever born, *half* were alive in the year 2000. If present projections hold, the Earth by the mid-to-late twenty-first century may be home to at least 10 billion people (Silver and DeFries 1990, 4). The world's population more than doubled in the 45 years between 1955 and 2000, to six billion, and may double again by about 2050 at current birth rates of roughly 225,000 a day.

Several factors are at work to enhance greenhouse-gas emissions in addition to a simple increase in human numbers. The average amount of fossil-fuel energy consumed per person is rising gradually worldwide as industrializing nations' middle classes adopt the comforts and conveniences of machine culture, including the basics of a middle-class living standard: an electrified home (often with heat and air conditioning), an automobile, and a variety of appliances, all of which consume electricity. People also live longer (the average life span in the United States did not pass 50 until shortly after the year 1900), providing each individual more time, on the average, to consume fossil-fuel-based energy.

Every time human population increases and a tree is cut down, the world balance of carbon dioxide changes a little. Compounded millions of times, the change is significant. Settlement of lands previously uninhabited by human beings (such as the rain forests of the Amazon and Indonesia) also introduces greenhouse gas-emitting animals and machinery (National Academy 1991, 5).

One index of material affluence (and greenhouse-gas production per capita) is the volume of trash and garbage generated. By about 1990, North Americans produced roughly four pounds of garbage per person per day, or almost a ton per year. Italians produced 1.5 pounds, and Nigerians one pound each per year (Gordon and Suzuki 1991, 184). The amount of garbage produced worldwide is another statistical curve which climbed slowly at first and then accelerated sharply in our own time, as material affluence has spread around the world. Population is increasing as the effluent per person also rises.

Timothy R. Barber and William M. Sackett find that human-induced greenhouse gases correlate strongly with global population growth, having risen about five-fold between 1950 and 1990. "Anthropogenic fossil sources are especially troublesome because they release previously buried carbon back into the atmosphere," they comment (Barber and Sackett 1993, 220).

Albert K. Bates writes,

Today [1989], the world must accommodate the equivalent of a new population of the United States and Canada every three years. If North American standards of consumption were universally achieved early in the twenty-first century, world carbon-dioxide production could reach 110 gigatons (billion tons) per year, 20 times present levels. The earth would experience a climate shock of profound proportions, beyond anything we have been able to estimate at the present time. (Bates 1990, 119)

"PREMEDITATED WASTE"

As human populations and per capita resource consumption rise, more products are being designed to be thrown away, as an increasing number of industries use "premeditated waste" to stimulate sales, income, and profit. Premeditated waste involves the manufacture of products designed to be used for only a year or two, until a new fashion or model replaces it. This practice originated in the high-fashion clothing industry and was embraced by automobile makers when they introduced annual model changes. By the end of the twentieth century, the same model of production had been adopted in many other industries. The most visible recent addition to the world garbage heap has been oceans of plastic waste generated by the rapid technological evolution of computer equipment.

According to Gordon and Suzuki, "Breaking the garbage habit means destroying the cultural myths that we have created as a society—myths that have allowed us to define ourselves by our possessions, that have led us to canonize time and convenience" (Gordon and Suzuki 1991, 186). Such a divorce may be very difficult because consumers love convenience, and industry loves selling them goods that will quickly become obsolete.

During the 1990s, the automobile industry came up with a new variation on an old sales gambit to maximize profits among those drivers who associate their personal identity with the shape, size, and power of their mode of transportation. The sports-utility vehicle allows automobile companies to sell a larger vehicle (along with more production capacity for greenhouse gases) and thereby increase profits. This strategy reflects a philosophy of "forced consumption," which was outlined during the middle 1950s in the *New York Journal of Retailing*:

Our enormously productive economy demands that we make consumption a way of life, that we convert the buying and use of goods into rituals, that we seek our spiritual satisfactions in consumption. . . . We need things consumed, burned up, worn out, replaced, and discarded at an ever-growing rate. (Gordon and Suzuki 1991, 186)

Some environmentalists have called upon consumers to boycott such high-pollution forms of personal transportation.

Probably the worst example of global warming abuse is the sports utility [vehicle]. . . . We are asking the public not to purchase vehicles that get poor gas mileage and therefore emit large quantities of carbon dioxide. . . . For example, owing to their poor gas mileage as determined by the Environmental Protection Agency, we are asking people not to buy 23 specific 1998 models of sports utility vehicles, which get 12 to 14 miles to the gallon (city). . . . A sports-utility [vehicle] that gets 12 miles to the gallon in the city will emit 800 pounds of carbon dioxide over a distance of 500 city miles. (Eco Bridge N.d.)

During May, 2000, the Ford Motor Company issued its first "corporate citizenship report" at the company's annual shareholders' meeting in Atlanta. The report acknowledged that sports-utility vehicles are environmental and safety hazards. William C. Ford, Jr., the company's chairman and a great-grandson of Henry Ford, said, according to an account in the *New York Times*, "that he worried automakers could wind up with reputations like those of big tobacco companies if they ignored sport utilities' problems" (Bradsher 2000, 1). Mr. Ford and Jacques Nasser, Ford's chief executive, said, however, that the company has no plans to stop making sports-utility vehicles, including the largest of all, the Ford Excursion, which weighs as much as two Jeep Grand Cherokees and gets 10 miles to a gallon of gas in the city and 13 on the highway. Sports-utility vehicles contributed about 20 percent of Ford's sales by the year 2000, up from five percent in 1990.

THE BILL WILL NOT ARRIVE UNTIL THE PRESENT GENERATION HAS LEFT THE TABLE

It will be decades, perhaps a century, "before the oceans and the atmosphere fully redistribute the absorbed energy and the currently 'committed' temperature rise is actually 'realized' " (National Academy 1991, 93). The "equilibrium temperature" is the level at which the system will come to rest once the full effects of greenhouse warming become evident. This feedback mechanism, to which various scientists assign a 10- to 100-year delay, makes it difficult to "prove" global warming simply through the use of present-day temperature readings, since they are reflecting a state of the atmosphere that reflects greenhouse-gas generation several decades ago. The Earth has already been committed to a 0.7 to 2.0 degree C. warming from carbon and other greenhouse-gas emissions during the late twentieth century. Only a fraction of this warming had become evident by the year 2000, due to feedback delays in the dynamics of the atmosphere. The "bill" for emission levels in the year 2000 probably will fall due about 2040 (Ramanathan 1991).

Author Jonathan Weiner compares the Earth's predicament to the cooking of a roast in an oven.

Suppose you preheat an oven to 450 degrees F. and put in a roast. It is easy enough to predict that the roast will eventually reach 450 degrees. But how fast will the meat's surface temperature rise in the first five minutes in the oven? To answer that question, you would need to know a lot about that piece of meat. You would have to make a study of what climate experts call the roast's "thermal inertia."

Now, suppose you did not preheat the oven. Instead, you turned up the dial very, very slowly toward 450 degrees, and someone else opened and closed the oven door a few times while you weren't looking. Predicting the meat's temperature five minutes ahead would be almost hopeless. . . . With Planet Earth . . . it is as if we have been turning up the dial slowly at first, then faster and faster. . . . How long will it take the planet to catch up with the dial? We do not know, but we know we are cooking the planet. (Weiner 1990, 261)

Global fossil-fuel consumption began to jump very rapidly during the 1950s and 1960s; it is possible that the surge in global warmth during the 1980s and 1990s can be traced, at least in part, to the arrival of the "bill" for that increase. Today's effluent, roughly double that of 1950, will be "billed" to future generations.

CAPITALISM'S PENCHANT FOR GROWTH

Edward Abbey has called capitalism the ideology of the cancer cell. The ecological implications of this metaphor are becoming more fully apparent as human numbers, material affluence, and global temperatures rise in tandem.

The Cold War has ended with the collapse of communism as an international state-level ideology, in no small part because the nineteenth-century ideology of Karl Marx paid little attention to the effects of environmental degradation. Capitalism is now the world's dominant economic system, as the drive to profit, along with human desires for wealth and comfort, are bringing millions of people into the carbon dioxide producing middle and upper classes. The "American Dream" (a family home, a car, appliances, and so forth) is spreading around the world, as greenhouse-gas levels accelerate. Brazil is "developing" the Amazon as Indonesia moves homesteaders to fire-cleared areas of Sumatra and Borneo.

Capitalism demands growth and, like Marxism, makes little economic allowance for environmental consequences. The atmosphere becomes the ultimate "free good," to be used and abused without cost or penalty. The Earth is a finite habitat, however, so growth (especially growth in fossil-fuel effluent) cannot continue indefinitely.

THE SPECIAL INTEREST STATE

Special interest republicanism, as it has evolved in the United States (and as something of a world model), will not be well-suited to a time when the due bills of the greenhouse effect may curtail acceptable ranges of individual choices. Ecologically inclined voters in the United States saw a preview of this problem during the 2000 election, when Al Gore, who has written and spoken extensively about global warming, ran for president and said very little about the issue. The presidential campaign (and other political contests) thus becomes a rhetorical ritual involving clichés and rituals calculated to win votes. Any coincidence of rhetoric with ecological reality may be accidental.

A sterling example of this problem was described by the national newspaper *Indian Country Today* during the early months of the 2000 presidential campaign.

Vice President Al Gore and former Senator Bill Bradley were debating health care in mid-January when suddenly demonstrators pushed open the door to their meeting hall and shouted, "Global warming is the issue!" Bradley didn't comment, but Gore, after disagreeing with the demonstrators' tactics, did say: "The issue is important. I've raised it, it's just nobody else wants to talk about it." (Earth Out 2000, A-4)

Molly Ivins captures the problem in a nutshell: "As Sherlock Holmes once explained to Watson, the dog that did not bark in the night is the key to the case. Our current [2000] presidential campaign is the sound of *no* dogs barking." Ivins suggests reading *Something New Under the Sun: An Environmental History of the Twentieth-century World* by J.R. McNeill. According to Ivins, McNeill says,

The world's population quadrupled during the last 100 years; the global economy expanded 14-fold; energy use increased 16 times, and industrial output went up by a factor of 40. Water use rose nine times, and carbon dioxide emissions went up 13 times. And what may be the most important overall is that humans in the 20th century used 10 times more energy than their forebears during the entire 1,000 years before 1900. (McNeill 2000, xv–xvi, 272, 360–361, cited in Ivins 2000, 39)

THE GASOLINE-POWERED AUTOMOBILE

The automobile is an exemplar of fossil-fuel culture—easy to use, convenient, suited to the individualism of our time, and generative of greenhouse gases in its manufacture, as well as its everyday use. Each automobile brings with it a bundle of carbon dioxide, carbon monoxide, methane, nitrous oxide, and waste heat. The tailpipes of motorized vehicles emit 47 percent of North America's nitrous oxides, a primary component of tropospheric ozone (Gordon and Suzuki 1991, 198). The automobile, in addition to burning fossil fuels, is also a little greenhouse factory of its own, as the windshield captures the sun's heat and cooks the air inside its steel and glass casing.

According to Bill McKibben, "The average American car, driven the average American distance—ten thousand miles—in an average . . . year releases its own weight in carbon into the atmosphere" (McKibben 1989, 6). Global warming, in a sense, is the exhaust pipe of the "American Dream," of the bigger and better, the new and improved, the mobile life in the fast lane. Cars and trucks used in the United States burn 15 percent of the world's oil production. Transportation alone consumes one-fourth of the energy and two-thirds of oil used in the United States (Cline 1992, 200). In the late 1980s, 500 million automobiles were registered around the world, each burning an average of almost two gallons of gasoline or diesel fuel per day.

Stephen Schneider recalls having shared a stage with John Denver, the singer, in Aspen, Colorado. "John said, in introducing me to this group," Schneider recalled, "that he had been unaware of his own impact on global nature until he learned about the global warming problem. 'As soon as I learned about that,' he confessed, 'I sold my Porsche.' Everybody applauded" (Schneider 2001).

Automobiles consume a third of the world's oil production. By the late 1980s and 1990s, the number of automobiles was increasing more quickly than population, especially in the Third World (Silver and Defries 1990, 49). New automobile registrations soared 30 to 40 percent between 1987 and 1992 in Africa, Latin America, and Asia, compared with about 7 percent in Japan, and 2 percent in Western Europe and the United States (Bates 1990, 119).

The "automotive age" took hold in the United States during the early twentieth century. Between 1900 and 1920, annual gasoline consumption in the United States rose from four thousand to four million gallons. By 1927, 85 percent of the automobiles in the world were plying a growing web of highways which laced the United States (Oppenheimer and Boyle 1990, 22). There were

26.7 million automobiles in the United States in 1929, traveling more than a million miles of paved roads by 1935.

As climate experts call for reductions in greenhouse-gas emissions around the world, the largest corporate automakers are doing their best to pave the same Earth with their products, to generate profits for shareholders. A feature story in the *Wall Street Journal* described this form of economic "globalization":

RAYONG, Thailand—On an old pineapple plantation here along the steamy Gulf of Thailand, the Big Three auto giants are building a piece of Detroit—bowling alleys and all—[as a] phalanx of Ford Ranger pickup trucks roars off an assembly line at the company's new $500 million plant. (Frank 1999, B-1)

With North American automobile markets saturated, United States car makers are looking to Asia for growth, selling the same pickup trucks familiar to North Americans, bringing United States culture along for the ride: "At the Bangkok International Motor Show last spring [1999], Ford flew in a country-western singer from Texas to cavort with a group of traditional Thai dancers, and sing a number called 'I'm Crazy About My Ford Truck' " (Frank 1999, B-5). Ford's automobile evangelizers realize that one-third of the world's automobiles are being driven in the United States, and another third of the total is on the road in Europe. Asia, with its enormous population, is home to only 15 percent of the world's automobiles. Africans own less than two percent of the world's cars (Jenney 1991, 285).

The growing numbers of sports-utility vehicles roaming the roads during the 1990s recall the words of Henry Ford: "Mini-cars make mini-profits" (Commoner 1990, 80). Some of the earliest automobile designs were electric; they were pushed aside by the development of powerful internal combustion engines during the 1930s. Ford has had workable designs for small, efficient, electric cars since the 1970s. Electric autos have not been developed nor promoted to the motoring public. The problem is less technological than a matter of driver preference and corporate profitability.

As respected students of the greenhouse effect were calling on the world to reduce its fossil-fuel emissions by at least half, the twentieth century rolled to a finish in a cavalcade of automobile traffic. A report by the American Highway Users Alliance released during November of 1999 described Americans as increasing the size of their cars, the number of trips taken in them, and the number of hours spent stalled in gridlock.

The same study found that population in the United States' largest urban areas had grown 22 percent during the previous 15 years, but that traffic congestion (defined as the amount of time spent stalled in traffic) had risen more than 150 percent (Bowles 1999, 1A). United States population grew 32 percent (203 million to 267 million) between 1970 and 1997, while the number of licensed drivers grew 65 percent, the number of registered vehicles rose 87 percent (111 million to 208 million), and the number of miles driven increased 127 percent.

People are driving further to work as the suburban hinterland of large cities grows. In Indianapolis, for example, population grew 17 percent between 1982 and 1997, while the number of miles driven—an indicator of greenhouse gases emitted—rose an estimated 100 percent (Bowles 1999, 2A).

FOSSIL-FUELED AIR TRAVEL

Airline projections that air traffic will double within two decades have compelled atmospheric scientists to ask whether growing air traffic is altering the chemistry of the stratosphere through which jets travel. The combustion of jet fuel releases into the atmosphere several chemicals which affect the balance of greenhouse gases: carbon dioxide, water vapor, nitrous oxides, sulfur oxides, and particulate matter (soot). Most work has concentrated on aircraft engines' nitrous oxide production and its relation to the ozone level of the atmosphere. Scientific scrutiny of jet aircraft's role in global warming is "considerably poorer than that of ozone chemical processes" (Friedl 1999, 58).

On arriving and departing, a Boeing 747 produces as much pollution as a car driven 5,600 miles, and as much nitrogen oxide as a car driven nearly 26,500 miles. Aircraft emissions are especially damaging because some of their pollution takes place in the upper atmosphere, at jet-stream level. At this altitude, in the stratosphere, nitrogen oxides in jet exhaust convert to ozone, a powerful greenhouse gas. Aviation's contribution to greenhouse-gas emissions may soon rise to half that of the world's automobile fleet. In the meantime, while alternatives to the fossil-fueled automobile are on some drawing boards, no one seriously proposes to introduce a jet aircraft powered by photovoltaic solar power, electricity, or hydrogen. No other technological alternative produces the sudden thrust required to get a large aircraft off the ground and keep it in the air at the speed allowed by carbon-based jet fuel.

During the 1990s, aviation was the fastest-growing mode of travel in the United States. To handle this growth, 32 of the nation's 50 busiest airports had plans to expand in 1999. Sixty of the 100 largest airports were proposing to build new runways. During the late 1990s, passenger and freight air transport mileage was doubling roughly every 10 years. By the late 1990s, more than two billion passengers and 42 million tons of cargo were being transported by air worldwide per year. The number of passengers is increasing by eight percent a year, on average, as the volume of cargo rises by 13 percent annually.

Airline travel, which was estimated to contribute five percent to the total greenhouse effect by about 1990, has been increasing at a much faster rate than overall fossil-fuel usage. Between 1969 and 1989, according to the International Energy Agency, worldwide airline passenger miles increased 400 percent (International Energy Agency 1993, 44). World air travel increased at about five percent a year during the 1990s. If this rate of increase is maintained, air travel may account for 15 percent of greenhouse-gas emissions by the year 2050 (Clover 1999). An explosion in commercial air traffic also may be triggering formation of high-altitude clouds that trap heat and could worsen global warming. Despite improvements in jet-aircraft fuel efficiency, as well as technological improvements, air travel's role in global warming will increase because the number of passengers and miles flown is increasing rapidly.

An IPCC report on air travel and global warming stimulated calls for environmental taxes and stricter emissions targets for airlines. The report was produced jointly by the United Nations and the World Meteorological Organization. "Transportation is the area of greatest growth (in terms of impact on the climate), and aviation is growing more rapidly than any other sector," said John Houghton of Britain, one of the report's co-chairs (Aircraft Pollution 1999). "Policy options include more stringent aircraft engine emissions regulations, removal of subsidies and incentives [and] market-based options such as environmental levies," the report said (Aircraft Pollution 1999).

Given present growth trends, global air traffic may increase six-fold by 2050. Even in 1999, in some air-traffic corridors above Europe, jet contrails sometimes covered as much as four percent of the sky at any given time. In northeast United States, jet contrails sometimes covered as much as six percent of the sky. Contrail coverage over Asia was increasing so quickly that contrails could multiply by a factor of ten in fifty to sixty years.

Steve McCrea, editor of the *Eco-Tourist Journal*, calls air travel "eco-tourism's hidden pollution" (McCrea 1996). Tourists who take the utmost ecological care when they visit exotic locales rarely give a thought to the greenhouse gases that they generate while reaching their destinations. According to McCrea,

One ton of carbon dioxide enters the atmosphere for every 4,000 miles that the typical eco-tourist flies. A round trip from New York to San Jose, Costa Rica (the world's leading eco-tourist destination) is 4,200 miles, so the typical eco-tourist generates roughly 2,100 pounds of carbon dioxide by traveling to a week of sleeping in the rainforest. (McCrea 1996)

To balance carbon dioxide generated by their air travel, McCrea suggests that eco-tourists plant three trees for every 4,000 miles, to compensate not only for the carbon dioxide, but also for other greenhouse gases created by the combustion of jet fuel.

During the 1990s, a rising number of people in the United States were commuting to work via the airlines, sometimes thousands of miles per week. The "suburbs" of Los Angeles have reached out and enveloped Santa Fe, New Mexico, where Professor Ken Lincoln leaves his home early each workweek for a flight to Los Angeles. Professor Lincoln holds a tenured appointment in the English Department on the University of California campus in Los Angeles. The *New York Times* has carried accounts of Manhattan jobholders who fly into the city from Rochester, New York. Silicon Valley, where the price of the average house was $400,000 early in the year 2000, draws weekly commuters, according to one *Times* account, from "Arizona, Idaho, Nevada, Oregon, and Utah" (Johnston 2000, 1G). Some people who work at New York's Lincoln Center commute from Florida. The typical schedule includes long workdays Tuesday through Thursday, travel Monday and Friday, and a weekend at home hundreds, sometimes thousands, of miles from the office.

The status vehicle in elite circles of the future won't be a sports-utility vehicle. According to one observer, it will be "the family plane" (Fallows 1999, 84). Such an aircraft, the Cirrus SR20, was being produced by the late 1990s, as the National Aeronautics and Space Administration (NASA) "is quietly advocating an Interstate Skyway Network . . ." (Fallows 1999, 88).

THE WORLD BANK'S PRIORITIES

The World Bank has been criticized by the United States Institute for Policy Studies and Friends of the Earth International for favoring projects which aggravate global warming. Between 1992 and 1998, according to a report, the World Bank Group spent $12.4 billion on gas, oil, and coal projects.

The report also indicated that "Greenhouse-gas emissions produced by current World Bank projects over their lifetimes (9.9 billion tons of carbon) will total more than three times the output of greenhouse gases from all OECD countries (2.7 billion tons of carbon) for 1995," and that the World Bank spends nearly 100 times more on promoting climate change than in averting it (Rich Nations 1998). The World Bank Group includes four institutions which provide loans, credit, equity guarantees, and risk insurance: the International Bank for Reconstruction and Development (IBRD), the International Development Association (IDA), the International Finance Corporation (IFC), and the Multilateral Investment Guarantee Agency (MIGA).

Additionally, according to the report,

Because there is almost a dollar-for-dollar relationship between contributions to the Bank and procurement contracts, nine-tenths of World Bank fossil-fuel money ends up with multinational companies such as Amoco, Chevron, Exxon and Mobil. These companies are leading lobbyists—through the Global Climate Coalition—in resisting official action on climate change. (Rich Nations 1998)

During the late 1990s, according to the report, one-third of all World Bank fossil-fuel funding was spent in China, which will require an extra 15 gigawatts of power annually to meet its planned economic-growth rate of eight percent. Much of this money is financing the construction of Chinese coal-fired power plants, which are contributing significantly to rising greenhouse-gas levels. The World Bank is also spending billions to open previously untapped coal, oil, and natural-gas reserves in Nigeria, the Caspian Sea, the Amazon Valley, and other Third World locations.

Daphne Wysham of the Institute for Policy Studies commented,

Our report shows that the World Bank is not part of the solution to climate change. It is part of the problem. The Bank is supposed to tackle poverty and promote sustainable development. In fact a good part of its money goes to help promote the single largest environmental threat to our planet. (Rich Nations 1998)

The World Bank Group's October 29, 1998 draft paper "Fuel for Thought: A New Environmental Strategy for the Energy Sector" has been debated within the World Bank's bureaucracy. This paper was part of a gradual recognition by the World Bank (among other human institutions) that an ecologically sustainable future begins by working the bias toward fossil fuels out of the world's energy mix, an idea that will be developed in this book in Chapter 10, "Possible Solutions."

Taken together, humankind's problems with greenhouse warming are rooted in the ways human economies and cultures conduct their day-to-day business. The problem is going to demand of humankind not only new definitions of how to generate the energy required by our machine culture, but also changes in some of our basic definitions of what is worth having, what is worth doing, and how these activities should be performed in the world of the future. The present course leads, in a matter of a few centuries, to a planetary habitat which would be scarcely recognizable to most people alive today.

REFERENCES

Abrahamson, Dean Edwin, ed. 1989. *The Challenge of Global Warming*. Washington, D.C.: Island Press.

Abrahamson, Dean Edwin, 1989. "Global Warming: The Issue, Impacts, Responses." In Dean Edwin Abrahamson, ed., *The Challenge of Global Warming*. Washington, D.C.: Island Press, 3–34.

Agarwal, Anil, and Sunita Narain. 1991. *Global Warming in an Unequal World: A Case of Environmental Colonialism*. New Delhi: Centre for Science and Environment.

"Aircraft Pollution Linked to Global Warming; Himalayan Glaciers are Melting, with Possibly Disastrous Consequences." 1999. Reuters. In *Baltimore Sun*, 13 June, 13A.

Baar, Hein J.W. de, and Michel H.C. Stoll. 1992. "Storage of Carbon Dioxide in the Oceans." In F. Stuart Chapin III, Robert L. Jefferies, James F. Reynolds, Gaius R. Shaver, and Josef Svoboda, eds., *Arctic Ecosystems in a Changing Climate: An Ecophysiological Perspective*. San Diego: Academic Press, 143–177.

Barber, Timothy R., and William M. Sackett. 1993. "Anthropogenic Fossil Carbon Sources of Atmospheric Methane." In Richard A. Geyer, ed., *A Global Warming Forum: Scientific, Economic, and Legal Overview*. Boca Raton, Fla.: CRC Press, 209–223.

Barbier, Edward B., Joanne C. Burgess, and David W. Pearce. 1991. "Technological Subsitution Options for Controlling Greenhouse-gas Emissions." In Rutiger Dornbusch and James M. Poterba, eds., *Global Warming: Economic Policy Reponses*. Cambridge, Mass.: MIT Press, 109–161.

Bates, Albert K., and Project Plenty. 1990. *Climate in Crisis: The Greenhouse Effect and What We Can Do*. Summertown, Tenn.: The Book Publishing Co.

Bolin, Bert, John T. Houghton, Gylvan Meira Filho, Robert T. Watson, M.C. Zinyowera, James Bruce, Hoesung Lee, Bruce Callander, Richard Moss, Erik Haites, Roberto Acosta Moreno, Tariq Banuri, Zhou Dadi, Bronson Gardner, J. Goldenberg, Jean-Charles Hourcade, Michael Jefferson, Jerry Melillo, Irving Mintzer, Richard Odingo, Martin Parry, Martha Perdomo, Cornelia Quennet-Thielen, Pier Vellinga, and Narasimhan Sundararaman. 1995. *Intergovernmental Panel on Climate Change. Second Assessment Synthesis of Scientific-Technical Information Relevant to Interpreting Article 2 of the United Nations Framework Convention on Climate Change*. Approved by the IPCC at its eleventh session, 11–15 December, Rome. [http://www.unep.ch/ipcc/pub/sarsyn.htm]

Bowles, Scott. 1999. "National Gridlock: Traffic Really *Is* Worse Than Ever—Here's Why." *USA Today*, 23 November, 1A–2A.

Bradsher, Keith. 2000. "Ford is Conceding S.U.V. Drawbacks." *New York Times*, May 12, Business Section, 1.

Brown, Paul. 1999. "Global Warming: Worse Than We Thought." *World Press Review*, February, 44.

Carrel, Chris. 1998. "Boeing Joins Fight Against Global Warming." *Seattle Weekly*, 17 September. [http://climatechangedebate.com/archive/09–18_10–27_1998.txt]

Christianson, Gale E. 1999. *Greenhouse: The 200-Year Story of Global Warming*. New York: Walker and Company.

Cline, William R. 1992. *The Economics of Global Warming*. Washington, D.C.: Institute for International Economics.

Clinton, Bill. 1997. Remarks by the President on Global Warming and Climate Change. Speech given at the National Geographic Society, Washington, D.C., 22 October. [http://uneco.org/Global_Warming.html]

Clover, Charles. 1999. "Air Travel is a Threat to Climate." *The Daily Telegraph* (London), 5 June.

Commoner, Barry. 1990. *Making Peace With the Planet*. New York: Pantheon.

Davidson, Keay. 1999. "Jet-bred Cirrus Clouds Could Pose Threat; Research Suggests Increase in Air Travel May Speed Global Warming." *Milwaukee Journal-Sentinel*, 16 March.

"Earth Out of Balance." 2000. *Indian Country Today*, 9 February, editorial, A-4.

Eco Bridge. N.d. What Can We Do About Global Warming? [http://www.ecobridge.org/content/g_wdo.htm]

Economists' Letter. 1997. Economists' Letter on Global Warming, 23 June. [http://uneco.org/Global_Warming.html]

Epstein, Paul, Georg Grabbher, Tom Karl, Ellen Mosley-Thompson, Kevin Trenberth, and George M. Woodwell. 1996. Current Effects: Global Climate Change. An Ozone Action Roundtable, 24 June, Washington, D.C. [http://www.ozone.org/curreff.html]

Fallows, James. 1999. "Turn Left at Cloud 109." *New York Times Sunday Magazine*, 21 November, 84–89.

Frank, Robert. 1999. "How Thailand Became the 'Detroit of the East.'" *Wall Street Journal*, 8 December, B-1, B-5.

Friedl, Randall R. 1999. "Perspectives: Atmospheric Chemistry, Unraveling Aircraft Impacts." *Science* 286 (1 October): 57–58.

Gelbspan, Ross. 1998. "Beyond Kyoto." *Amicus Journal*, Winter. [http://www.nrdc.org/nrdc/nrdc/nrdc/nrdc/eamicus/98win/toc.html]

———. 1997a. *The Heat is On: The High Stakes Battle Over Earth's Threatened Climate*. Reading, Mass.: Addison-Wesley Publishing Co.

———. 1995. "The Heat is On: The Warming of the World's Climate Sparks a Blaze of Denial." *Harper's Magazine*, December. [http://www.dieoff.com/page82.htm]

———. 1997b. "A Global Warming." *American Prospect* 31 (March/April). [http://www.prospect.org/archives/31/31gelbfs.html]

Gordon, Anita, and David Suzuki. 1991. *It's a Matter of Survival*. Cambridge: Harvard University Press.

Hardin, Garrett, and John Baden, eds. 1977. *Managing the Commons*. San Francisco: W.H. Freeman.

Hariss, Robert C., Terry Bensel, and Denise Blaha. 1993. "Methane Emissions to the Global Atmosphere From Coal Mining." In Richard A. Geyer, ed., *A Global Warming Forum: Scientific, Economic, and Legal Overview*. Boca Raton, Fla.: CRC Press, 339–346.

Hashimoto, Ryutaro, Prime Minister of Japan. 1997. Excerpts from Statements on Climate Change by Foreign Leaders at Earth Summit, 23 June. [http://uneco.org/Global_Warming.html]

Hegerl, Gabriele. 1998. "Climate Change: the Past as Guide to the Future." *Nature* 392 (23 April): 758–759.

Hertsgaard, Mark. 1999. "Will We Run Out of Gas? No, We'll Have Plenty of Carbon-based Fuel to See Us Through the Next Century. That's the Problem." *Time*, 5 November, 110–111.

Houghton, John. 1997. *Global Warming: The Complete Briefing*. Cambridge, England: Cambridge University Press.

Huntley, Brian. 1990. "Lessons from the Climates of the Past." in Jeremy Leggett, ed., *Global Warming: The Greenpeace Report*. New York: Oxford University Press, 133–148.

Indermuhle, A., T.F. Stocker, F. Joos, H. Fischer, H.J. Smith, M. Wahlen, B. Deck, D. Mastroianni, J. Tschumi, T. Blunier, R. Meyer, and B. Stauffer. 1999. "Holocane Carbon-cycle Dynamics Based on CO2 Trapped in Ice at Taylor Dome, Alaska." *Nature* 398 (11 March): 121–126.

International Energy Agency. 1993. *Cars and Climate Change*. Paris: International Energy Agency.

Ivins, Molly. 2000. "Environment is Ticking, but Will We Hear It?" *Chicago Sun-Times*, 13 July, 39.

Jager, J., and H.L. Ferguson. 1991. *Climate Change: Science, Impacts, and Policy. Proceedings of the Second World Climate Conference*. Cambridge: Cambridge University Press.

"Japan Agency Warns CO2 Emissions May Soar." 1997. United Press International, 6 November. [http://benetton.dkrz.de:3688/homepages/georg/kimo/0254.html]

Jenney, L.L. 1991. "Reducing Greenhouse-gas Emissions From the Transportation Sector." In G.I. Pearman., ed., *Limiting Greenhouse Effects: Controlling Carbon-dioxide Emissions*. Report of the Dahlem Workshop on Limiting the Greenhouse Effect, Berlin, December 9–14, 1990. New York: John Wiley & Sons, 283–302.

Johnston, David Cay. 2000. "Some Need Hours to Start Another Day at the Office." *New York Times*. In *Omaha World-Herald*, 6 February, 1G.

Kane, R.L., and D.W. South. 1991. "The Likely Roles of Fossil Fuels in the Next 15, 50, and 100 Years, with or without Active Controls on Greenhouse-gas Emissions." In G.I. Pearman, ed., *Limiting Greenhouse Effects: Controlling Carbon-dioxide Emissions*. Report of the Dahlem Workshop on Limiting the Greenhouse Effect. Berlin, December 9–14, 1990. New York: John Wiley & Sons, 189–227.

Kok, Wim, Prime Minister, Netherlands. 1997. On Behalf of the European Union. Statements on Climate Change by Foreign Leaders at Earth Summit, 23 June. [http://uneco.org/Global_Warming.html]

Leggett, Jeremy, ed. 1990. *Global Warming: The Greenpeace Report*. New York: Oxford University Press.

Mann, Michael E., Raymond S. Bradley, and Michael K. Hughes. 1998. "Global-scale Temperature Patterns and Climate Forcing over the Past Six Centuries." *Nature* 392 (23 April): 779–787.

McCrea, Steve. 1996. "Air Travel: Eco-tourism's Hidden Pollution." *San Diego Earth Times*. August. [http://www.sdearthtimes.com/et0896/et0896s13.html]

McKibben, Bill. 1989. *The End of Nature*. New York: Random House.

McNeill, J.R. 2000. *Something New Under the Sun: An Environmental History of the Twentieth-century World*. New York: W.W. Norton.

National Academy of Sciences. 1991. *Policy Implications of Greenhouse Warming*. Washington, D.C.: National Academy Press.

Newman, Cathy. 2000. "Prescott Warns U.S. Over Climate." *Financial Times* (London), 27 April, 7.

Oppenheimer, Michael, and Robert H. Boyle. 1990. *Dead Heat: The Race Against the Greenhouse Effect*. New York: Basic Books.

Pew Center. 1999. "Experts Say Global Warming More Than Predicted." A New Study Released by the Pew Center on Global Climate Change Foresees Greater Global Warming than Previously Predicted, Along with Greater Extremes of Weather and Faster Sea-level Rise, 10 July. [http://www.gsreport.com/articles/art000175.html]

Ramanathan, V. 1991. "Trace Gas Trends and Change." Testimony before the Senate Subcommittee on Environmental Protection, 23 January 1987. Cited in Lynne T. Edgerton and the Natural Resources Defense Council, *The Rising Tide: Global Warming and World Sea Levels*. Washington, D.C.: Island Press, 11, 108.

Rao, P.K., ed. 2000. *The Economics of Global Warming*. Armonk, N.Y.: M.E. Sharp.

Rich Nations, Multinationals Profit From Fossil Fuel Funding. 1998. Global Exchange, 15 May. [http://www.globalexchange.org/economy/rulemakers/ClimateChange.html]

Ridenour, David. 1998. "Hypocrisy in Buenos Aires: Millions of Gallons of Fuel to Be Burned by Those Seeking Curbs on Fuel Use." *National Policy Analysis*: A Publication of the National Center for Public Policy Research, no. 217 (October). [http://nationalcenter.org/NPA217.html]

Rolfe, Christopher. 1996. "Comments on the British Columbia Greenhouse Gas Action Plan." West Coast Environmental Law Association. A Presentation to the Air and Water Management Association, 17 April. [http://www.wcel.org/wcelpub/11026.html]

Rowlands, Ian H. 1995. *The Politics of Global Atmospheric Change*. Manchester, U.K.: Manchester University Press.

Schneider, Stephen H. 1989. *Global Warming: Are We Entering the Greenhouse Century?* San Francisco: Sierra Club Books.

Schneider, Stephen H. 2001. "No Therapy for the Earth: When Personal Denial Goes Global." In Michael Aleksiuk and Thomas Nelson, eds., *Nature, Environment & Me: Explorations of Self In A Deteriorating World.* Montreal: McGill-Queens University Press.

Silver, Cheryl Simon, and Ruth S. DeFries. 1990. *One Earth, One Future: Our Changing Global Environment.* Washington, D.C.: National Academy Press.

Weiner, Jonathan. 1990. *The Next One Hundred Years: Shaping the Fate of Our Living Earth.* New York: Bantam Books.

Whalley, John, and Randall Wigle. 1990. "The International Incidence of Carbon Taxes." Paper presented at a Conference on Economic Policy Responses to Global Warming, September, Rome, Italy. Cited in William R. Cline, *The Economics of Global Warming.* Washington D.C.: Institute for International Economics, 1992, 342, 392.

Whalley, John, and Randall Wigle. 1991. "The International Incidence of Carbon Taxes." In Rudiger Dornbusch and James M. Poterba, eds., *Global Warming: Economic Policy Reponses.* Cambridge, Mass.: MIT Press, 233–263.

Yellen, Janet, Chair, White House Council of Economic Advisors. 1998. Statement on the Economics of the Kyoto Protocol before the Committee on Agriculture, Nutrition, and Forestry, U.S. Senate, 5 March, Washington, D.C. [http://www.fetc.doe.gov/products/gcc/research/economic.html]

CHAPTER 2

The General Consensus on Global Warming

While a lively debate in political circles and the press questions whether human activity is significantly warming the Earth, scientific evidence has been accumulating steadily in support of the idea. Much of this evidence, unobscured by special economic interests which sometimes cloud popular debate, is not at all ambiguous. With the exception of a minority (see Chapter 3), the human role in a rapid warming of the Earth has become nearly incontrovertible.

Kathy Maskell and Irving M. Mintzer, writing in the British medical journal *Lancet*, describe the carbon cycle's natural balancing act:

Over the past 10,000 years, the concentration of CO2 and other greenhouse gases has remained fairly constant, and this represents a remarkable balancing act of nature. Every year natural processes on the land and in the oceans release to and remove from the atmosphere huge amounts of carbon, about 200 billion tons (gigatons) in each direction. Since the atmosphere contains about 700 gigatons of carbon, small changes in natural fluxes could easily produce large swings in atmospheric concentrations of CO2 and CH4 [methane]. Yet for ten millennia natural fluxes have remained in remarkably close balance. (Maskell and Mintzer 1993, 1027)

Human activities are disturbing this long-standing balance in a fundamental manner, adding to concentrations of several greenhouse gases, including, according to Maskell and Mintzer, "Water vapor, the principal greenhouse gas . . . [which] is expected to increase in response to higher temperatures that would further enhance the greenhouse effect" (Maskell and Mintzer 1993, 1027).

Scientists who study the future potential of human-induced warming also point to several other natural mechanisms which could cause the pace of change to accelerate, through "biotic feedbacks," such as release of carbon dioxide and methane from permafrost and continental shelves in the oceans. The possibility of a "runaway" greenhouse effect by the year 2050 is raised in the literature, often with a palpable sense of urgency. This sense of urgency intensified at the turn of the millennium, as scientists traced the relationship between warming of the lower atmosphere and stratospheric cooling, a major factor in depletion of the ozone layer over the poles.

EVIDENCE OF RISING TEMPERATURES

By the late 1990s, scientific debates regarding global warming were taking place against a climatic background of rapidly rising surface global temperatures. According to records gleaned from ice cores, tree rings, and other "proxy indicators" of temperature, the 1990s was the warmest decade of the second millennium on the Christian calendar, with 1998 the warmest single year on record. Each of the 12 months from August 1997 through September 1998 set a new all-time worldwide monthly high-temperature record. Seven of the 10 warmest years in the past 130 years occurred during the 1990s. By 1998, the Earth had sustained 20 consecutive years above the 1961 to 1990 average temperature, with the upward curve steepening toward the end of the period. (Hansen et al., GISS Analysis, 1999).

Also, during 1998,

[A]t least 56 countries suffered severe floods, while 45 baked in droughts that saw normally unburnable tropical forests go up in smoke from Mexico to Malaysia and from the Amazon to Florida. . . . Spring in the Northern Hemisphere is coming a week earlier [and] the altitude at which the atmosphere chills to freezing is rising by nearly 15 feet a year. (Augenbraun et al. N.d.)

The warmest temperatures (compared to averages) during 1998 occurred in North America, in a pattern that commonly occurs during El Niño years. Almost the entire world was much warmer than millennial averages during the year, as well. According to the National Aeronautics and Space Administration (NASA),

The El Niño, by itself, cannot account for either the observed long-term global warming trend or the extreme warmth of 1998. Because the Pacific Ocean temperature has returned to a more normal level, it is anticipated that the global temperature in 1999 will be less warm than during 1998, but will remain well above the long-term mean for the period of climatology, 1951–1980. The rapid global warming since the mid 1970s exceeds that of any previous period of equal length in the time of instrumental data. (Goddard Institute 1999)

The National Climatic Data Center said that the spring of 2000 was the warmest in 106 years of record keeping in the United States, averaging 55.5 degrees F., 0.4 degrees F. above the previous record, set in 1910. (January through May also was the warmest on record. Temperatures in the United States averaged 48.5 degrees F.; the old record, set in 1986, was 47.4 degrees F.) In the United States, all 50 states reported above-average temperatures, which is very unusual.

Worldwide, 2000 was the twenty-fourth consecutive March-through-May period in which temperatures were higher than average in the Northern Hemisphere. The effects of La Niña seem restricted to the tropics, where temperatures were slightly below average. Large areas of the tropical Pacific Ocean also were below average for the same reason (Spring 2000, 1). "It's quite amazing," said Kelly Redmond, Deputy Director and Regional Climatologist at the Western Regional Climate Center in Reno (Hall 2000, A-1). Kelly said that La Niña, which continued into the year 2000 in the equatorial Pacific, produces shifts in sea surface temperatures and wind patterns that usually translate into cooler-than-normal temperatures for large areas of the country, especially northern latitudes. That pattern was broken during the first half of the year 2000. "For a La Niña year, it's been rather out of character," Redmond said (Hall 2000, A-1).

An analysis of the climate of the last 1,000 years suggests that human activity is the dominant force behind the sharp global warming trend seen during the late-twentieth century. The analysis, by Thomas J. Crowley, a geologist at Texas A & M University, found that natural factors, such as fluctuations in sunshine or volcanic activity, were the most powerful influences on temperatures until 1900. After 1900, however, natural forces have accounted for only about a quarter of observed worldwide warming, Crowley estimates (Revkin 2000, A-12). "These twin lines of evidence provide further support for the idea that the greenhouse effect is already here," Dr. Crowley wrote in the July 14, 2000 issue *Science* (Revkin 2000, A-12; Crowley 2000, 276).

Crowley fed past variations in the solar radiance, bursts of volcanic activity, and other natural variations into a computer model simulating the flow of energy to and from the earth. This model produced temperatures which matched most actual climate variations from the year 1000 to the middle 1800s. Crowley found that the same simulation broke down completely during the twentieth century. The only climatic "forcing" which remotely matched the jump in temperatures seen in the last half of the century was the rise in emissions of greenhouse gases.

Michael E. Schlesinger, a climatologist at the University of Illinois, said that policy makers and the public should not assume that temperature trends will follow a smooth course. He said the relationship between the oceans and the atmosphere is so complex that alterations in it could easily cause temporary cool periods or other unpredictable and sometimes abrupt changes that could create confusion and paralyze work to attack global warming (Revkin 2000, A-12).

A steady warming of the Earth during the twentieth century has been confirmed by readings taken in more than 600 holes drilled into Earth's surface, mainly at mining sites. Shaopeng Huang, Henry N. Pollack, and Po-Yu Shen took temperature readings from 616 previously drilled boreholes on six continents. The authors of this study write, "subsurface temperatures comprise an independent archive of past surface temperature changes that is complementary to both the instrumental record and . . . climate proxies," such as tree rings (Huang et al. 2000, 756). Underground measurements are more apt to detect long-term trends than many surface proxies, the authors of the borehole study assert.

The borehole data confirm that the warming trend accelerated in the latter half of the twentieth century. "Some 80 percent of that warming corresponds with the growth of industrialization," said Henry Pollack, a geology professor at the University of Michigan in Ann Arbor and co-author of the study in the February 17, 2000 issue of *Nature* (Borehole 2000). The study confirms others, and this indicates that the warming trend in the latter half of the twentieth century is without precedent in the past 400 to 1,000 years.

"We do not know of any combination of natural mechanisms that can explain this phenomenon," writes Jonathan Overpeck, a geoscientist at the University of Arizona, in *Nature*. "So we are left with the likelihood that human-induced global warming is underway, caused by emissions of 'greenhouse gases' such as carbon dioxide, and that the next century is going to see even greater warming," he added (Borehole 2000).

"The upper 500 meters of the crust is an archive of what has taken place over the past 1,000 years," said Pollack. "We send a thermometer down and take the temperature at a number of depths along the way, establishing a profile of temperature down the borehole" (Borehole 2000). Borehole temperatures in Pollack's study also show greater warming over the past 500 years than temperatures calculated by other means. "My hunch is boreholes do a better job of establishing long-term trends," he said (Borehole 2000).

To a consensus of scientists who make up the Intergovernmental Panel on Climate Change (IPCC), a United Nations committee, global warming is no longer a matter of whether, but how much, how soon, and with how much damage to the Earth's flora and fauna, including humankind. Many of the scientists who are closest to the evidence are urging policy makers to decide what is to be done, and, failing remedies on a global scale, what price will be paid by future generations.

In 1997, a group of prominent scientists issued this warning:

Our familiarity with the scale, severity, and costs to human welfare of the disruptions that the climatic changes threaten leads us to introduce this note of urgency and to call for early domestic action to reduce U.S. emissions via the most cost-effective means. We encourage other nations to join in similar actions with the purpose of producing a substantial and progressive global reduction in greenhouse-gas emissions, beginning immediately. We call attention to the fact that there are financial as well as environmental advantages to reducing emissions. More than 2000 economists recently observed that there are many potential policies to reduce greenhouse-gas emissions for which total benefits outweigh the total costs. (Scientists' Statement 1997)

This statement was signed by George M. Woodwell, founder and director of Woods Hole Research Center; Dr. John P. Holdren; Teresa Heinz; John Heinz, Professor of Environmental Policy and Director, Program in Science, Technology, and Public Policy, John F. Kennedy School of Government; Dr. William H. Schlesinger; James B. Duke, Professor of Botany at Duke University; Jane Lubchenco, Ecology Professor at Oregon State University and chair of the American Association for the Advancement of Science; Dr. Harold Mooney, Paul S. Achilles Professor of Environmental Biology, Stanford University; Dr. Peter Raven, Director of the Missouri Botanical Garden; and F. Sherwood Rowland, professor of chemistry at the University of California at Irvine and recipient of the 1995 Nobel Prize in Chemistry.

Global warming has been no secret to atmospheric scientists. Articles on the subject began to appear occasionally in the scientific literature of meteorology and other fields during the middle 1970s. A scientific consensus was forming as early as October, 1985, at a conference in Villach, Austria, where a consensus statement began

As a result of the increasing concentrations of greenhouse gases, it is now believed that in the first half of the next century a rise in global mean temperature could occur which is greater than any in man's history. (Scientific Consensus 1989, 63)

Other conferences on global warming were convened in Villach (October, 1987) and Bellagio, Italy (November, 1987). An executive summary of these meetings said that "It is now generally agreed that if present trends of greenhouse-gas emissions continue during the next hundred years, a rise in global mean temperature could occur that is larger than any experienced in human history" (Jager 1989, 97). This summary also projected that the most extreme increases would come in the northern latitudes during the winter, a forecast that was being borne out by observations a decade later.

HISTORY OF THE GREENHOUSE EFFECT AS AN IDEA

The fossil-fueled industrial revolution was born in England. As coal-fired industry (as well as home heating and cooking) filled English skies with acrid smoke, some English homeowners protested coal's use as a fuel. Others described the horrors of coal mines, into which children as young as six years of age were sent to work. Queen Elizabeth sometimes forbade the burning of coal in London while Parliament was in session. In 1661, John Evelyn wrote a book that complained about the noxious nature of coal smoke.

The coal-burning steam engine was invented by Thomas Newcomen in 1712 and refined into a form which was widely adaptable for industrial processes by James Watt, beginning in 1769. Within a century of industrialism's first stirrings, during the 1820s, Jean Baptiste Joseph Fourier, a Frenchman, compared the atmosphere to a greenhouse. During the 1860s, John Tyndall, an Irishman, developed the idea of an "atmospheric envelope," suggesting that water vapor and carbon dioxide in the atmosphere are responsible for retaining heat radiated from the sun. Tyndall also wrote that climate might warm or cool based on the amount of carbon dioxide and other gases in the atmosphere. Tyndall, who speculated in 1861 that a fall in carbon dioxide levels could have accounted for the ice ages, was the first person to make quantitative, spectroscopic measurements showing that water vapor and carbon dioxide absorb thermal radiation and could therefore trap solar heat in the atmosphere.

In 1896, Svante August Arrhenius, a Swedish chemist, published a paper in *The London, Edinburgh, and Dublin Philosophical Magazine and Journal of Science* titled: "On the Influence of Carbonic Acid in the Air upon the Temperature of the Ground." In his paper, Arrhenius theorized that a rise in the atmospheric level of carbon dioxide could raise the temperature of the air. He was not the only person thinking along these lines at the time; Swedish geologist Arvid Hogbom had delivered a lecture on the same idea three years earlier, which Arrhenius incorporated into his article. Arrhenius was a well-known scientist in his own time, not for his theories describing the greenhouse effect, but for his work in electrical conductivity, for which he was awarded a Nobel Prize in 1903. Later in his life, Arrhenius directed the Nobel Institute in Stockholm. His work in global-warming theory was not much discussed during his own life. Arrhenius, using the available measurements of absorption and transmission by water vapor and carbon dioxide, developed the first quantitative mathematical model of the Earth's greenhouse effect and obtained results of acceptable accuracy by today's standards for equilibrium climate sensitivity to carbon dioxide changes.

Arrhenius developed his theory through the use of equations, by which he calculated that a doubling of carbon dioxide in the atmosphere would raise air temperatures about 10 degrees F. Arrhenius thought 3,000 years would have to pass before human-generated carbon dioxide levels would double, a miscalculation which shares something with Benjamin Franklin's belief, late in the eighteenth century, that European-American expansion across North America would take a thousand years.

Arrhenius applauded the possibility of global warming, telling audiences that a warmer world "would allow all our descendants, even if they only be those of a distant future, to live under a warmer sky and in a less harsh environment than we were granted" (Christianson 1999, 115). In 1908, in his book *Worlds in the Making*, Arrhenius wrote, "By the influence of the increasing percentage of carbonic acid in the atmosphere, we may hope to enjoy ages with more equable and better climates, especially as regards the colder regions of the Earth, ages when the Earth will bring forth much more abundant crops than at present for the benefit of rapidly propagating mankind" (Christianson 1999, 115).

Arrhenius' ideas were not widely discussed, but they did not completely die during the early twentieth century. Alfred J. Lotka, an American physicist, warned in 1924,

Economically we are living on our capital; biologically, we are changing radically the complexion of our share in the carbon cycle by throwing into the atmosphere, from coal fires and metallurgical furnaces, ten times as much carbon dioxide as in the process of breathing. (Oppenheimer and Boyle 1990, 35)

Calculating on the basis of fossil fuel use in 1920, at the beginning of the automotive age, Lotka ventured that the level of carbon dioxide in the atmosphere would double in 500 years because of human activities, one-sixth of the time period forecast by Arrhenius.

By the late 1930s, the prospect of global warming was catching the eye of G.S. Callendar, a British meteorologist, who gathered records from more than 200 weather stations around the world. He argued that the Earth had warmed 0.4 degrees C. between the 1880s and the 1930s because of carbon dioxide emissions by industry (Callendar 1938). While Callendar's assertions were met with skepticism by many English scientists at the time, he was laying the foundation for modern day efforts to make more precise measurements of atmospheric trace-gas trends and radiative properties and to design more capable climate models to simulate climate change.

Two decades after Callendar, in 1956, Gilbert Plass, a scientist at Johns Hopkins University in Baltimore, suggested that carbon dioxide is an important climate-control mechanism. He also projected that burning of fossil fuels would raise the global temperature 1.1 degrees C. (2.0 degrees F.) by the end of the century, very close to the actual worldwide increase.

ROGER REVELLE, "GRANDFATHER OF THE GREENHOUSE EFFECT"

In 1957, Roger Revelle and Hans Suess warned, as part of the International Geophysical Year,

Human beings are now carrying out a large-scale geophysical experiment of a kind that could not have happened in the past nor be reproduced in the future. Within a few centuries we are returning to the atmosphere and oceans the concentrated organic carbon stored in sedimentary rocks over hundreds of millions of years. (Christianson 1999, 155–156)

Revelle would become known to the world some years later as the mentor of a graduate student, Albert Gore, who, in 1992 (a year after Revelle died), was elected vice president of the United States. The same year, Gore published a book, *Earth in the Balance*, which argued for mitigation of the greenhouse effect. Until the late 1950s, scientists had no reliable records of carbon dioxide and other greenhouse-gas levels in the atmosphere. At about the same time that Revelle and Suess issued their warning, Charles David Keeling began to assemble documentation indicating that the worldwide level of carbon dioxide had risen to about 315 parts per million (p.p.m.), compared to about 280 p.p.m. at the end of the previous century. Keeling's calculations also indicated that the level was continuing to rise. It reached 370 p.p.m. by the year 2000.

When Keeling decided to measure the concentration of carbon dioxide in the atmosphere, he first had to construct a machine to obtain readings in parts per million. No such machine existed at the time. Keeling worked on his "manometer" for a year, building the machine from an old blueprint at the California Institute of Technology in Pasadena. Keeling's first readings, on the roof of a Caltech laboratory, showed an atmospheric concentration of 310 parts per million. Keeling next took his manometer on family vacations, recording readings. What he found contradicted scientific assumptions of the time, which held that carbon dioxide levels would vary widely, depending on local sources of the gas. Instead, he found that carbon dioxide readings were very similar in different places.

"I decided all the data in the literature was wrong," Keeling later told William K. Stevens, a science reporter for the *New York Times* (Stevens, *The Change*, 1999, 140). After Keeling and his associates had taken a number of readings for more than a year, Keeling made a discovery which confirmed Revelle and Suess' assertions. The readings on his manometer were rising steadily, year by year. The resulting graph which plots the level of carbon dioxide in the atmosphere has come to be known among scientists as "The Keeling Curve."

Keeling also was discovering that atmospheric carbon dioxide levels vary annually, with lower readings in the spring and summer (when plants are respiring oxygen on the large land masses of the Northern Hemisphere) and higher levels during fall and winter, when many plants are dormant and more carbon is being released due to vegetative decay. This annual cycle can vary by as much as three percent in the Northern Hemisphere, compared to one percent in the Southern Hemisphere, where the dominance of oceans mutes seasonal cycles and restricts variability in the carbon dioxide level.

Until the work of Revelle, Suess, and Keeling, most scientists who studied carbon dioxide levels in the atmosphere assumed that the oceans absorbed all of the extra carbon that human activities were injecting into the air. The readings of Keeling and his associates showed that human activity was steadily raising the proportion of carbon dioxide in the atmosphere more quickly than the oceans or other "sinks" of the gas could absorb it.

By the early 1960s, a growing number of scientists were watching as Keeling's carbon dioxide readings from the Mauna Loa observatory climbed steadily. In 1963, the Conservation Foundation issued a warning in a report titled *Implications of Rising Carbon Dioxide Content of the Atmosphere*. The report concluded: "It is estimated that a doubling of carbon-dioxide content in the atmosphere would produce a temperature rise of 3.8 degrees [C.]" (Kellogg 1990, 99).

Revelle, who became known as the "grandfather" of the greenhouse effect, was born in Seattle on March 7, 1909, into a family of Huguenot descent on his father's side and Irish descent on his mother's side. His parents, William Roger Revelle, an attorney and schoolteacher, and Ella Robena Dougan Revelle, also a schoolteacher, both graduated from the University of Washington.

After he was admitted to Pomona College at the age of 16, Revelle entertained thoughts of a career in journalism. However, under the influence of a charismatic professor, Alfred Woodford, Revelle became interested in geology and, after receiving his B.A. in 1929, spent a year of additional study with Woodford. Revelle entered graduate studies at the University of California, Berkeley, in 1930, under the tutelage of geologist George Louderback, who stimulated his interest in marine sedimentation. In 1936, Revelle completed his doctorate at the University of California.

Called to active duty in the U.S. Navy six months before the attack on Pearl Harbor, Revelle was assigned to oceanographic research applied to wartime needs as an officer in the U.S. Navy. Near the end of World War II, Revelle was assigned to Joint Task Force One, the military command supervising the first postwar atomic test on Bikini Atoll (Operation Crossroads). He organized the Crossroads scientific program, which included a study of the diffusion of radionuclides in the atoll, as well as radioactivity's impact on marine life. What he learned through these studies, and others, made Revelle a lifelong opponent of nuclear weapons. After the war, in 1951, Revelle became director of the Scripps Institute of Oceanography.

In 1964, Revelle accepted an appointment as Richard Saltonstall Professor of Population Policy at Harvard University, where he served as Director of the Center for Population Studies until 1974. Revelle held the endowed chair at Harvard until 1978. Revelle also was sympathetic to the 1962 book *Silent Spring* by Rachel Carson, which argued that ecosystems were being disrupted by the use of halogenated hydrocarbon pesticides such as DDT.

Revelle played a key role in the creation, during 1970, of the Scientific Committee on Problems of the Environment of the International Council of Scientific Unions (ICSU). He also suggested the objective for the ICSU's International Geosphere-Biosphere Program in 1986: "To describe and understand the interaction of the great global physical, chemical, and biological systems regulating planet Earth's favorable environment for life, and the influence of human activity on that environment" (Malone et al. N.d.).

Revelle served as president of the American Association for the Advancement of Science during 1974. In 1976, he returned to the University of California, at San Diego, to become a professor of science and public policy. He received the National Medal of Science in 1991 "for his pioneering work in the areas of carbon dioxide and climate modifications, oceanographic exploration presaging plate tectonics, and the biological effects of radiation in the marine environment, and studies for population growth and global food supplies." To a reporter asking why he received the medal, Revelle said, "I got it for being the grandfather of the greenhouse effect" (Malone et al. N.d.).

GLOBAL WARMING BECOMES A POLITICAL ISSUE

The modern debate over whether the earth is warming due to human activity began in policy circles during 1979, after a small number of well-known scientists reported to the Council on Environmental Quality, "Man is setting in motion a series of events that seem certain to cause a significant warming of world climates unless mitigating steps are taken immediately" (Pomerance 1989, 260). At the same time, the National Academy of Sciences initiated a study of the greenhouse effect. Also during 1979, in the United States, the President's Council on Environmental Quality mentioned global warming: "The possibility of global climate change induced by an increase of carbon dioxide in the atmosphere is the subject of intense discussion and controversy among scientists" (Anderson 1990).

An alarm regarding global warming was sounded in the British journal *Nature*, May 3, 1979: "The release of carbon dioxide to the atmosphere by the burning of fossil fuels is, conceivably, the most important environmental issue in the world today" (Bernard 1993, 6). At about the same time, a study conducted by a scientific team chaired by meteorologist Jule Charney estimated that doubling the carbon dioxide level in the atmosphere would raise the average global temperatures by about three degrees C., plus or minus 1.5 degrees C. Four years later, the United States Environmental Protection Agency released a report, "Can We Delay a Greenhouse Warming." A National Academy of Sciences report, also issued in 1983, stated, "We do not believe that the evidence at hand about CO_2-induced climate change would support steps to change current fuel-use patterns away from fossil fuels" (Pomerance 1989, 261).

Until the early 1980s, those who argued that infrared forcing had (or would) raise the temperature of the atmosphere near the Earth's surface had a statistical problem. Starting in 1940, until about 1975, average global temperatures actually fell slightly. After 1975, temperatures began a steady, accelerating rise, which made global warming a notable political issue by the late 1980s.

During 1985, Veerabhaadran Ramanathan, Ralph Cicerone, and their colleagues at the National Center for Atmospheric Research argued that trace gases other than carbon dioxide could be as dangerous vis-à-vis greenhouse warming as carbon dioxide. Also, by 1985, Claude Lorius was beginning to demonstrate that lower carbon dioxide levels in the atmosphere strongly correlated with low temperatures during the last ice age.

The potential impact of warming due to infrared forcing was raised again at United Nations sponsored conferences in Villach, Austria, during the middle 1980s. Shortly after these conferences, Senators David Durenberger of Minnesota and Albert Gore of Tennessee called for an international "Year of the Greenhouse" to raise the issue in public consciousness. Gore already had played a role in congressional hearings on the issue in 1982 and 1984, when he was serving in the House of Representatives.

As a political issue in the United States of America, global warming came of age during the notably hot summer of 1988. The year 1988 provided something of a wake-up call in the debate over global warming because it was the warmest year since reliable records had been kept in the middle of the nineteenth century. During 1988, 400 electrical transformers in Los Angeles blew out on a single day as temperatures rose to 110 degrees F. Two thousand daily temperature records were set that year in the United States. Widespread heat and drought caused some crop yields in the U.S. Midwest to fall between 30 and 40 percent. In Moscow, Russians escaping their hottest summer in a century flocked to rivers and lakes, and drowned in record numbers (Christianson 1999, 197).

During 1988, Colorado Senator Timothy E. Wirth, whose hearings on global warming the previous winter had drawn little attention, played the weather card. He called another hearing, this time during the summer. As it happened, the hearing convened on a particularly hot, humid day in Washington, D.C. during which the high temperature reached a record 101 degrees F. At Wirth's hearing, James E. Hansen, head of the federal government's Goddard Institute of Space Studies, testified that the unusually warm temperatures of the 1980s were an early portent of global warming caused by the burning of fossil fuels and not solely a result of natural variation. Hansen's remarks became front-page news nationwide within hours.

Hansen also continued a running battle throughout the 1980s, during the Reagan and Bush administrations, to call the science of global warming as he saw it, despite repeated threats to the funding of the Goddard Institute. The Office of Management and Budget forced Hansen to censor the severity of his findings several times. The pressure was so intense that Hansen sometimes asked to testify as a private citizen rather than as a federal employee (Hansen 1989).

In the meantime, the scientific debate over global warming was intensifying. By the end of 1988, the United Nations General Assembly had approved the creation of the Intergovernmental Panel on Climate Change (IPCC). A year later, Hansen said that it was "time to cry wolf":

When is the proper time to cry wolf? Must we wait until the prey, in this case the world's environment, is mangled by the wolf's grip? The danger of crying too soon, which much of the scientific community fears, is that a few cool years may discredit the whole issue. But I believe that decision-makers and the man-in-the-street can be educated about natural climate variability. . . . A greater danger is to wait too long. The climate system has great inertia, so as yet we have realized only a part of the climate change which will be caused by gases we have already added to the atmosphere. Add to this the inertia of the world's energy, economic, and political systems, which will affect any plans to reduce greenhouse gas emissions. Although I am optimistic that we can still avoid the worst-case climate scenarios, the time to cry wolf is here. (Nance 1991, 267–268)

Hansen elaborated,

I said three things [in 1988]. The first was that I believed the earth was getting warmer and I could say that with 99 percent confidence. The second was that with a high degree of confidence we could associate the warming and the greenhouse effect. The third was that, in our climate model, by the late 1980s and early 1990s, there's already a noticeable increase in the frequency of drought. (Parsons 1995, 7)

Between June 27 and 30, 1988, as the Earth's warmest summer on record (to that time) was getting under way, more than 300 leaders in science, politics, law, and environmental studies gathered in Toronto at the invitation of Canada's government to address problems related to climate change, including prospects of global warming. A scientific consensus was forming around the idea that human activity already was altering the Earth's atmosphere at an unprecedented rate. A consensus statement issued by the Montreal climate conference asserted that "There can be a time lag of the order of decades between the emission of gases into the atmosphere and their full manifestation in atmospheric and bio-logical consequences. Past emissions have already committed planet Earth to a significant warming" (Ferguson 1989, 48).

The climate system is never actually in thermodynamic equilibrium. Rather, it is forever playing catch-up with the daily and seasonal variations of incident sunlight, as the ground tries to come into thermal equilibrium with changes in solar radiation. This is where the heat capacity of the ground, the atmosphere, and the ocean come into play, as well as the thermal opacity of the atmosphere, which regulates how readily heat energy from the ground can be radiated to space. The time it takes the system to get to a new equilibrium is expressed in terms of a time constant, or "e-folding" time, in other words, the length of time that it takes the system to reach approximately 63 percent of its final equilibrium temperature. Mathematically, it takes forever to reach "true" equilibrium; but practically, after a few e-folding times, the system can be said to be in effective equilibrium. The e-folding time of the atmosphere is a few months; the ocean mixed layer, a few years; and the total (deep) ocean, a few hundred years. The present climate has accumulated about 0.5 watts per meter squared of unrealized warming.

In 1989, Veerabhaadran Ramanathan stated,

The rate of decadal increase of the total radiative heating of the planet is now about five times greater than the mean rate from the early part of this century. Non-CO2 trace gases in the atmosphere are now adding to the greenhouse effect by an amount comparable to the effect of CO2 increase. . . . The cumulative increase in the greenhouse forcing until 1985 has committed the planet to an equilibrium warming of about 1 to 2.5 degrees C. (Ramanathan 1989, 241)

Ramanathan continued,

The climate system cannot restore the equilibrium instantaneously, and hence the surface warming and other changes will lag behind the trace-gas increase. Current models in-dicate that this lag will range [from] several decades to a century. However, analyses of temperature records of the last 100 years as well as proxy records . . . [of] paleoclimate changes indicate that climate changes can also occur abruptly instead of a gradual return to equilibrium as estimated by models. The timing of the warming is one of the most uncertain aspects of the theory. (Ramanathan 1989, 245)

At about the same time, Peter Ciborowski projected, "[W]ithin 50 years, we will be committed to a mean global temperature rise of 1.5 degrees C. to 5 degrees C. And if no attempt is made to slow the rate of increase, we could be committed to another 1.5 degrees C. to 5 degrees C. in another 40 years" (Ciborowski 1989, 227–228).

Debate regarding the greenhouse effect intensified when temperature readings in 1990 eclipsed the record warmth of 1988. Later in the decade, 1995 became the warmest year, followed by 1997, and 1998. During most of these years, The Southern Oscillation (El Niño) weather pattern (called "ENSO" in climate-change literature) was an important factor in world weather. The ENSO involves a marked warming of the tropical ocean off the west coast of South America. The pattern occurred more often during the 1990s than at any time for the century and a half during which detailed worldwide weather records have been available.

Some atmospheric scientists have asserted that the ENSO was associated with a gradual warming of the lower atmosphere during most of the twentieth century. In 1996, Kevin E. Trenberth and Timothy J. Hoar of the National Center for Atmospheric Research pointed out that El Niño periods had occurred more frequently in the 1980s and 1990s than during the previous century (Christianson 1999, 224).

During the 1992 presidential campaign in the United States, candidate Bill Clinton criticized the Bush administration's refusal to join in worldwide diplomatic efforts to reduce emissions of greenhouse gases. Clinton's first budget proposed a carbon tax, a measure that was quickly dropped under pressure from Republicans in Congress. In Congress, the carbon tax died in committee. Meanwhile, a Climate Convention signed by 161 countries at the Conference on Environment and Development in Rio de Janeiro, during 1992, contained a directive, in Article 2, favoring stabilization of greenhouse gases "at levels and on a time scale that do not produce unacceptable damage to ecosystems and that allow for sustainable economic development" (Woodwell, *Biotic Feedbacks in the Global Climate System*, 1995, v).

At about the same time several studies (examples being Easterling et al. 1997 and Karl et al. 1993) indicated that daily minimum temperatures had increased more rapidly during much of the twentieth century than daily maxima. Easterling reports that between 1950 and the mid-1990s, daily minima have increased at a rate of 1.86 degrees C. per century, while maxima have increased 0.88 degrees C. (less than half as much) during the same period. A study by Henry F. Diaz and Raymond S. Bradley supported climate-model forecasts that daily minimum temperatures will rise more rapidly than maxima in a warmer world. Diaz and Bradley, who studied temperature changes during the twentieth century at high-elevation sites, wrote, "The signal appears to be more closely related to increases in daily minimum temperatures than changes in the daily maximum. The changes in surface temperature vary spatially, with Europe (particularly Western Europe) and parts of Asia displaying the strongest high-altitude warming during the period of record" (Diaz and Bradley 1997, 253).

A number of legislative bodies in different parts of the world took initiatives to limit greenhouse-gas emissions soon after the memorably hot summer of 1988. During 1989, the Netherlands passed a National Environmental Policy Plan, which required a freeze on carbon dioxide emissions at 1989–1990 levels. The parliament of Norway decided to limit carbon dioxide emissions in that country to 1989 levels by the year 2000, with a decline in emissions mandated after that. The Vermont State Legislature enacted a law outlawing automobile air conditioners after 1993 unless a substitute was developed to replace chlorofluorocarbons (CFCs). On June 13, 1990, just before Germany was reunified, the West German Cabinet committed the country to a 25 percent reduction in greenhouse gases, based on 1987 levels, by the year 2005. Later, the unified government stood behind these limits, but added allowances for energy-inefficient industries in what had been East Germany.

Stephen H. Schneider's *Global Warming: Are We Entering the Greenhouse Century?* (1989) was one of the first popular treatments of global warming in book form. During the next several years, Schneider became a leading scientific voice in public debates over the issue. By the late 1990s, Schneider was a professor in the Department of Biological Sciences and a senior fellow at the Institute for International Studies, at Stanford University. He was honored in 1992 with a MacArthur Fellowship for his ability to integrate and interpret the results of global climate research through public lectures, seminars, classroom teaching, environmental assessment committees, media appearances, Congressional testimony, and research collaboration with colleagues. Schneider served as a consultant to several federal agencies as well as White House staff in the Nixon, Carter, Reagan, Bush, and Clinton administrations.

Schneider received his Ph.D. in Mechanical Engineering and Plasma Physics from Columbia University in 1971. He studied the role of greenhouse gases and suspended particulate material on climate as a postdoctoral fellow at NASA's Goddard Institute for Space Studies. Schneider was awarded a postdoctoral fellowship at the National Center for Atmospheric Research (NCAR) and was a member of the scientific staff of NCAR from 1973 to 1996, where he co-founded the Climate Project. In 1975, he founded the interdisciplinary journal, *Climatic Change*, and served as its editor. Schneider also edited *The Encyclopedia of Climate and Weather* (1996); *Global Warming: Are We Entering the Greenhouse Century?* (1989); and *Laboratory Earth: The Planetary Gamble We Can't Afford to Lose* (1997).

Schneider's *Global Warming: Are We Entering the Greenhouse Century?* ends with a passage that sounded rather prescient in the year 2000:

"Are we now entering the Greenhouse Century?" I asked in the subtitle of this book. It should be clear by now that I believe we've been in it for a while already, but admit that it will take a decade or so more of record heat, forest fires, intense hurricanes, or droughts to convince the substantial number of skeptics that still abound. Unfortunately, while the antagonists debate, the greenhouse gases keep building up in the atmosphere. I wonder what we will say to our children when they eventually ask what we did—or didn't do—to create the Greenhouse Century they will inherit. (Schneider 1989, 285)

Schneider wrote, "I strongly suspect that by the year 2000 increasing numbers of people will point to the 1980s as the time the global warming signal emerged from the natural background of climatic noise" (Schneider 1989, 32). When he was asked whether human activities had assumed a dominant role in climate change, Schneider said, "I'm not 99 percent sure, but I am 90 percent sure. Why do we need 99 percent certainty when nothing else is that certain? If there were only a 5 percent chance the chef slipped some poison in your dessert, would you eat it?" (Landsea 1999). Schneider argues that everyone will be 100 percent certain that scientists' beliefs about climate change are accurate only when they observe dramatic, adverse changes in climate. By then, he asserts, it will be too late to take remedial measures.

In 1991, author John J. Nance, in *What Goes Up: The Global Assault on Our Atmosphere*, quoted Dr. Susan Soloman, an atmospheric chemist:

I can't go home and dump my garbage in my neighbor's backyard. The police would arrest me in five minutes. But [until they were banned] I could take a tank of chloro-fluorocarbons, put it in my backyard, turn it on and let it go into the atmosphere all day long, and no one [could] stop me. Somehow that's very wrong. (Nance 1991, 9)

By the year 2000, global warming was becoming a national political issue in some countries. One was Norway, where, on March 9, Prime Minister Kjell Magne Bondevik announced the resignation of his government after losing a vote of confidence in parliament. Bondevik's became the first government in the world to fall as a result of issues related to global warming. The issue was his objection to the building of several natural gas fired power stations. The minority three-party coalition government, with just 42 seats in the 165-member parliament, was bitterly at odds with the parliamentary majority over whether to build the gas fired power plants to burn some of Norway's large stocks of natural gas. The government argued that the plants would release far too much carbon dioxide linked to global warming. The government wanted to put the gas power plants on hold until more efficient, cleaner technology is developed to make natural gas fired power plants pollution free.

The opposition, an alliance of conservatives and labor, wanted to build the plants anyway, on the grounds that there is no alternative to meet the demand for electricity. Norway currently generates almost all its electricity from hydro-power, which emits no greenhouse gases. Further hydropower development is not popular in Norway because of its effects on the landscape, according to the Environment News Service (Norwegian Government 2000).

BRITAIN'S ACTIVIST ROLE IN CLIMATE-CHANGE DIPLOMACY

Britain's government has diplomatically nudged the United States on several occasions on the global-warming issue. The British government's involvement in global warming as a political issue dates to the tenure of Tory Prime Minister Margaret Thatcher, who had earned an undergraduate degree in chemistry. In 1997, as 160 nations prepared to negotiate the Kyoto Protocol, British Deputy Prime Minister John Prescott traveled to Washington, D.C. to advocate a stronger response to the problem by the United States. According to a report by the British Broadcasting Corporation,

Mr. Prescott . . . avoided direct criticism of the Clinton administration. He said simply that he had told his hosts that cutting greenhouse gases was presently the most important issue facing the world and that acceptable targets had to be agreed upon. But the British Minister of the Environment, Michael Meacher, who also recently visited Washington, was less cautious. He said it as now time for the Americans to show greater leadership. (Britain Urges 1997)

Britain's environmental secretary, John Gummer, called for an emissions-reduction target of 50 percent, beginning with an end to all subsidies for oil and coal use, as Britain's *Financial Times* pinned the ineffectiveness of the Kyoto Protocol on resistance from established interests in the United States:

Right now the final protocol promises to be a sad affair given the power of the oil and coal lobby in the United States, which has still failed to meet its obligations under the Rio Treaty. . . . It is therefore a shame that the U.S. government lacks the courage to admit that while some jobs in some industries may be lost, new jobs will more than compensate. (Global Warming 1997)

During March, 1999, the European Union called for a 15 percent reduction in carbon dioxide, methane, and nitrous oxide emissions (using 1990 levels) by 2010. At its December 15, 1999 meeting, the European Parliament passed a resolution supporting speedy approval of the Kyoto Climate Protocols. The resolution called for a carbon tax, as well as for emission controls. The lack of legislative action on global warming in the United States has been, by contrast, notable by its absence. According to United Nations figures, carbon dioxide emissions in Britain fell roughly eight percent between 1990 and 1998, while emissions rose 10.7 percent in the United States, 9.5 percent in Japan, and 12 percent in Australia (Meacher 2000).

SCIENTIFIC CONSENSUS

The public policy debate regarding global warming has often conveyed an impression that scientists are hopelessly divided over the issue of whether human activities are warming the lower atmosphere. In actuality, a high degree of agreement has existed since the IPCC's *First Assessment* was published in 1990. The IPCC's first major report forecast widely varying temperature rises by region with an assumed doubling of carbon dioxide in the atmosphere. The largest increases (six to seven degrees C.) were forecast in the interiors of northern North America and Asia during the winter; increases in the summer for the same regions were forecast at between three and four degrees C. The largest summer temperature increase (4.8 degrees C.) was forecast for interior southern Asia. The smallest increases year-round were forecast for the tropics, especially areas near large bodies of water.

An IPCC conference during November, 1990, at Geneva, Switzerland, issued a "ministerial declaration" representing 137 countries which agreed that while climate had varied in the past, "[t]he rate of climate change predicted by the IPCC to occur over the next century [due to greenhouse warming] is unprecedented." The ministers declared, "[C]limate change is a global problem of unique character" (Jager and Ferguson 1991, 525). The ministers also declared that the eventual goal should be "to stabilize greenhouse gas concentrations at a level that would prevent dangerous anthropogenic interference with climate" (Jager and Ferguson 1991, 536).

The question of whether the Earth is becoming unnaturally warmer because of human activities was largely settled in scientific circles by 1995, with publication of the *Second Assessment* of the Intergovernmental Panel on Climate Change (IPCC), a worldwide group of about 2,500 experts. The panel concluded that the earth's temperature had increased between 0.5 and 1.1 degrees F. (0.3 to 0.6 degrees C.) since reliable worldwide records became available between 1850 and 1900. The IPCC noted that warming accelerated as measurements approached the present day (Bolin et al. 1995).

The IPCC's *Second Assessment* concluded that human activity—increased generation of carbon dioxide and other greenhouse gases—is at least partially responsible for the accelerating rise in global temperatures. The amount of carbon dioxide in the atmosphere has been rising nearly every year due to increased use of fossil fuels by ever-larger human populations experiencing higher living standards. The IPCC's *Second Assessment*, according to one observer, "makes an unprecedented, though qualified, attribution of the observed climate change to human causes. Though the human signal is still building and somewhat masked within natural variation, and while there are key uncertainties to be resolved, the Panel concludes that 'the balance of evidence suggests that there is a discernible human influence on global climate' " (Landsea 1999).

In its *Second Assessment*, the IPCC declared,

During the past few decades, two important factors regarding the relationship between humans and the Earth's climate have become apparent. First, human activities, including the burning of fossil fuels, land-use change and agriculture, are increasing the atmospheric concentrations of greenhouse gases (which tend to warm the atmosphere) and, in some regions, aerosols (microscopic airborne particles, which tend to cool the atmosphere). These changes in greenhouse gases and aerosols, taken together, are projected to change regional and global climate and climate-related parameters such as temperature, precipitation, soil moisture and sea level. Second, some human communities have become more vulnerable to hazards such as storms, floods and droughts as a result of increasing population density in sensitive areas such as river basins and coastal plains. Potentially serious changes have been identified, including an increase in some regions in the incidence of extreme high-temperature events, floods and droughts, with resultant consequences for fires, pest outbreaks, and ecosystem composition, structure and functioning, including primary productivity. (Bolin et al. 1995)

The *Second Assessment* indicated that rising atmospheric concentrations of greenhouse gases would cause interference with the climate system to grow in magnitude, as "the likelihood of adverse impacts from climate change that could be judged dangerous will become greater" (Bolin et al. 1995). Because of these dangers, the IPCC called upon the governments of the world to "take precautionary measures to anticipate, prevent or minimize the causes of climate change and mitigate its adverse effects" (Bolin et al. 1995).

The IPCC linked expected climate changes to rising levels of several greenhouse gases in the atmosphere following the rapid spread of fossil-fueled industry. These gases included carbon dioxide (from about 280 to almost 360 parts per million as of 1992); methane (from 700 to 1720 parts per billion); and nitrous oxide (from about 275 to about 310 parts per billion). While aerosols may have a short-term impact in some areas, the IPCC's *Second Assessment* said that their short-lived nature does little to mitigate the effects of the greenhouse gases, which are resident in the atmosphere for hundreds of years.

THE COMPLEXITY OF ATMOSPHERIC CHEMISTRY AND DYNAMICS

Determining the net climatic effect of a given mix of greenhouse gases is no simple matter, in part because the mix is so complex. Some gases (an example is carbon dioxide) may continue to increase in the atmosphere, while others (an example during the 1990s was methane) may stabilize or even decline. Other gases (an example being nitrous oxides) may counteract some of the effects of others.

After doubling from pre-industrial times to about 1980, the proportion of methane in the atmosphere parted paths with carbon dioxide during the 1990s, as its rate of accumulation slowed markedly, and, during 1992 and 1993, actually declined. According to a June 1999 report by E.J. Dlugokencky and associates in the British scientific journal *Nature*, methane levels may have declined during

1992 and 1993 for several reasons. Mount Pinatubo's eruption in 1991 caused a cooling of global temperatures; cooler temperatures caused less methane than usual to be released from boreal wetlands.

At the same time, the countries comprising the former Soviet Union continued a decade-long decline in methane emissions from oil and natural gas production due to social and political collapse. In flusher times, the Soviet Union's oil and gas drillers vented immense amounts of methane from their operations. Decreased biomass burning in the tropics also may have contributed to the decline in atmospheric methane. The scientists who disclosed the news that methane levels have stabilized speculate that atmospheric levels will rise slowly again for a few decades, then stabilize (Dlugokencky et al. 1999, 449). While scientists can measure the level of methane in the atmosphere, they have difficulty accounting for increases or decreases in that level because methane sources are many, varied, and often too small to detect in global-scale calculations.

The interactions of various greenhouse gases in the atmosphere are so complex, in some cases, that scientists are only now beginning to understand them. Mark G. Lawrence and Paul J. Crutzen surveyed the role of nitrogen oxides in the troposphere, or lower atmosphere, for example, and found that nitrous oxide "would be expected to reduce the atmospheric lifetimes of greenhouse gases— such as methane—as well as to increase aerosol production rates and cloud reflectivities, therefore exerting a cooling influence on the climate" (Lawrence and Crutzen 1999, 168). According to Lawrence and his co-author, human activities account for about half the atmosphere's nitrogen oxides. Their investigation centered on nitrogen oxide emissions by ships, which eject enough of the chemicals into the air of frequently used shipping lanes to raise nitrogen oxide levels to as much as 100 times natural "background" levels (Lawrence and Crutzen 1999, 168).

WHITHER THE MISSING CARBON DIOXIDE?

The term "missing sink" with reference to the global carbon budget was coined by ocean modelers W.S. Broecker, U. Siegenthaler, and H. Oeschger during the late 1970s when they were unable to account for all the carbon released into the atmosphere from fossil-fuel combustion and land-use changes in oceanic sinks. They were uncomfortable attributing the unaccounted-for carbon to a terrestrial sink without having a detailed model to explain terrestrial biosphere carbon storage mechanisms (Broecker, Fate of Fossil Fuel, 1979; Siegenthaler and Oeschger 1978). Scientists do not yet completely know how much of the carbon dioxide that human activity produces is stored in the ocean. There is about 60 times as much carbon dioxide in the oceans as in the atmosphere, so the interaction of the gas in the atmosphere and in the ocean is a very important element on a worldwide scale (Silver and DeFries 1990, 38).

Of the more than seven billion tons of carbon dioxide being pumped into the atmosphere per year by human activity at the dawn of the third millennium on the Christian calendar, roughly half remained in the air. Scientists can account for about two gigatons (billion tons), which is absorbed by various carbon "sinks," mainly the oceans. A major mystery in this debate is the "missing carbon," roughly 1.5 gigatons (ggt) a year, which has evaded source detection. Plants (mainly trees) may be absorbing some of it, but with deforestation, that "sink" should be diminishing. "The mystery of the missing carbon . . . is cause for concern," writes S. George Philander. "How can we be sure that the unidentified sink is not becoming saturated? If that should happen, [the] atmospheric carbon-dioxide level will soon increase more rapidly" (Philander 1998, 197).

Some speculation has arisen that a "missing sink" of carbon exists somewhere in the terrestrial ecosystem. The size and location of such a carbon "sponge" is important to climate diplomats because, under the Kyoto Protocol, countries that harbor such carbon "sinks" could use them to balance their emissions. By 1998, some scientific argument was tending toward acceptance of a very large carbon sink over inland North America which indicates, according to Jocelyn Kaiser, that "North America may have drawn the winning card in the carbon-sink sweepstakes" (Kaiser 1998, 386).

Nature may not intend irony, but its imprint here is visible to anyone who studies the politics of climate change. The proposed North American carbon sink, which is posited by S. Fan and associates (1998), is large enough, according to Kaiser, "to suck up every ton of carbon dioxide discharged annually by fossil-fuel burning in the United States and Canada" (Kaiser 1998, 386; Fan et al. 1998). The sink proposed by Fan and colleagues is large enough to soak up nearly a quarter of the more than six billion tons of carbon generated each year by human activity worldwide. The missing sink appears to center around the Great Lakes area (Fan et al 1998). A later study by David Schimel and colleagues suggests that "processes such as regrowth on abandoned agricultural land or in forests harvested before 1980 have effects as large or larger than the direct effects of CO2 and climate" (Schimel et al. 2000, 2004). This study of net carbon storage in terrestrial ecosystems of the United States between 1895 and 1993 also suggests that the contributions of these factors may change by 100 percent year to year as a result of climate variability. While this study supports the general thrust of the Fan group, the size of the carbon sink it projects is "an order of magnitude" (e.g., one-tenth) less than that projected by Fan and associates (Schimel et al. 2000, 2006).

No general consensus exists regarding what might be causing massive carbon absorption around the Great Lakes; some of it could be caused by new growth on formerly forested lands, or previously unexamined absorption by wetlands and soils. Carbon uptake also may be increasing because of intensive application of nitrogen fertilizers in an area that is intensely farmed. The Fan group also may have overstated its results through errors in methodology. The results also may have been skewed by the eruption of Mt. Pinatubo in 1991.

The North American carbon sink is far from a scientific done deal, however. None of the proposed causes account for its size, and even the authors of this controversial idea admit that "its magnitude remains uncertain and its cause unknown" (Fan et al. 1998, 445). Several other scientists have criticized this study's methodology, and another study (Phillips et al. 1998) indicates another, very large, carbon sink over tropical South America. This study indicates that biomass is being generated in tropical forests more quickly than it is being consumed by deforestation, thus "reducing the rate of increase in atmospheric carbon dioxide" (Phillips et al. 1998, 439). About 40 percent of the Earth's biomass exists as tropical forests, so Phillips and associates believe that "a small perturbation in this biome could result in a significant change in the global carbon cycle" (Phillips et al. 1998, 439).

The closer scientists look at the carbon cycle, the more reasons some find to doubt previous assumptions. For example, most models of forest response to increased greenhouse-gas levels assume that increasing temperatures will provoke the decomposition of organic matter which will raise the carbon dioxide levels in the lower atmosphere, eventually raising temperatures. This assumption is reflected in research cited in previous pages of this volume. In April, 2000, however, Christian P. Giardina and colleagues made a case in *Nature* that this assumption may be mistaken.

In two separate studies (Giardina and Ryan 2000; Valentini et al. 2000), new measurements of carbon flux from individual forests indicated that "decomposition of organic matter is not very sensitive to temperature" (Grace and Rayment 2000, 820) and "[soil] respiration is [an] important component of the carbon balance in northerly latitudes despite the cold temperatures there" (Grace and Rayment 2000, 820). Giardina and Ryan compiled measurements from 82 sites on five continents, while Valentini and colleagues obtained their measurements from 15 European forests. If these results are confirmed by other studies, climate modelers may be forced to revise their work. "Does this mean that the doomsday view of runaway global warming now seems unlikely? We hope so," write John Grace and Mark Rayment (Grace and Rayment 2000, 321).

Roger M. Gifford and colleagues assert that increasing temperatures and carbon dioxide levels in the atmosphere will, up to a point, cause creation of greater plant mass, which will absorb a significant amount of the additional carbon. They contend that the ability of plants to absorb additional carbon in this manner "is now well below saturation" (Gifford et al. 2000, 88). They add:

The modeled magnitude of the CO2 fertilizing effect on net C [carbon] storage by the biosphere . . . is approximately the amount needed to account for the missing sink in the global C budget. Given this fact, the option must remain open that the terrestrial biosphere is responding to the increasing global atmospheric CO2 concentration, with support in some areas from the deposition of anthropogenic N [nitrogen]. (Gifford et al. 2000, 88)

THE RELATIONSHIP BETWEEN A GIVEN RISE IN ATMOSPHERIC GREENHOUSE GASES AND A GIVEN RISE IN TEMPERATURE

Atmospheric scientists know that the Earth's surface temperatures often have varied more or less in tandem with the atmosphere's level of carbon dioxide and other greenhouse gases. What they do not know precisely is *how much* of a temperature increase may be triggered by a given rise in the level of carbon dioxide and other greenhouse gases. Atmospheric modeling is at once very simple, because the general relationship between heat-absorbing gases and temperature is known, and very complex, because climate can be influenced by many other factors as well.

Although the level of carbon dioxide in the atmosphere usually rises and falls with temperature, paleoclimatologists have found some specific epochs in the Earth's history during which glaciation occurred when carbon dioxide levels were several times higher than today, even at human-enhanced levels. One such example is the Late Ordovician (Hirnantian) glaciation during the early Paleozoic, a rare spike of cold weather during an otherwise balmy period when atmospheric carbon dioxide levels sometimes reached 14 to 16 times pre-industrial levels. Mark T. Gibbs and colleagues assert that positioning of the continents at that time may have contributed to this most unusual of ice ages (Gibbs et al. 2000, 386).

Adding another intriguing wrinkle to the paleoclimatic puzzle, investigators have found evidence that sometimes appears to decouple carbon dioxide levels in the atmosphere from climate change. For example, M. Pagani, M.A. Arthur, and K.H. Freeman (1999) report in *Paleoceanography* that during the Miocene Climatic Optimum (roughly 14.5 to 17 million years ago), the Earth experienced its warmest climate in 35 million years when temperatures averaged about six degrees C. higher than today. This level of warmth was achieved with atmospheric carbon dioxide levels between 180 and 290 parts per million, compared to the present day level of about 370 p.p.m. The same researchers also found that rising, not falling, carbon dioxide levels accompanied growth in the East Antarctic Ice Sheet between 12.5 and 14 million years ago. Such findings do not necessarily contradict the infrared-forcing capacities of carbon dioxide itself but indicate that other factors (including the levels of other gases, such as methane, and effects of changes in oceanic circulation patterns) compete with carbon dioxide levels to shape the climate of any particular place on the Earth at any given time.

By the late 1990s, scientists lacking solid theoretical models which establish a direct causal link between carbon dioxide level (by itself) and temperature had largely stopped forecasting that a particular carbon dioxide level will cause a specific temperature rise. Instead, increasingly sophisticated studies are attempting to describe what role greenhouse gases will play in the context of other influences, or "forcings," in the atmosphere.

As more scientists examine the dance of climate change, they find more possible influences which complicate the identification of a specific, quantifiable role for infrared forcing on its own. For example, in the April 11, 2000 edition of the *Proceedings of the National Academy of Sciences of the United States of America*, Charles D. Keeling and Timothy P. Whorf propose that a 1,800-year oceanic tidal cycle is influencing climate. "We propose that such abrupt millennial changes, as seen in ice and sedimentary core records, were produced at least in part by well-characterized, almost periodic variations in the strength of the global oceanic tide-raising forces caused by resonances in the periodic motions of the Earth and Moon" (Keeling and Whorf 2000, 3814). Increased tidal turbulence influences the rate of "vertical mixing" in the oceans and, thus, their surface temperature. Greater oceanic mixing cools the surface of the oceans, and exerts an influence on global average temperatures. The last peak in this cycle is said, by Keeling and Whorf, to have coincided with the "Little Ice Age" which climaxed about 1600 A.D. (Keeling and Whorf 2000, 3814).

ENDURING IMPACTS OF GLOBAL WARMING

When considering the steady rise in atmospheric carbon-dioxide levels, it is crucial (and sobering) to realize that these increases are "essentially irreversible over periods of hundreds of years" (Cline 1992, 17). Moreover, the IPCC estimated in its *Second Assessment* (1995) "an immediate reduction [in emissions] of 50 to 70 percent and further reductions thereafter" to keep carbon dioxide levels in the atmosphere from rising above present levels (Bolin et al. 1995). The IPCC presented similar scenarios for other greenhouse gases, as it warned, "The stabilization of greenhouse-gas concentrations does not imply that there will be no further climate change. After stabilization is achieved, global mean surface temperature would continue to rise for some centuries and sea level for many centuries" (Bolin et al. 1995).

By the year 2000, levels of carbon dioxide and methane in the Earth's atmosphere had risen higher than the upper range of natural concentrations during warm spells between glaciations. Until the year 2000, the "Keeling Curve" resided on uncharted atmospheric ground as far back in time as human measurement extends. By early in the year 2000, that record, from Antarctic ice cores, extended to 420,000 years before the present.

As if to illustrate how quickly baseline scientific knowledge has been evolving in this area, Paul Pearson of the University of Bristol and Martin Palmer of Imperial College, London, reported in *Nature*, August 17, 2000, that they have developed proxy methods for measuring the atmosphere's carbon dioxide level to 60 million years before the present. Their records considerably extend the 400,000-plus year record of ice cores. The upshot of Pearson and Palmer's studies is their conclusion that carbon dioxide levels at the year 2000 were as high as they have been in at least the last 20 million years. According to their records, however, carbon dioxide levels reached the vicinity of 2,000 p.p.m.

during "the late Palaeocene and earliest Eocene periods (from about 60 to 52 million years ago)" (Pearson and Palmer 2000, 695).

Pearson and Palmer used plankton shells drilled from the seabed to estimate the acidity (and thus the carbon content) of sea water over a span of time back almost to the era of the dinosaurs. By 2100, at present rates of increase, their figures indicate that the carbon dioxide level of the Earth's atmosphere may match the level last seen in the Eocene, about 50 million years ago. At that time the Earth had no permanent pack ice, and (as characterized by one English newspaper) "London was a steaming mangrove swamp" (Radford 2000, 9).

As levels of greenhouse gases have risen, concern has been expressed by a number scientists and policy makers that the Earth may be entering a period of rapid, human-induced warming which may damage animal (including human) and plant life. The IPCC's *Second Assessment* indicated that deserts "are likely to become more extreme—in that, with few exceptions, they are projected to become hotter but not significantly wetter" (Bolin et al. 1995). Temperature increases could threaten organisms, such as oceanic corals, which already live near their heat tolerance limits. Large areas of land may become deserts, a process which often takes the land involved to an ecological dead-end: "Land degradation in arid, semi-arid and dry sub-humid areas resulting from various factors, including climatic variations and human activities, is more likely to become irreversible if the environment becomes drier and the soil becomes further degraded through erosion and compaction" (Bolin et al. 1995).

In aquatic and coastal ecosystems, such as lakes and streams, warming would, according to the IPCC,

have the greatest biological effects at high latitudes, where biological productivity would increase, and at the low-latitude boundaries of cold- and cool-water species ranges, where extinctions would be greatest. The geographical distribution of wetlands is likely to shift with changes in temperature and precipitation. . . . Some coastal ecosystems are particularly at risk, including saltwater marshes, mangrove ecosystems, coastal wetlands, sandy beaches, coral reefs, coral atolls and river deltas. (Bolin et al. 1995)

REGIONAL VARIATIONS OF WARMING MODELED

The IPCC's *Second Assessment* (Bolin et al. 1995) suggests that a doubling of the carbon dioxide level in the atmosphere could raise temperatures an average of 1.9 degrees C. to 5.2 degrees C. These figures are global averages which leave a great deal of room for regional variations. In general, temperatures are expected to rise most dramatically in the Arctic and Antarctic, with smaller increases nearer the equator. Temperatures also are expected to rise more in winter than in summer, and more at night than during the day.

In the North American Arctic, according to IPCC models, temperature rises may range from seven degrees C. in the lower Mackenzie Valley to more than nine degrees C. over the Arctic islands during winter, with an average summer rise of three to four degrees C. "The northward movement of permafrost could further increase the emissions of greenhouse gases into the atmosphere through the release of CO_2 from soils that are highly organic, such as in Siberia," Barrie Maxwell comments (Maxwell 1992, 23–24). "If the soils thaw more deeply, however, methane emissions might also be reduced" because microbes will oxidize some of the methane. Maxwell notes that none of these biochemical feedbacks had been factored into existing climate models by the early 1990s (Maxwell 1992, 29).

During the late 1980s, climate researchers at England's University of East Anglia attempted to sketch conditions in a warmer world by compiling the records of a set of warm years and a set of cold years during the twentieth century. The scientists at East Anglia subjected 115 years of global temperature data to statistical analysis, and compared the result to computer climate models. They found that the correlations were far stronger for the actual temperature data than for the simulations taken from the two models. The implication of their results, the authors say, is that this century's anthropogenic greenhouse-gas-induced warming trend has overpowered natural variability.

The results of the East Anglia study corresponded rather closely to those obtained by the IPCC's models. The greatest warming was expected in the polar regions, with a few areas, including India, the Middle East, much of Mesoamerica, and parts of Southeast Asia, expected to experience a slight cooling trend as the rest of the world warms. Precipitation may change in a more haphazard fashion, according to the East Anglia models—increasing along the United States coasts, and in northern North America, Scandinavia and the Baltic States, northeastern Africa and the Middle East, India, and parts of China. Decreases are expected by this model over most of the rest of Europe and Asia, the center of North America, and Mexico (Gribben, *Hothouse Earth*, 1990, 218).

Volcanic eruptions were found by the East Anglia models to be too infrequent to affect long-term climate significantly. The models also suggested that variations in solar output during the last century could have been large enough to influence some long-term temperature trends, however. While variations in the sun's energy output has a role in this natural equation, the East Anglia study concluded, "Solar forcing alone is insufficient to explain the behavior of the observed temperature data" (Kirby, Climate Change, 1998). The study indicated that combining solar input with changes in greenhouse gas levels produced generally credible results. Wigley said that the results "strengthen yet further our confidence that there has been a discernible human influence on climate. . . . Furthermore, they provide additional evidence that the models used to make projections of future climate change are realistic" (Kirby, Climate Change, 1998).

Changes in landscape may modify local climates significantly. Scientists have begun to model the effects on climate of landscape changes, such as the expected northward movement of boreal forests with warming temperatures. "Because boreal forests absorb much more solar radiation than tundra does, poleward shifts in the location of the forest-tundra boundary during a period of warming can amplify climate changes by as much as fifty percent" (Melillo 1999, 138). Other studies have found that over-grazed land south of the Mexican border with Arizona experiences temperatures as much as four degrees C. warmer than nearby land on the United States side of the border which had not been grazed. The spread of irrigated farming on the eastern slope of the Rocky Mountains in Colorado has caused temperatures in some locations to fall by as much as two degrees C. Deforested areas of the Amazon Valley create "hot spots" on satellite photographs, with temperatures averaging one degree C. higher (with 30 percent less rainfall) than nearby areas which have not been logged and replaced by grassland. One observer reported in *Science* that some deforested areas of the Amazon resembled "a lunar landscape" (Couzin 1999, 317). In south-central Asia, irrigation has drained the Aral Sea, making summers hotter, winters colder, and the entire annual cycle drier.

Studies by R.S. Cerveny and Robert Balling support the idea that human-induced urban warming (along with increases in anthropogenic water vapor) influence downwind storminess. Cerveny and Balling examined 16 years of storm data for the eastern seaboard of the United States, and found that 20 percent more precipitation falls on weekends than week days. They contend that pollution generated in urban areas during the work week creates condensation nuclei for precipitation the following weekend (Cerveny and Balling, 1998; Couzin 1999, 318–319).

BIOTIC FEEDBACKS:
WILL GLOBAL WARMING FEED UPON ITSELF?

Human-induced warming of the Earth may be working in tandem with several natural feedback mechanisms to accelerate climate change through biotic feed-backs. The possibility that human-induced warming way feed upon itself pro-duces a special sense of urgency in many climate scientists' public statements. Along with a sense of urgency, there is the possibility of biotic "surprises," which infuses a high degree of uncertainty into all forecasts of global warming's possible effects.

Biotic feedbacks can enhance warming (as when deforestation decreases the Earth's ability to produce oxygen) or not (as when a jungle is replaced by a higher-albedo desert). The most important biotic feedback factors involve water vapor, the amount and nature of cloudiness, and changes in the Earth's reflec-tivity (albedo), most notably from high-reflection ice and snow vis-à-vis open ocean water (with, for example, a decrease of ice cover in the Arctic), which ab-sorbs a higher proportion of incoming solar radiation. Feedbacks magnify the tem-perature change that would be produced by a radiative forcing, such as a carbon dioxide increase, in the absence of feedbacks (Hansen et al. 1984; Boyle 1999).

George M. Woodwell, one of the world's most respected experts on biotic-feedback mechanisms, has written,

A significant body of experience . . . suggests that there are mechanisms entrained by a change in global climate that tend to increase the trend of temperature change. . . . [T]here is a possibility that the warming itself may cause a series of further changes in the earth that will speed the warming. . . . The most serious questions have to do with the potential for . . . surprises, especially surprises which lead to positive feedbacks. The potential appears significant. (Woodwell et al. 1995, 393, 406)

Woodwell explains how global warming could feed powerfully upon itself:

Disruptions of forests globally, especially in the higher latitudes of the northern hemisphere, will lead to a significant increase in the release of carbon into the atmosphere. That release can easily be in the range of 1–2 billion tons per year. It means that allowing the warming to progress leads to a potential for surprises. That's only one of the surprises.

If that were to occur it would mean that correcting the problem would be even more difficult than it is at the moment. The releases from the combustion of fossil fuels at the moment (1996) are about 6 billion tons a year. The annual accumulation is 3–4 billion tons a year. Stabilizing the composition of the atmosphere would require removing from current releases something of the order of 3 billion tons, perhaps a little more, immediately. That's a half or more of the current releases of fossil fuels—a very important challenge.

Warmer temperatures speed the decay of organic matter in soils. Forests . . . and tundra soils of the Northern Hemisphere contain sufficient carbon to add significantly to the annual emissions, thereby speeding the accumulation of carbon dioxide in the atmosphere. Such a positive feedback has not been incorporated into current estimates of the warming. It is one of several potential surprises lurking in the wings as warming proceeds. (Epstein et al. 1996)

In cold water, for example, methane clathrates form crystal structures which are somewhat similar to water ice. Warming temperatures could destabilize the clathrates, and release some of their stored methane. Roughly 10 trillion tons of methane is trapped under pressure in crystal structures in permafrost or on the edges of the oceans' continental shelves, "the Earth's largest fossil-fuel reservoir," according to Gerald Dickens, a geologist at James Cook University in Townsville, Australia (Pearce 1998). The greenhouse potential of all the methane stored in clathrates on the continental shelves and in permafrost worldwide is roughly equal to that of all the world's coal reserves (Cline 1992, 34). Some of the land masses which host these deposits already have warmed two to four degrees C. (four to seven degrees F.) during the twentieth century (Lachenbruch and Marshall 1986).

Atmospheric scientist Roger Revelle has estimated that, with a three degree C. rise in global average temperature, methane emissions from clathrates would increase half a gigaton per year worldwide. Over a century, this rate could be enough to double the amount of methane in the atmosphere. Add to this another 12 gigatons of methane that could be released by clathrates liberated from ocean bottoms under the Arctic Ocean once the ice cap now covering them melts. "It is possible," writes Jonathan Weiner, "that [this] . . . feedback effect is already underway and the rise in Earth temperatures in the last hundred years has already sprung many gigatons of methane from their molecular prisons at the bottom of the sea" (Weiner 1990, 118).

Woodwell writes,

If, for instance, a sufficient decline in the water table occurs in the boreal and tundra peatlands, subterranean fires could speed oxidation of the peat in the vast, remote peat-lands of Canada and Russia, spewing forth smoke, CO2, and CH4 [methane], throughout the northern hemisphere for years. . . . [I]f water tables remain high, these peatlands might shift toward the production of CH4 at high rates. (Woodwell 1995, 406–407)

Woodwell and colleagues, writing in *Climatic Change*, contend that while terrestrial ecosystems may have absorbed some of the increased carbon gener-ated by human activity during most of the twentieth century, "The recent rate of increase in temperature . . . leads to concern that we are entering a new phase in climate, one in which the enhanced greenhouse effect is emerging as the dominant influence on the temperature of the Earth" (Woodwell et al. 1998, 495).

Biotic feedbacks were discussed in Paris, during early December 1998, at a conference on climate variability organized by the World Meteorological Or-ganization. At this conference, Stephen Schneider warned that the permafrost of Siberia and Arctic North America may already be melting and releasing methane into the air because global warming is occurring as quickly in Siberia as any-where else on the planet.

Scientists who study biotic feedbacks sometimes remind themselves that while models are linear, nature can behave in random ways which confound linear analysis. The speed and geographic variability of ozone depletion surprised many scientists who had studied the anticipated effects of chlorofluorocarbons (CFCs) in theory. It is believed that the role of biotic feedbacks in global warm-ing could be similarly surprising. In its *Second Assessment* (Bolin et al. 1995), the IPCC stated that nonlinear systems, "when rapidly forced," are particularly subject to unexpected behavior ("surprises"). Examples of such "surprises," ac-cording to Schneider, may "include rapid decrease in the thermohaline circula-tion in the North Atlantic Ocean, excitation of certain dynamical modes of response of the climate system, rapid decarbonization of terrestrial ecosystems (e.g., forest die-back in fires or insect outbreaks), [and] catastrophic deglaciation of ice shelves in the West Antarctic" (Schneider N.d.).

Atmospheric levels of carbon dioxide, which increased between 1.5 and 2.4 parts per million per year during the later years of the twentieth century, build like a bank account compounding interest. Every five years, the total from which the range of increase is calculated rises by about eight percent. Add to this the fact that soils tend to release more carbon dioxide and methane naturally under warmer conditions. In scientific language, "Any changes that increase temperature or reduce pressure may liberate CH4 from hydrate. . . . The major potential dangers include massive emission from Arctic hydrate, especially in western Siberia" (Nisbit and Ingham 1995, 193, 212). Higher temperatures accelerate the oxidation rates of sulfur dioxide and nitrogen oxide to sulfuric and nitric acids, the precursors of acid rain.

Warming temperatures may change the behavior of the Earth's hydrological cycle. Warmer ocean water removes less carbon dioxide from the atmosphere than cooler water, so warming of the oceans may feed upon itself in coming years. Water vapor is also a potent absorber of heat in the atmosphere. It has been estimated that a doubling of carbon dioxide in the atmosphere would increase its water content by about 30 percent, raising temperatures an additional 1.4 degrees C. (Hansen et al. 1981, 957) Many models project a rise in cloudiness, and attendant atmospheric moisture, in a warmer, more humid world. George M. Woodwell raises the possibility of a rapid surge in global warming beyond any possibility of human control:

The possibility exists that the warming will proceed to the point where biotic releases from the warming will exceed in magnitude those controlled directly by human activity. If so, the warming will be beyond control by any steps now considered reasonable. We don't know how far we are from that point because we do not know sufficient detail about the circulation of carbon dioxide among the pools of the carbon cycle. We are not going to be able to resolve those questions definitely soon. Meanwhile, the concentration of heat-trapping gases in the atmosphere rises. (Woodwell 1990, 130)

Given Woodwell's expectations, the peoples of the Earth in the year 2000 are approaching a point of no return with regard to biotic feedbacks. Deforestation is accelerating around the world due to growing populations and levels of material affluence. Use of fossil fuels, which has increased at an annual rate of roughly five percent during most of this [20th] century, shows no signs of stabilizing, much less falling by half in the next 30 years. China, alone, projects burning enough fossil fuel (mainly coal) by 2025 to account for about half the present consumption of fossil fuels by everyone on Earth (Leggett 1990, 27).

THE "METHANE BURP" HYPOTHESIS

Researchers who have drilled into sediment layers near the east coast of Florida found evidence that melting methane clathrates thawed suddenly (over the course of a few thousand years) about 55 million years ago, initiating a sudden episode of global warming which ended with crocodiles and palm trees in the Arctic. At the peak of this episode, greenhouse-gas levels in the atmosphere were between two and six times as high as at present. Lisa Sloan, a paleoclimatologist at the University of California (Santa Cruz), and Gerald Dickens, a paleoceanographer at James Cook University in Queensland, Australia (two of several scientists who conducted the study), reported the findings at a meeting of the American Geophysical Union late in 1999.

The study of methane clathrates has become more popular in recent years, as evidence accumulates that their release, especially from oceans, may be a major driving force in Earth's climate cycles. James P. Kennett and colleagues studied climate records for the last 60,000 years off Santa Barbara, California, and parts of Greenland, finding that "surface and bottom temperatures change in concert" (Kennett et al. 2000; Blunier 2000, 68). This finding supports assertions by E.G. Nisbit (1990) that massive release of oceanic methane from clathrates have played a significant role in rapid warmings during the past, even without added forcings by human industry.

Scientists have yet to reach any sort of consensus on causes of the Earth's "methane burps." No one yet knows why a trillion tons of methane may be released so suddenly from solid methane hydrates around the world. This chemical reaction provoked a sudden (in geologic time) global warming of four to eight degrees C. James Cook and Gerald Dickens theorize, "The methane probably oxidized to form carbon dioxide. which eventually reached the atmosphere, driving greenhouse warming" (Kerr, Smoking Gun, 1999, 1465).

The sediment cores drilled by Katz and associates contained remnants of small marine organisms called *foraminifera*, which preserve a record in their shells of carbon levels in the ocean. The shells tell a story of an extreme warming (possibly more than 10 degrees F.) in the ocean over a short time, which killed more than half of the *foraminifera*. The sediment core also contains evidence of an underwater landslide which scientists believe took place as melting methane clathrates "warmed dramatically, breaking apart into water and methane gas, and bubbl[ed] ferociously out of the sea floor" (Witze 2000, 4).

This line of reasoning was supported by Richard Norris of the Woods Hole Research Center, and Ursula Rohl of Germany's University of Bremen, who wrote in *Nature* that the "methane burp" occurred when an as-yet-unknown natural provocation pumped greenhouse gases into the atmosphere, causing a sudden bout of global warming: "Our results suggest that large natural perturbations to the global carbon cycle have occurred in the past . . . at rates that are similar to those induced today by human activity" (Norris and Rohl 1999, 775–778). Miriam E. Katz, Dorothy K. Pak, Gerald R. Dickens, and Kenneth G. Miller assert that this surge in global temperatures may have played a crucial role in the evolution of warm-blooded mammals as the Earth's dominant species 10 million years after a cataclysmic event (probably the impact, on the Earth, of a very large asteroid) ended dominance by the dinosaurs. Katz and colleagues contend that "elevated temperatures quickly opened high latitude migration routes for the widespread dispersal of mammals" (Katz et al. 1999, 1531)

Stephen P. Hesselbo and colleagues reported that roughly 140 to 200 million years ago, large quantities of methane were liberated from ocean floors, possibly because of warming global temperatures. This "methane pulse"—a "voluminous and extremely rapid release of methane from gas hydrate contained in marine continental-margin sediments" (Hesselbo et al. 2000, 392)—combined with oxygen in the oceans to form carbon dioxide, accelerating the worldwide warming. Along the way, a large proportion of oceanic animal life (perhaps 80 percent) died for lack of oxygen. "One of the important questions that is debated a lot today is the stability of this methane hydrate reservoir, and how easy it is to release the methane," said Stephen P. Hesselbo, lead author of the paper. "The extinction and the association with lack of oxygen has been fairly well established, but the association with methane release is something that hadn't been realized before," he said (Prehistoric 2000, 9).

NEW FORECASTS OF TEMPERATURE RISE

A study issued by the Pew Center on Global Climate Change in 1999 asserts that temperatures will rise somewhat more by the year 2100 than forecast in 1995 by the Intergovernmental Panel on Climate Change's *Second Assessment*. The study, titled *The Science of Climate Change: Global and U.S. Perspectives*, was researched and written by Tom Wigley of the National Center for Atmospheric Research for the Pew Center.

The Pew Center study projects global-mean temperature increases ranging from 1.3 to 4.0 degrees C. (2.3 to 7.2 degrees F.), compared with the IPCC's projections of 0.8 to 3.5 degrees C. (1.4 to 6.3 degrees F.). The Pew Center study also forecasts a sea-level rise of 17 to 99 centimeters (7 to 39 inches) by the year 2100, compared to a previous IPCC projection of 13 to 94 centimeters (5 to 37 inches).

The study suggests that the rate of warming in the United States may be "noticeably faster than the global-mean rate" (Pew Center 1999). The study expects temperatures in the southeastern and southwestern sections of the United States to warm slightly less than the global mean. The northernmost states, from North Dakota eastward to Maine, are expected to warm as much as twice the global mean during winter months, according to Wigley's projections. The study also forecasts, "The frequency of high-precipitation events is likely to increase, bringing increased chances of flooding" (Pew Center 1999). The Pew Center study also estimates that about one-third of the expected global warming during the next century may be attributed to changes in the sun's radiative output.

A report by the United Kingdom's Hadley Center for Climate Change, incorporating improved representations of ocean currents into models of the climate system, suggests that a "runaway" greenhouse effect is possible by the end of the twenty-first century (Brown 1998). The study contends that as lack of rainfall turns large swaths of the Amazon, the Eastern United States, Southern Europe and other areas into near-deserts, the ability of plants and trees to absorb greater amounts of carbon dioxide will be reduced, resulting in higher atmospheric concentrations and rapid global temperature increases.

By 2050, this report projects that agricultural output in central and southern Africa will be severely reduced, and North America's agricultural heartland could see wheat and corn yields fall by as much as 10 percent. Extreme water shortages will affect 170 million people, according to this study, which was presented at the Buenos Aires climate conference in 1998. The study forecasts that temperatures on land will rise an average of 6 degrees C. by the year 2100, subjecting about 100 million people to annual hazards of coastal flooding from rising sea levels.

The Hadley Center study also projects that the Gulf Stream, which is an important warming influence on much of Europe during the winter, will be 20 percent less strong in the future, but that Europe still will warm considerably. This study, unlike some others (See Chapter 4, "Icemelt"), does not foresee a weakening of the Gulf Stream as portending colder winters for Europe while most of the rest of the world warms. Instead, the Hadley Center model forecasts that Western Europe, including Scotland, will gain the ability to grow extra grain and that European storms will become more severe, especially during the winter.

The Hadley Center study anticipates that the benefits for plants of a carbon-enhanced atmosphere will be outweighed by lack of rainfall in many important agricultural areas. The study also asserts that many tropical grasslands will be transformed into deserts, leading to widespread extinction of wildlife. Michael Meacher, British environment minister, told *The Guardian* that "These are sobering findings. Millions of people will have life made miserable by climate change, with increased risk of hunger, water shortages and extreme events like flooding. Combating climate change is the greatest challenge of human history" (Brown 1998). The Hadley Center study also anticipates that much of central and southern Africa will experience a reduced ability to grow staple crops. While the agricultural heartland of the United States may suffer production reductions because of drought and heat, the study projects that Canada will experience a wheat production increase of about 2.5 percent.

Temperature readings during the late 1990s indicated that a steep rise in temperatures seemed to be underway. On March 9, 2000 the National Oceanic and Atmospheric Administration (NOAA) said the winter of 1999–2000 was the warmest such season in the United States since the government began keeping records 105 years earlier. This marked the third year in a row that record warmth was recorded in the United States during the winter months. Since 1980, more than two-thirds of U.S. winters have been warmer than average, NOAA said.

The average temperature in the United States between December, 1999 and February, 2000 was 38.4 degrees F., six-tenths of a degree warmer than the record set the previous year. Scientists at NOAA reported that every state in the continental United States was warmer than its long-term average, with 21 states from California to the Midwest ranked as much above average. As the report was being released, a winter carnival was being canceled near Wausau, Wisconsin, for lack of snow. Other casualties of the warmth included North America's largest cross-county ski race and an ice fishing derby in International Falls, Minnesota.

A week later, the Great Lakes, the world's largest body of fresh water, were measured at their lowest level in recorded history, because of scant snowfall during the winter, after several years of declining water levels. Consequences have included dry wells, landlocked docks, obstacle courses for commercial shipping and pleasure boaters, and smelly drinking water in some areas. Many docks have become useless, while emergency dredging has been required for others. At the same time, the temperature of the lakes' water has been climbing. During 2000, Buffalo, New York reported that its harbor's water temperature at the end of March had equaled the record warmest (39 degrees F.) set in 1998.

During 1990, Congress commissioned a study of how global warming would affect various regions by the year 2100. The report ("Climate Change Impacts on the United States" [www.nacc.usgcrp.gov]) was issued in draft for public comment during the summer of 2000. The report, which involved 5,000 people in nine federal agencies, projected that average temperatures will rise 5 to 10 degrees F. by the end of the century. This report analyzed possible climate changes in eight regions of the United States, "based on a pair of state-of-the-art climate models—one from the Canadian Climate Center and one from the United Kingdom Hadley Centre for Climate Research and Prediction" (Kerr, Dueling Models, 2000, 2113). The entire study was coordinated by Thomas Karl, director of the NOAA National Climate Center in Asheville, North Carolina, who said that the report illustrated a "range of our uncertainties." To Karl, the report also indicated that "The past isn't going to be a very good guide to future climate" (Kerr, Dueling Models, 2000, 2113).

The two climate models used in the study sometimes contrast sharply. The Canadian model, for example, indicates frequent severe drought in the United States' agricultural heartland, while the Hadley model suggests plentiful rainfall in the same area. Generally, however, the report supports a 5 to 10 degree F. rise in temperatures during the century, in line with the models of the Intergovernmental Panel on Climate Change. The report was largely a compilation of data from existing sources because when Congress mandated the study no funds were provided to pay for it.

According to this report, water levels in the Great Lakes are expected to drop five feet by century's end. By the year 2000, Lakes Erie, Michigan, and Huron had dropped three feet in three years, while Lakes Superior and Ontario were down about 18 inches during the same period. Todd Thompson of the Indiana Geological Survey was quoted as saying that the levels of the Great Lakes ebb and flow in 30-year cycles (Flesher 2000, B-1). The lakes receded during the drought years of the 1930s, then again in the 1960s. Coming years will reveal whether present lake-level declines are merely cyclical, or part of a new trend related to global warming. In the meantime, the Toledo Beach Marina was spending $1 million during the year 2000 to add three and a half feet of draft to its docking facilities (Flesher 2000, B-3). Lake Erie averages only 70 feet deep, and in some places a few inches makes a big difference for cargo shipping. At about the same time, the National Environmental Trust released a report describing global warming's anticipated effects on the Great Lakes. Philip Clapp, president of the organization, said dredging of shipping lanes caused by declining lake levels could cost billions of dollars (Fauber and Brook, Global Warming, 2000, 1A).

The same federal report also anticipates some beneficial effects of warming, including reduced costs for snow removal in many midwestern cities, and an opportunity for farmers to profit by planting more than one crop a year (Fialka 2000, A-24).

As news of recent temperature rises arrived during the year 2000, the IPCC strengthened its previous statement affirming human modification of the Earth's climate. In 1995 the IPCC had concluded that *"the balance of evidence suggests a discernible human influence."* In its 2000 assessment, this language reads: *"There has been a discernible human influence* on global climate" (Kerr, Global Warming, 2000, 590, emphasis added). The temperature records of the last 1,000 years now leave very little doubt that the upward spike in temperatures since 1980 has been influenced in large part by human greenhouse-gas emissions. The IPCC's new forecasts for global temperatures in 2100 changed little in 2000 from the 1990 or 1995 assessments—one degree C. to five degrees C., according to several sets of assumptions about how human numbers, societies, economies, and technologies may change during that time.

THE POSSIBLE SPEED OF CLIMATE CHANGE

During May of 1997, twenty-one nationally prominent ecologists warned President Clinton that rapid climate change due to global warming could ruin ecosystems on which human societies depend. The signers, including Stephen H. Schneider and three colleagues from Stanford University, urged Clinton to take a "prudent course" in the then-upcoming global climate-change negotiations in Kyoto, Japan (Basu 1997). The scientists warned that the warming would happen so quickly that many plant and animal species will not be able to adapt. The resulting breakdown of ecosystems could lead to disturbances with major effects on human populations, the scientists warned. These may include increasing numbers of fires, floods, droughts, and storms, as well as erosion and outbreaks of pests and pathogens. The letter said that if present (1997) levels of greenhouse-gas emissions continue to rise, the climate will change more quickly during the coming century than at any time during the past 10,000 years.

"The signers include the leading international experts on many particular dimensions of this problem," said Harold Mooney, Stanford professor of biological sciences and the organizer of the effort. "As you will read in the letter, they all have deep concerns about the ecological consequences of rapid climatic change" (Basu 1997). Among the signers are Mooney, as well as Paul Ehrlich of Stanford (an international leader in ecological research), and Jane Lubchenco of Oregon State University, a past president of the American Association for the Advancement of Science. Seven of the signers are members of the National Academy of Sciences and five are past presidents of the Ecological Society of America.

The scientists said, in the United States

[R]apid climate change could mean the widespread death of trees, followed by wildfires and . . . replacement of forests by grasslands. National parks and forests could become inhospitable to the rare plants and animals that are preserved there—and where the parks are close to developed or agricultural land, the species themselves may disappear for lack of another safe haven. Worldwide, fast-rising sea levels could inundate the marshes and mangrove forests that protect coastlines from erosion and serve as filters for pollutants and nurseries for ocean fisheries. "The more rapid the rate [of change] the more vulnerable to damage ecosystems will be," the scientists told the president. "We are performing a global experiment [with] little information to guide us. (Basu 1997)

The ecologists warned that in some United States temperate-zone forests, rapid climate change could lead to "widespread tree mortality, wildfires and replacement of the forests by grasslands. Species that are long-lived, rare, or endangered will be severely disadvantaged" (Basu 1997). "It would be difficult to imagine, for example," the scientists wrote, "how the imperiled species of Everglades National Park, such as the Cape Sable Sparrow and American Crocodile, could migrate north into the urban and agricultural landscapes of coastal and central Florida and successfully re-establish themselves" (Basu 1997).

The scientists' letter seemed to have an impact at the White House, judging from presidential rhetoric. In his 1999 State of the Union speech, President Clinton called global warming "our most fateful new challenge," as he recalled 1998 as the warmest year ever recorded, with heat waves, floods, and storms which "are but a hint of what future generations may endure if we do not act now." Clinton proposed creation of a new "clean air fund to help communities reduce greenhouse and other pollutions." Clinton also said he "want[ed] to work with members of Congress in both parties to reward companies that take early, voluntary action to reduce greenhouse gases" (Clinton 1999). Many of Clinton's proposals were repeated a year later in his 2000 (and last) State of the Union speech (Clinton 1999).

"We know from ice-core records and deep-sea sediment records that the earth's climate is capable of changing much more quickly than we had previously thought," said Jeff Severinghaus of the University of California (Webb, World Temperatures, 1998). "In some cases," said Severinghaus, "the climate warmed abruptly in less than 10 years . . . up to possibly 10 degrees centigrade" (Webb, World Temperatures, 1998). Severinghaus' findings were presented at the 1998 climate-change conference in Buenos Aires.

Severinghaus continued,

It is possible that by increasing greenhouse gases, we will induce such a change and that, instead of the smooth warming that's being anticipated over the next 50 years, we'll instead go along for a while with very little warming and then all of a sudden in a matter of three or five or ten years we'll have a very large catastrophic warming. (Webb, World Temperatures, 1998)

At the National Ice Core Laboratory in Denver, thousands of meter-long tubes are arrayed on shelves, holding ice cores from the Arctic and Antarctic at minus 36 degrees C. These ice cores contain records of the Earth's changing climate for the last 420,000 years. From studies of ice cores taken in Greenland, scientists have assembled a climatic record which indicates that during the last 8,000 years the earth's climate has been relatively mild and stable. At the end of ice ages (the most recent one, which ended about 12,000 years ago is an example), temperatures tend to swing wildly in cycles of five to twenty years. According to Gale E. Christianson, "Temperatures rose by an astonishing 10 degrees C. within the lifespan of a Paleolithic hunter, and some scientists now think that even that figure is too low by half" (Christianson 1999, 128).

Richard B. Alley, also writing in the *Proceedings of the National Academy of Sciences*, reviewed ice-core evidence from the last 110,000 years which indicates that climate may vary very little over long periods, then undergo changes as large as those between glacial and interglacial conditions, sometimes within a few years or decades. Alley points out that the development of complex human civilization has taken place during a period without such rapid changes (Alley 2000, 1331).

Thomas V. Lowell of the University of Cincinnati's Geology Department, also writing in the *Proceedings of the National Academy of Sciences*, used changes in glacial mass to track climatic change. Using such measures, Lowell provides graphic evidence (from glacial samples near coastal Alaska's Prince William Sound and Western Greenland) that average temperatures declined slowly from about 1450 A.D. to almost 1900 A.D. (the so-called "Little Ice Age"). At that time, temperatures began to climb rapidly, at a pace of about 0.8 degrees C. per century, four times the rate of change during the previous 900 years. The temperature curve rises at an increasingly steep angle toward the end of the century (Lowell 2000, 1351).

Jonathan Overpeck, director of the paleoclimatology program for the National Oceanic and Atmospheric Administration (NOAA), said at the end of 1998, "There is no period that we can recognize in the last 1,200 years that was as warm on a global basis [as the present]." Overpeck presented his findings at a meeting of the American Geophysical Union in San Francisco. "That makes what we're now seeing more unusual, and more difficult to explain without turning to a 'greenhouse gas' mechanism," said Overpeck (Warrick 1998). By the 1990s, according to the IPCC's *Second Assessment* the temperature was rising at the most rapid rate in at least 10,000 years (Bolin et al. 1995).

Until the 1990s, many climate scientists believed that the Earth had warmed dramatically during the period of time which Europe called the Middle Ages, roughly 900 to 1400 A.D. New research, based on tree rings, glaciers and other "proxy" measurements of past climate around the world, indicates that this warming was limited mainly to northern latitudes in Europe and North America. Evidence of a rapid warm-up during the Middle Ages has been used as "proof" by some global-warming skeptics that natural variations may explain rapid temperature increases worldwide during the last quarter of the twentieth century. "Our study of the Medieval Warm Period supports the likelihood that no known natural phenomenon can explain the record twentieth-century warmth," Overpeck said. "Twentieth century global warming is a reality and should be taken seriously" (Warrick 1998).

Dean Edwin Abrahamson confirms Overpeck's analysis:

One must go back in time 5 to 15 million years to the late Tertiary to find a time that was 3 or 4 degrees C. warmer than now. During periods when there was no permanent pack ice in the Arctic, climatic and vegetational region and boundaries were displaced as much as 1,000 to 2,000 kilometers north of their present position (a displacement which we may replicate during the next 100 years). (Abrahamson, Global Warming, 1989, 15)

During the period Abrahamson describes, intense aridity was the norm from present day North and South Dakota to Missouri and Alabama, as well as throughout Central and Southern Africa. These changes may be similar to those which will be experienced by generations to come. As Abrahamson comments,

Beyond the year 2050. . . . we could be committed to a far larger warming—probably on the order of 6 to 10 degrees C. The climatic conditions that might be associated with such a warming are, with few exceptions, pure mystery. . . . Today's climate models have little, if any, validity for such extreme warming. . . . There can be no planned adaptation under these conditions. (Abrahamson, Global Warming, 1989, 21)

By early in the year 2000, scientists working for NOAA released compilations of global temperatures for the last half of the twentieth century which reveal a speed of warming that most climatologists had not expected until late in the twenty-first century. The rate of warming (one degree F. over the entire century) increased to a rate of four degrees F. during the century's last quarter, according to calculations of Tom Karl and associates, published in the March 1, 2000 edition of *Geophysical Research Letters*. This is roughly the rate of increase which several climate models had forecast for the second half of the twenty-first century. "The next few years could be very interesting," Karl told the *Los Angeles Times*. "It could be the beginning of a new increase in temperatures" (Analysis 2000, 12). Tom Wigley, a senior scientist at the National Center for Atmospheric Research in Boulder, Colorado, said that warming was strengthened by frequent El Niño events, which he said are not human-induced. "Those months were unusual," he said, "but they weren't unusual due to human influences" (Analysis 2000, 12).

Karl and colleagues begin a statistical analysis of recent global temperature trends with the observation that between May of 1997 and September of 1998, sixteen months in a row, global temperatures set observational (e.g., century-scale) monthly records. Their analysis of a century-plus of records (roughly 1880 to 2000) indicates that the rate of warming tends to surge upward, then relent a little, and then surge again. "The increase in global mean temperatures is by no means constant" (Karl et al. 2000, 719). Karl and colleagues conclude, "The warming rate over the past few decades [since the mid-1970s] is already comparable to that projected during the twenty-first century based on IPCC business-as-usual scenarios of anthropogenic climate change" (Karl et al. 2000, 719).

We interpret the results to indicate that the mean rate of warming since 1976 is clearly greater than the mean rate of warming averaged over the late nineteenth and twentieth centuries. It is less certain whether the rate of temperature change has been constant since 1976 or whether the recent string of record-breaking temperatures represents yet another increase in the rate of temperature change. . . . Moreover, these results imply that if the climate continues to warm at present rates of change, more events like the 1997 and 1998 record warmth can be expected. (Karl et al. 2000, 720–721)

OZONE DEPLETION'S RELATIONSHIP TO THE GREENHOUSE EFFECT

Chlorofluorocarbons (CFCs) initially raised no environmental questions when they were first marketed by Dupont Chemical during the 1930s under the trade name Freon. Freon was introduced at a time when such questions usually were not askèd. At about the same time, asbestos was being proposed as a high-fashion material for clothing, and radioactive radium was being built into time-pieces so that they would glow in the dark.

By 1976, manufacturers in the United States were producing 750 million pounds of CFCs a year, and finding all sorts of uses for them, from propellants in aerosol sprays, to solvents used to clean silicon chips, to automobile air conditioning, and as blowing agents for polystyrene cups, egg cartons, and containers for fast food. "They were amazingly useful," wrote Anita Gordon and Peter Suzuki. "Cheap to manufacture, non-toxic, non-inflammable, and chemically stable" (Gordon and Suzuki 1991, 24). By the time scientists discovered, during the 1980s, that CFCs were thinning the ozone layer over the Antarctic, they found themselves taking on a $28 billion-a-year industry. The ozone shield protects plant and animal life on land from the sun's ultraviolet rays, which can cause skin cancer, cataracts, and damage to the immune systems of human beings and other animals. Thinning of the ozone layer also may alter the DNA of plants and animals.

By the time their manufacture was banned internationally during the late

1980s, CFCs had been used in roughly 90 million car and truck air conditioners, 100 million refrigerators, 30 million freezers, and 45 million air conditioners in homes and other buildings. Because CFCs remain in the stratosphere for up to 100 years, they will deplete ozone long after industrial production of the chemicals ceased.

These human-created chemicals do more than destroy stratospheric ozone. They also act as greenhouse gases, with several thousand times the per-molecule greenhouse potential of carbon dioxide. What's more, the warming of the near-surface atmosphere (the lower troposphere) seems to be related to the cooling of the stratosphere, which accelerates depletion of ozone at that level. An increasing level of carbon dioxide near the Earth's surface "acts as a blanket," said NASA research scientist Katja Drdla. "It is trapping the heat. If the heat stays near the surface, it is not getting up to these higher levels" (Borenstein 2000).

At about the same time, scientists were looking for reasons why the ozone layers over the Arctic and Antarctic were failing to repair themselves as expected following the international ban on production of CFCs. They began to suspect that global warming near the surface might be related to ozone depletion in the stratosphere.

During the middle 1990s, scientists were beginning to model a relationship between global warming and ozone depletion. A team led by Drew Shindell at the Goddard Institute for Space Studies created the first atmospheric simulation to include ozone chemistry. The team found that the greenhouse effect was responsible not only for heating the lower atmosphere, but also for cooling the upper atmosphere. The cooling poses problems for ozone molecules, which are most unstable at low temperatures. Based on the team's model, the buildup of greenhouse gases could chill the high atmosphere near the poles by as much as 8 degrees C. to 10 degrees C. The model predicted that maximum ozone loss would occur between the years 2010 and 2019 (Shindell et al. 1998, 589).

In 1998, the Antarctic ozone hole reached a new record size roughly the size of the continental United States. Some researchers came to the conclusion that, as Richard A. Kerr describes in *Science*,

Unprecedented stratospheric cold is driving the extreme ozone destruction. . . . Some of the high-altitude chill . . . may be a counterintuitive effect of the accumulating greenhouse gases that seem to be warming the lower atmosphere. The colder the stratosphere, the greater the destruction of ozone by CFCs. (Kerr, Deep Chill, 1998, 291)

"The chemical reactions responsible for stratospheric ozone depletion are extremely sensitive to temperature," Shindell and colleagues wrote in *Nature*. "Greenhouse gases warm the Earth's surface but cool the stratosphere radiatively, and therefore affect ozone depletion" (Shindell et al. 1998, 589). By the decade 2010 to 2019, Shindell and colleagues expect ozone loses in the Arctic to peak at two-thirds of the "ozone column," or roughly the same ozone loss observed in Antarctica during the early 1990s. "The severity and duration of the Antarctic ozone hole are also expected to increase because of greenhouse-gas-induced stratospheric cooling over the coming decades" (Shindell et al. 1998, 589).

During the middle 1990s, scientists began to detect ozone depletion in the Arctic after a decade of measuring a growing ozone "hole" over the Antarctic. By the year 2000 during March and April, the ozone shield over the Arctic had thinned to about half its previous density. Ozone depletion over the Arctic reaches its height in late winter and early spring, as the sun rises after the midwinter night. Solar radiation triggers reactions between ozone in the stratosphere and chemicals containing chlorine or bromine. These chemical reactions occur most quickly on the surface of ice particles in clouds, at temperatures less than minus 80 degrees C. (minus 107 degrees F.).

Space-based temperature measurements of the Earth's lower stratosphere, a layer of the atmosphere from about 17 kilometers to 22 kilometers (roughly 10 to 14 miles) above the surface, indicate record cold at that level as record surface warmth has been reported during the 1990s. Roy Spencer of NASA and John Christy of the University of Alabama at Huntsville and the Global Hydrology and Climate Center obtained temperature measurements of layers within the entire atmosphere of the Earth from space, using microwave sensors aboard several polar-orbiting weather satellites. They found that, despite significant, short-lived warming following the eruptions of El Chichon in Mexico in 1982 and Mt. Pinatubo in the Philippines in 1991, the stratosphere as a whole has been cooling steadily during the past 15 years.

Steve Hipskind, atmospheric and chemistry dynamics branch chief at NASA's Ames Research Center, Moffett Field, California, has been quoted as saying that chlorine atoms use clouds as "a platform" to destroy stratospheric ozone (Arctic Region 2000, 4). Clouds form more frequently over the Arctic at lower temperatures. Ice crystals, which form as part of polar stratospheric clouds, assist the chemical process by which ozone is destroyed. CFCs' appetite for ozone molecules rises notably below minus 80 degrees C. (minus 107 degrees F.), a level that was reached in the Arctic only rarely until the 1990s. During the winter of 1999–2000, temperatures in the stratosphere over the Arctic were recorded at minus 118 degrees F. or lower (the lowest on record), forming the necessary clouds to allow accelerated ozone depletion.

As Dennis L. Hartmann and colleagues explain,

The pattern of climate trends during the past few decades is marked by rapid cooling and ozone depletion in the polar lower stratosphere of both hemispheres, coupled with an increasing strength of the wintertime westerly polar vortex and a poleward shift of the westerly wind belt at the Earth's surface. . . . [I]nternal dynamical feedbacks within the climate system . . . can show a large response to rather modest external forcing. . . . Strong synergistic interactions between stratospheric ozone depletion and greenhouse warming are possible. These interactions may be responsible for the pronounced changes in tropospheric and stratospheric climate observed during the past few decades. If these trends continue, they could have important implications for the climate of the twenty-first century. (Hartmann et al. 2000, 1412)

Ozone depletion has been measured only for a few decades, so these re-searchers caution that they are not entirely certain that rapid warming at the surface is not caused by natural variations in climate, which is powerfully in-fluenced by the interactions of oceans and atmosphere. "However," they con-clude, "it seems quite likely that they are at least in part human-induced." Hartmann and associates also raise the possibility that the poleward shift in westerly winds may be accelerating melting of the Arctic ice cap, part of what they contend may be a "transition of the Arctic Ocean to an ice-free state during the twenty-first century." A continued northward shift in these winds also could portend additional warming over the land masses of North America and Eurasia, they write (Hartmann et al. 2000, 1416).

The connection between global warming, a cooling stratosphere, and depletion of stratospheric ozone was confirmed in April, 2000, with release of a lengthy report by more than 300 NASA researchers as well as several European, Japa-nese, and Canadian scientists. The report found that while ozone depletion may have stabilized over the Antarctic, ozone levels north of the Arctic circle were still falling, in large part because the stratosphere has cooled as the troposphere has warmed. The ozone level over parts of the Arctic was 60 percent lower during the winter of 2000 than during the winter of 1999, measured year over year (Scientists Report, 2000).

In addition, scientists learned that as winter ends, the ozone-depleted atmos-phere tends to migrate southward over heavily populated areas of North America and Eurasia. "The largest ultraviolet increases from all of this are predicted to be in the mid-latitudes of the United States," said University of Colorado at-mospheric scientist Brian Toon. "It affects us much more than the Antarctic [ozone "hole"]" (Borenstein 2000).

Ross Salawitch, a research scientist at NASA's Jet Propulsion Laboratory in Pasadena, California, said that if the pattern of extended cold temperatures in the Arctic stratosphere continues, ozone loss over the region could become "pretty disastrous" (Scientists Report 2000, 3A). Salawitch said that the new data has "really solidified our view" that the ozone layer is sensitive not only to ozone-destroying chemicals, but also to temperature (Stevens, New Survey, 2000, A-19). "The temperature of the stratosphere is controlled by the weather that will come up from the lower atmosphere," said Paul Newman, another scientist who took part in the Arctic ozone project. "If we have a very active stratosphere we tend to have warm years, when stratosphere weather is quiescent we have cold years" (Connor, Ozone Layer, 2000, 5). New research indicates that global warming will continue to cool the stratosphere, making ozone de-struction more prevalent even as the volume of CFCs in the stratosphere is slowly reduced. "One year does not prove a case," said Paul Newman of

NASA's Goddard Space Flight Center in Greenbelt, Maryland. "But we have seen quite a few years lately in which the stratosphere has been colder than normal" (Aldhous 2000, 531).

"We do know that if the temperatures in the stratosphere are lower, more clouds will form and persist, and these conditions will lead to more ozone loss," said Michelle Santee, an atmospheric scientist at NASA's Jet Propulsion Laboratory in Pasadena and co-author of a study on the subject in the May 26, 2000 issue of *Science* (McFarling 2000, A-20). The anticipated increase in cloudiness over the Arctic could itself become a factor in ozone depletion. The clouds, formed from condensed nitric acid and water, tend to increase snowfall, which accelerates depletion of stratospheric nitrogen. The nitrogen (which would have acted to stem some of the ozone loss had it remained in the stratosphere) is carried to the surface as snow.

REFERENCES

Abrahamson, Dean Edwin, ed. 1989. *The Challenge of Global Warming*. Washington, D.C.: Island Press.

Abrahamson, Dean Edwin, 1989. "Global Warming: The Issue, Impacts, Responses." In Abrahamson, Dean Edwin, ed., *The Challenge of Global Warming*. Washington, D.C.: Island Press, 3–34.

Aldhous, Peter. 2000. "Global Warming Could Be Bad News for Arctic Ozone Layer." *Nature* 404 (6 April): 531.

Alley, Richard B. 2000. "Ice-core Evidence of Abrupt Climate Changes." *Proceedings of the National Academy of Sciences of the United States of America* 97:4 (15 February): 1331–1334.

"Analysis: Climate Warming at Steep Rate." 2000. *Los Angeles Times*. In *Omaha World-Herald*, 23 February, 12.

Anderson, J.W. 1990. The History of Climate Change as a Political Issue. The Weathervane: A Global Forum on Climate Policy Presented by Resources for the Future, August. [http://www.weathervane.rff.org/features/feature005.html]

"Arctic Region Quickly Losing Ozone Layer." 2000. Knight-Ridder News Service. In *Omaha World-Herald*, 6 April, 4.

Arrhenius, Svante. 1896. "On the Influence of Carbonic Acid in the Air Upon the Temperature of the Ground." *The London, Edinburgh, and Dublin Philosophical Magazine and Journal of Science*, 5th ser. (April): 237–276.

Augenbraun, Harvey, Elaine Matthews, and David Sarma. N.d. The Greenhouse Effect, Greenhouse Gases, and Global Warming. [http://icp.giss.nasa.gov/research/methane/greenhouse.html]

Basu, Janet. 1997. Ecologists' Statement on the Consequences of Rapid Climatic Change: 20 May. [http://www.dieoff.com/page104.htm]

Bernard, Harold W., Jr. 1993. *Global Warming: Signs to Watch For*. Bloomington: Indiana University Press.

Blunier, Thomas. 2000. " 'Frozen' Methane Escapes from the Sea Floor." *Science* 288 (7 April): 68–69.

Bolin, Bert, John T. Houghton, Gylvan Meira Filho, Robert T. Watson, M.C. Zinyowera,

James Bruce, Hoesung Lee, Bruce Callander, Richard Moss, Erik Haites, Roberto Acosta Moreno, Tariq Banuri, Zhou Dadi, Bronson Gardner, J. Goldenberg, Jean-Charles Hourcade, Michael Jefferson, Jerry Melillo, Irving Mintzer, Richard Odingo, Martin Parry, Martha Perdomo, Cornelia Quennet-Thielen, Pier Vellinga, and Narasimhan Sundararaman. 1995. *Intergovernmental Panel on Climate Change. Second Assessment Synthesis of Scientific-Technical Information Relevant to Interpreting Article 2 of the United Nations Framework Convention on Climate Change.* Approved by the IPCC at its eleventh session, 11–15 December, Rome. [http://www.unep.ch/ipcc/pub/sarsyn.htm]

Borehole Temperatures Confirm Global Warming. 2000. *Nature*, 17 February. [http://www.cnn.com/2000/NATURE/02/17/boreholes.enn/]

Borenstein, Seth. 2000. "Arctic Lost 60 percent of Ozone Layer; Global Warming Suspected." Knight-Ridder News Service, 6 April, in LEXIS.

Boyle, Robert H. 1999. "Global Warming: You're Getting Warmer." *Audubon*, November–December, 80–87.

Britain Urges U.S. to Get Tough on Global Warming. 1997. British Broadcasting Corp. On-line, 11 June. [http://benetton.dkrz.de:3688/homepages/georg/kimo/0254.html]

Broecker, W.S. 1979. "Fate of Fossil Fuel Carbon Dioxide and the Global Carbon Budget." *Science*, 206:409–418.

Brown, Paul. 1998. "World's Biggest Super-computer Predicts Runaway Greenhouse Effect that will Bring Drought, Deserts, and Disease in its Wake." *The Guardian* (London), 3 November. [http://bonanza.lter.uaf.edu/~davev/nrm304/glbxnews.htm]

Callendar, G.S. 1938. "The Artificial Production of Carbon Dioxide and its Influence on Temperature." *Quarterly Journal of the Royal Meteorological Society* 64:223–240.

Cerveny, R.S., and R.C. Balling, Jr. 1998. "Weekly Cycles of Air Pollutants, Precipitation and Tropical Cyclones in the Coastal NW Atlantic Region." *Nature* 394:561–563.

Charlson, Robert J. 1999. "Giants' Footprints in the Greenhouse: The Seeds of Our Understanding of Global Warming Were Sown by Early Heroes." *Nature* 401 (21 October): 741–742.

Christianson, Gale E. 1999. *Greenhouse: The 200-Year Story of Global Warming.* New York: Walker and Company.

Ciborowski, Peter. 1989. "Sources, Sinks, Trends, and Opportunities." In Dean Edwin Abrahamson, ed., *The Challenge of Global Warming.* Washington, D.C.: Island Press, 213–230.

Cline, William R. 1992. *The Economics of Global Warming.* Washington, D.C.: Institute for International Economics.

Clinton, Bill. 1999. "President Clinton's State of the Union Address." *New York Times*, 20 January. [http://geography.rutgers.edu/courses/99spring/370sp99/news01_20_99.html#anchor33828]

Connor, Steve. 2000. "Ozone Layer Over Northern Hemisphere is Being Destroyed at 'Unprecedented Rate.' " *The Independent* (London), 5 March, 5.

Couzin, Jennifer. 1999. "Landscape Changes Make Regional Climate Run Hot and Cold." *Science* 283 (15 January): 317–318.

Crowley, Thomas J. 2000. "Causes of Climate Change Over the Past 1000 Years." *Science* 289 (14 July): 270–277.

December 1997 is Coldest Month on Record in the Stratosphere. 1998. 20 January. [http: //science.msfc.nasa.gov/newhome/headlines/essd20jan98_1.htm]

Diaz, Henry F., and Raymond S. Bradley. 1997. "Temperature Variations During the Last Century at High-elevation Sites." *Climatic Change* 36:253–279.

Dlugokencky, E.J., K.A. Masrie, P.M. Lang, and P.P. Tans. 1999. "Continuing Decline in the Growth Rate of the Atmospheric Methane Burden." *Nature* 393 (4 June): 447–450.

Easterling, David R., Briony Horton, Phillip D. Jones, Thomas C. Peterson, Thomas R. Karl, David E. Parker, M. James Salinger, Vyacheslav Razuvayev, Neil Plummer, Paul Jamason, and Christopher K. Folland. 1997. "Maximum and Minimum Temperature Trends for the Globe." *Science* 277:364–366.

Eco Bridge: What Can We Do About Global Warming? N.d. [http://www.ecobridge.org/ content/g_wdo.htm]

Edgerton, Lynne T., and the Natural Resources Defense Council. 1991. *The Rising Tide: Global Warming and World Sea Levels.* Washington, D.C.: Island Press.

Epstein, P.R., H.F. Diaz, S. Elias, G. Grabherr, N.E. Graham, W.J.M. Martens, E. Mosley-Thompson, and J. Susskind. 1998. "Biological and Physical Signs of Climate Change: Focus on Mosquito-borne Diseases." *Bulletin of the American Meteorological Society* 79, no. 3 (March): 409–417.

Epstein, Paul, Georg Grabbher, Tom Karl, Ellen Mosley-Thompson, Kevin Trenberth, and George M. Woodwell. 1996. Current Effects: Global Climate Change. An Ozone Action Roundtable, 24 June, Washington, D.C. [http://www.ozone.org/ curreff.html]

Fan, S., M. Gloor, J. Mahlman, S. Pacala, J. Sarmiento, T. Takahashi, P. Tans. 1998. "A Large Terrestrial Carbon Sink in North America Implied by Atmospheric and Oceanic Carbon Dioxide Data and Models." *Science* 282 (16 October): 442–446.

Fauber, John, and Tom Vanden Brook. 2000. "Global Warming May Take Great Lakes Gulp; Plunge in Coming Century Would Have Significant Ripple Effect, Reports Say." *Milwaukee Journal-Sentinel,* 14 June, 1A.

Ferguson, H.L. 1989. "The Changing Atmosphere: Implications for Global Security." In Dean Edwin Abrahamson, ed., *The Challenge of Global Warming.* Washington, D.C.: Island Press, 48–62.

Fialka, John J. 2000. "U.S. Study on Global Warming may Overplay Dire Side." *Wall Street Journal,* 26 May, A-24.

Flesher, John. 2000. "The Great Loss: Lakes' Water Drop Incites Debate on Cause, Concern about Impact." *Toledo Blade,* 21 May, B-1, B-3.

Flower, Benjamin P. 1999. "Warming Without High CO2?" *Nature* 399 (27 May): 313–314.

Fourier, Jean-Baptiste. 1824. "Les Temperatures du Globe Terrestre et des espaces planetaires." *Memoires de L'Academe Royale des Sciences de L'Institut de France* 7:569–604.

Giardina, Christian P., and Michael G. Ryan. 2000. "Evidence That Decomposition Rates of Organic Carbon in Mineral Soil Do Not Vary with Temperature." *Nature* 404 (20 April): 858–861.

Gibbs, Mark T., Karen L. Bice, Eric J. Barron, and Lee R. Kump. 2000. "Glaciation in the Early Paleozoic 'Greenhouse': The Roles of Paleo-geography and Atmospheric CO2." In Brian T. Huber, Kenneth G. MacLeod, and Scott L. Wing, eds.,

Warm Climates in Earth History. Cambridge, U.K.: Cambridge University Press, 386–422.

Gifford, Roger M., Damian J. Barrett, Jason L. Lutze, and Ananda B. Samarakoon. 2000. "The CO2 Fertilizing Effect: Relevance to the Global Carbon Cycle." In T.M.L. Wigley and D.S. Schimel, eds., *The Carbon Cycle.* Cambridge, U.K.: Cambridge University Press, 77–92.

"Global Warming." 1997. *The Financial Times,* 11 March, editorial. [http://benetton.dkrz.de:3688/homepages/georg/kimo/0254.html]

Goddard Institute for Space Science. 1999. Global Temperature Trends: 1998 Global Surface Temperature Smashes Record. [http://www.giss.nasa.gov/research/observe/surftemp]

Gordon, Anita, and David Suzuki. 1991. *It's a Matter of Survival.* Cambridge: Harvard University Press.

Grace, John, and Mark Rayment. 2000. "Respiration in the Balance." *Nature* 404 (20 April): 819–820.

Gribben, John. 1990. *Hothouse Earth: The Greenhouse Effect and Gaia.* London: Bantam Press.

Hall, Carl T. 2000. "Spring Scorches the Record Books; It was the Hottest in U.S. History. Study Rekindles Global Warming Debate." *San Francisco Chronicle,* 17 June, A-1.

Hansen, James E. 1989. The Greenhouse, the White House, and Our House. Typescript of a speech delivered at the International Platform Association, 3 August, Washington, D.C.

Hansen, J.E., D. Johnson, A. Lacis, S. Lebedeff, P. Lee, D. Rind, and G. Russell. 1981. "Climate Impact of Increasing Atmospheric Carbon Dioxide." *Science* 213:957–956.

Hansen, James E., A. Lacis, D. Rind, G. Russell, P. Stone, I. Fung, R. Ruedy, and J. Lerner. 1984. "Climate Sensitivity: Analysis of Feedback Mechanisms." *Geophysical Monograph 29 Maurice Ewing.* Vol. 5. American Geophysical Union, 130–163.

Hansen, J.E., R. Ruedy, J. Glascoe, and M. Sato. 1999. "GISS Analysis of Surface Temperature Change." *Journal of Geophysical Research* 104 (27 December): 30,997–31,022.

Hartmann, Dennis L., John M. Wallace, Varavut Limpasuvan, David W.J. Thompson, and James R. Holton. 2000. "Can Ozone Depletion and Global Warming Interact to Produce Rapid Climate Change?" *Proceedings of the National Academy of Sciences of the United States of America* 97:4 (15 February): 1412–1417.

Hesselbo, Stephen P., Darren R. Grocke, Hugh C. Jenkyns, Christian J. Bjerrum, Paul Farrimond, Helen S. Morgans Bell, and Owen R. Green. 2000. "Massive Dissociation of Gas Hydrate During a Jurassic Oceanic Anoxic Event." *Nature* 406 (27 July): 392–395.

Huang, Shaopeng, Henry N. Pollack, and Po-Yu Shen. 2000. "Temperature Trends Over the Past Five Centuries Reconstructed from Borehole Temperatures." *Nature* 403 (17 February): 756–758.

Intergovernmental Panel on Climate Change. 1990. *Scientific Assessment of Climate Change: Report Prepared for IPCC by Working Group I.* New York: World Meteorological Organization and United Nations Environmental Programme.

Jager, Jill. 1989. "Developing Policies for Responding to Climate Change." In Dean

Edwin Abrahamson, ed., *The Challenge of Global Warming*. Washington, D.C.: Island Press, 96–109.

Jager, J., and H.L. Ferguson. 1991. *Climate Change: Science, Impacts, and Policy. Proceedings of the Second World Climate Conference*. Cambridge, U.K.: Cambridge University Press.

Kaiser, Jocelyn. 1998. "Possibly Vast Greenhouse Gas Sponge Ignites Controversy." *Science* 282 (16 October): 386–387.

Karl, T. R., P.D. Jones, R.W. Knight, G. Kukla, N. Plummer, V. Razuvayev, K.P. Gallo, J. Lindsay, R.J. Charlson, and T. C. Peterson. 1993. "A New Perspective on Recent Global Warming: Asymmetric Trends of Daily Maximum and Minimum Temperature." *Bulletin of the American Meteorological Society* 74:1007–1023.

Karl, Thomas R., Richard W. Knight, and Bruce Baker. 2000. "The Record-breaking Global Temperatures of 1997 and 1998: Evidence for an Increase in the Rate of Global Warming." *Geophysical Research Letters* 27 (1 March): 719–722.

Katz, Miriam E., Dorothy K. Pak, Gerald R. Dickens, and Kenneth G. Miller. 1999. "The Source and Fate of Massive Carbon Input During the Latest Paleocene Thermal Maximum." *Science* 286 (19 November): 1531–1533.

Keeling, Charles D., and Timothy P. Whorf. 2000. "The 1,800-year Oceanic Tidal Cycle: A Possible Cause of Rapid Climate Change." *Proceedings of the National Academy of Sciences of the United States of America* 97, no. 8 (11 April): 3814–3819.

Kellogg, William W. 1990. "Theory of Climate Transition from Academic Challenge to Global Imperative." In Terrell J. Minger, ed., *Greenhouse Glasnost: The Crisis of Global Warming*. New York: Ecco Press, 99.

Kennett, James P., Kevin G. Cannariato, Ingrid L. Hendy, and Richard J. Behl. 2000. "Carbon Isotopic Evidence for Methane Hydrate Instability During Quaternary Interstadials." *Science* 288 (7 April): 128–133.

Kerr, Richard A. 1998. "Deep Chill Triggers Record Ozone Hole." *Science* 282 (16 October): 391.

———. 1999. "A Smoking Gun for an Ancient Methane Discharge." *Science* 286 (19 November): 1465.

———. 2000. "Global Warming: Draft Report Affirms Human Influence." *Science* 288 (28 April): 589–590.

———. 2000. "Dueling Models: Future U.S. Climate Uncertain." *Science* 288 (23 June): 2113.

Kirby, Alex. 1998. "Climate Change: It's the Sun and Us." British Broadcasting Corp. News, 26 November. [http://news.bbc.co.uk/hi/english/sci/tech/newsid_222000/222437.stm]

Lachenbruch, A.H., and B.V. Marshall. 1986. "Changing Climate: Geothermal Evidence from Permafrost in the Alaskan Arctic." *Science* 234:689–696.

Lacis, Andrew A. 2000. Personal communication, 5 September.

Landsea, Christopher. 1999. NOAA: Report on Intensity of Tropical Cyclones, 12 August, Miami, Fla. [http://www.aoml.noaa.gov/hrd/tcfaq/tcfaqG.html#G3]

Lawrence, Mark G., and Paul J. Crutzen. 1999. "Influence of Nitrous Oxide Emissions from Ships on Tropospheric Photochemistry and Climate." *Nature* 402 (11 November): 167–168.

Leggett, Jeremy, ed. 1990. *Global Warming: The Greenpeace Report*. New York: Oxford University Press.

Lowell, Thomas V. 2000. "As Climate Changes, So Do Glaciers." *Proceedings of the*

National Academy of Sciences of the United States of America. 97:4 (15 February): 1351–1354.

Malone, Thomas F., Edward D. Goldberg, and Walter H. Munk. N.d. Roger Randall Dougan Revelle, 1909–1991. [http://www.nap.edu/readingroom/books/biomems/rrevelle.html]

Maskell, Kathy, and Irving M. Mintzer. 1993. "Basic Science of Climate Change." *Lancet* 342:1027–1032.

Maxwell, Barrie. 1992. "Arctic Climate: Potential for Change Under Global Warming." In F. Stuart Chapin III, Robert L. Jefferies, James F. Reynolds, Gaius R. Shaver, and Josef Svoboda, eds., *Arctic Ecosystems in a Changing Climate: An Ecophysiological Perspective.* San Diego: Academic Press, 11–34.

McFarling, Usha Lee. 2000. "Scientists Warn of Losses in Ozone Layer over Arctic." *Los Angeles Times,* 27 May, A-20.

Meacher, Michael. 2000. "This is the World's Chance to Tackle Global Warming." *London Times,* 3 September, in LEXIS.

Melillo, Jerry M. 1999. "Warm, Warm on the Range." *Science* 283 (8 January): 183.

Nance, John J. 1991. *What Goes Up: The Global Assault on Our Atmosphere.* New York: William Morrow and Co.

Nisbit, E.G. 1990. Quoted in *Canadian Journal of Earth Sciences* 27:148.

Nisbit, E.G., and B. Ingham. 1995. "Methane Output from Natural and Quasinatural Sources: A Review of the Potential for Change and for Biotic and Abiotic Feedbacks." In George M. Woodwell and Fred T. MacKenzie, eds., *Biotic Feedbacks in the Global Climate System: Will the Warming Feed the Warming?* New York: Oxford University Press, 188–218.

Norris, Richard D., and Ursula Rohl. 1999. "Carbon Cycling and Chronology of Climate Warming During the Palaeocene/Eocene Transition." *Nature* 401 (21 October): 775–778.

Norwegian Government Falls on Global Warming Issue. 2000. Environment News Service, 9 March. [http://ens.lycos.com/ens/mar2000/2000L-03-09-05.html]

Oppenheimer, Michael, and Robert H. Boyle. 1990. *Dead Heat: The Race Against the Greenhouse Effect.* New York: Basic Books.

Overpeck, Jonathan T. 2000. "Climate Change: The Hole Record." *Nature* 403 (17 February): 714–715.

Pagani, M., M.A. Arthur, and K.H. Freeman. 1999. *Paleoceanography* 14:273–292.

Parsons, Michael L. 1995. *Global Warming: The Truth Behind the Myth.* New York: Plenum Press/Insight.

Pearce, Fred. 1998. "Nature Plants Doomsday Devices." *The Guardian* (England), 25 November. [http://go2.guardian.co.uk/science/912000568-disast.html]

Pearson, Paul N., and Martin R. Palmer. 2000. "Atmospheric Carbon Dioxide Concentrations Over the Past 60 Million Years." *Nature* 406 (17 August): 695–699.

Pew Center. 1999. Experts Say Global Warming More Than Predicted: A New Study Released by the Pew Center on Global Climate Change, 10 July, Foresees Greater Global Warming than Previously Predicted, Along with Greater Extremes of Weather and Faster Sea-level Rise. [http://www.gsreport.com/articles/art000175.html]

Philander, S. George. 1998. *Is the Temperature Rising? The Uncertain Science of Global Warming.* Princeton, N.J.: Princeton University Press.

Phillips, Oliver L., Yadvinder Malhi, Niro Higuchi, William F. Laurance, Percy V. Nunez, Rodolfo M. Vasquez, Susan G. Laurance, Leandro V. Ferreira, Margaret Stern, Sandra Brown, and John Grace. 1998. "Changes in the Carbon Balance of Tropical Forests: Evidence from Long-term Plots." *Science* 282 (16 October): 439–442.

Pomerance, Rafe. 1998. "The Dangers From Climate Warming: A Public Awakening." In Dean Edwin Abrahamson, ed., *The Challenge of Global Warming*. Washington, D.C.: Island Press, 259–269.

"Prehistoric Extinction Linked to Methane." 2000. Associated Press. In *Omaha World-Herald*, 27 July, 9.

Radford, Tim. 2000. "Greenhouse Buildup Worst for 20m [million] Years." *The Guardian* (London), 17 August, 9.

Ramanathan, V. 1989. "Observed Increases in Greenhouse Gases and Predicted Climatic Changes." In Dean Edwin Abrahamson, ed., *The Challenge of Global Warming*. Washington, D.C.: Island Press, 239–247.

Revelle, R., and H.E. Suess. 1957. "Carbon Dioxide Exchange Between Atmosphere and Ocean and the Question of an Increase of Atmospheric CO2 During the Past Decades." *Tellus* 9:18–27.

Revkin, Andrew C. 2000. "Study Faults Humans for Large Share of Global Warming." *New York Times*, 14 July, A-12.

Schimel, David, Jerry Melillo, Hanqin Tian, A. David McGuire, David Kicklighter, Timothy Kittel, Nan Rosenbloom, Steven Running, Peter Thornton, Dennis Ojima, Willam Parton, Robin Kelly, Martin Sykes, Ron Neilson, and Brian Rizzo. 2000. "Contribution of Increasing CO2 and Climate to Carbon Storage by Ecosystems in the United States." *Science* 287 (17 March): 2004–2006.

Schneider, Stephen H. 1994. "Detecting Climatic Change Signals: Are There Any Fingerprints?" *Science* 263 (21 January): 341–347.

———. 1989. *Global Warming: Are We Entering the Greenhouse Century?* San Francisco: Sierra Club Books.

———. N.d. Talk Abstract: Surprises and Scaling Connections between Climatology and Ecology. Institute for Mathematics and Its Applications. [http://www.ima.umn.edu/biology/wkshp_abstracts/schneider1.html]

"Scientific Consensus: Villach (Austria) Conference." 1989. In Dean Edwin Abrahamson, ed., *The Challenge of Global Warming*. Washington, D.C.: Island Press, 63–67.

"Scientists Report Large Ozone Loss." 2000. *USA Today*, 6 April, 3A.

Scientists' Statement on Global Climatic Disruption. 1997. Statements on Climate Change by Foreign Leaders at Earth Summit, 23 June. [http://uneco.org/Global_Warming.html]

Shindell, Drew T., David Rind, and Patrick Lonergan. 1998. "Increased Polar Stratospheric Ozone Losses and Delayed Eventual Recovery Owing to Increasing Greenhouse-gas Concentrations." *Nature* 392 (9 April): 589–592.

Siegenthaler, U., and H. Oeschger. 1978. "Predicting Future Atmospheric Carbon Dioxide Levels." *Science* 199:388–395.

Silver, Cheryl Simon, and Ruth S. DeFries. 1990. *One Earth, One Future: Our Changing Global Environment*. Washington, D.C.: National Academy Press.

"Spring 2000 is Warmest on Record." 2000. Associated Press. In *Omaha World-Herald*, 17 June, 1.

Stevens, William K. 1999. *The Change in the Weather: People, Weather, and the Science of Climate*. New York: Delacorte Press.

———. 2000. "New Survey Shows Growing Loss of Arctic Atmosphere's Ozone." *New York Times*, 6 April, A-19.

Today's Science: Global Warming and Ozone Hole Linked. 1998. Facts on File. Today's Science on File, June. [http://facts.com/cd/s70026.htm]

Tyndall, John. 1861. "On the Absorption and Radiation of Heat by Gases and Vapours, and on the Physical Connexion of Radiation, Absorption, and Conduction." *The London, Edinburgh, and Dublin Philosophical Magazine and Journal of Science*. 4th ser. (September): 169–194.

———. 1863. "On Radiation Through the Earth's Atmosphere." *The London, Edinburgh, and Dublin Philosophical Magazine and Journal of Science*. 4:200–207.

Valentini, R., G. Matteucci, A.J. Dolman, E.D. Schulze, C. Rebmann, E.J. Moors, A. Granier, P. Gross, N.O. Jensen, K. Pilegaard, A. Lindroth, A. Grelle, C. Bernhofer, T. Grunwald, M. Aubinet, R. Ceulmans, A.S. Kowalski, T. Vesala, U. Rannik, P. Berbigler, D. Loustau, J. Gudmundsson, H. Thorgiersson, A. Ibrom, K. Morgenstern, R. Clement, J. Moncrieff, L. Montagnani, S. Minerbi, and P.G. Jarvis. 2000. "Respiration as the Main Determinant of Carbon Balance in European Forests." *Nature* 404 (20 April): 861–865.

"Warming Affects Ocean Algae." 1999. ABC News, 14 January. [http://geography.rutgers.edu/courses/99spring/370sp99/news01_20_99.html#anchor33828]

Warrick, Jody. 1998. "Earth at Its Warmest In Past 12 Centuries; Scientist Says Data Suggest Human Causes." *Washington Post*, 8 December. [http://www.asoc.org/currentpress/1208post.htm]

Webb, Jason. 1998. "World Temperatures Could Jump Suddenly." Reuters, 4 November. [http://bonanza.lter.uaf.edu/~davev/nrm304/glbxnews.htm]

Weiner, Jonathan. 1990. *The Next One Hundred Years: Shaping the Fate of Our Living Earth*. New York: Bantam Books.

Weissert, Helmut. 2000. "Global Change: Deciphering Methane's Fingerprint." *Nature* 406 (27 July): 356–357.

Witze, Alexandra. 2000. "Evidence Supports Warming Theory." *Dallas Morning News*. In *Omaha World-Herald*, 12 January, Metro extra, 4.

Woodwell, George M. 1990. "The Effects of Global Warming." In Jeremy Leggett, ed., *Global Warming: The Greenpeace Report*. New York: Oxford University Press, 116–132.

———. 1995. "Biotic Feedbacks from the Warming of the Earth." In George M. Woodwell and Fred T. MacKenzie, eds., *Biotic Feedbacks in the Global Climate System: Will the Warming Feed the Warming?* New York: Oxford University Press, 3–21.

Woodwell, George M., and Fred T. MacKenzie, eds. 1995. *Biotic Feedbacks in the Global Climate System: Will the Warming Feed the Warming?* New York: Oxford University Press.

Woodwell, George M., Fred T. MacKenzie, R.A. Houghton, Michael J. Apps, Eville Gorham, and Eric A. Davidson. 1995. "Will the Warming Speed the Warming?" In George M. Woodwell and Fred T. MacKenzie, eds., *Biotic Feedbacks in the*

Global Climate System: Will the Warming Feed the Warming? New York: Oxford University Press, 393–411.

Woodwell, G.M., F.T. MacKenzie, R.A. Houghton, M. Apps, E. Gorham, and E. Davidson. 1998. "Biotic Feedbacks in the Warming of the Earth." *Climatic Change* 40: 495–518.

CHAPTER 3

Warmer Is Better; Richer Is Healthier: Global-warming Skeptics

At the end of the nineteenth century, the notion of the earth as a finite system of relationships between its resident flora and fauna was in its infancy among a tiny minority of European and European-American thinkers. While ecological perspectives are much more widespread a century and more later, they are hardly universal. Today, a number of scientific and popular skeptics scoff at the predictions of global warming shared by a majority of the world's scientists. These critics tend to distrust climate models, and in so doing deny the evidence portending rapid warming during the twenty-first century and beyond.

The anthropomorphic bias of the skeptics is reflected in the ways some of them propose to manipulate nature to adapt to global warming generated by humankind's use of fossil fuels. Proposals have been floated to inject sulfur into the atmosphere, so that its cooling effects can counteract those of greenhouse gases. The skeptics leave unsaid the fact that atmospheric sulfur is a primary ingredient of acid rain. Bombing the stratosphere with sulfur dioxide also could turn the sky an opaque, dirty white and accelerate ozone depletion. Other proposals rely on humankind's biochemical innovations to breed food plants which will survive a warmer, more humid world. One can almost hear the skeptics enthusiastically describing to one another the size of the corn that they imagine will grow around Hudson's Bay.

When they point out that the earth has survived carbon dioxide levels six to eight times those of the present, very few of the greenhouse skeptics mention that the weather during such times was rather miserable for warm-blooded animals, as warm as 20 degrees F. higher than today. Such temperatures may have been a balm for the dinosaurs, but tacking that much heat onto the daily summer averages where most people live would make much of humanity's present range nearly uninhabitable for several months a year. Omaha's average summer high, for example, could approach 110 degrees F., or roughly today's averages in the Central Sahara. The climate conservatives also do not dwell much on the fact that the earth had no permanent ice when the dinosaurs were its dominant animal species. Sea levels were roughly 280 feet higher than today.

When proponents of global warming cite a growing list of anecdotes indicating that temperatures are rising, skeptics respond with their own, shorter, list of exceptional recent coolings. Australia, for example, experienced its coldest winter in 132 years of recorded history during 1992. The same month, Jerusalem experienced its deepest snow in four decades (an even deeper snow, 15 inches, fell there in early January, 2000). During the record-warm winter of 1999–2000, Mongolia was beset by killing freezes and blizzards, as more than 60 people died in a cold wave that swept over Bangladesh. While reports indicate that glaciers are melting worldwide and ice sheets are thinning in parts of the Arctic and Antarctic, skeptics find a few places where the temperatures are colder or snowfall heavier.

Generally, the skeptics believe that warming will be a balm for human health. A skeptic's journal, the *World Climate Report* (*WCP*), replied to "a recent editorial in the prominent British medical journal *Lancet* [which] continued spreading the propaganda that malaria and other mosquito-borne diseases will become a scourge on the United States as the tropics creep northward from global warming" (Global Warming Pests 1995–2000). According to the *WCP* (echoing general skeptic thinking), stopping malaria has very little to do with temperature or humidity, and more to do with medical technology and air conditioning. The *WCP* points to statistics indicating that epidemics of malaria were common in most of the United States before the 1950s. In 1878, 100,000 Americans were infected and one-quarter of them died. Commented *WCP*, "The late-nineteenth century had the coldest summers in the entire period of record. Malaria was wiped out during the decades when temperatures were much higher" (Global Warming Pests 1995–2000). Skeptics point out that yellow fever broke out in Philadelphia during 1793, six years after the Constitutional Convention was held there, killing thousands of people. Yellow fever took a similar toll in Memphis almost a century later, also without global warming to blame.

THOMAS GALE MOORE: WARMING AS BALM FOR HUMANKIND

One prominent vehicle of global-warming skepticism has been the Hoover Institute, a conservative think tank, under whose banner Thomas Gale Moore has coined a signature slogan for the skeptics: "Global change is inevitable— warmer is better, richer is healthier" (Moore 1997). For pure evangelistic fervor in the face of "global warmists," few can excel Moore, a senior fellow at the Hoover Institute. Moore's 1998 book *A Politically Incorrect View of Global Warming: Foreign Aid Masquerading as Climate Policy* was published by the Cato Institute.

Moore believes, "Global warming, if it were to occur, would probably benefit most Americans" (Moore 1997). If global climate models indicate that a rising in the level of greenhouse gases in the atmosphere will cause temperatures to rise more at night than during the day, so much the better, according to Moore. Moore asserts that 90 percent of human deaths occur in categories that are more common in winter than summer (Moore 1996). Left unmentioned by Moore is the Intergovernmental Panel on Climate Change's (IPCC) estimate that a doubling of carbon dioxide levels could lead to approximately 10,000 estimated additional deaths per year for the current population of the United States from higher summer temperatures, even after factoring in the beneficial effects of warmer winters and assuming that people in a warmer world will become somewhat acclimatized to their environment. Moore argues, to the contrary, that human civilization has flourished during warm periods of history, and declined when climate cooled. Therefore, Moore argues that a warmer world will benefit human society and economy. In addition, he enthuses,

Less snow and ice would reduce transportation delays and accidents. A warmer winter would cut heating costs, more than offsetting any increase in air conditioning expenses in the summer. Manufacturing, mining and most services would be unaffected. Longer growing seasons, more rainfall and higher concentrations of carbon dioxide would benefit plant growth. (Moore 1997)

Virtually any attempt to ameliorate global warming, according to Moore, would entail "a huge price for virtually no benefit" (Moore 1997). The best way to deal with potential climate change, says Moore, is not to embark on a futile attempt to prevent it, but to promote growth and prosperity so that people will have the resources to deal with it: "[G]lobal warming is likely to be good for most of mankind. The additional carbon, rain and warmth should promote the plant growth necessary to sustain an expanding world population" (Moore 1997). Contrary to some scientists, who project an intensification of storms in a warmer world, Moore believes, "Warmer periods bring benign rather than more violent weather" (Moore 1995). Moore, like most greenhouse skeptics, celebrates humankind's dominance of nature.

A rise in worldwide temperatures will go virtually unnoticed by inhabitants of the industrial countries. As modern societies have developed a larger industrial base and become more service-oriented, they have grown less dependent on farming, thus boosting their immunity to temperature variations. (Moore 1995)

Moore relishes the prospect that the polar regions are projected to warm most rapidly in many climate models:

A warmer climate would produce the greatest gain in temperatures at northern latitudes, with less change near the equator. Not only would this foster a longer growing season and open up new territory for farming, but it would mitigate harsh weather. The contrast between the extreme cold near the poles and the warm atmosphere on the equator drives storms and much of the earth's climate. This difference propels air flows; if the disparity is reduced, the strength of winds driven by equatorial highs and arctic lows will be diminished. (Moore 1995)

Patrick J. Michaels, a climatologist and professor of environmental sciences at the University of Virginia, agrees with Moore, writing, "[M]oderate climate change would be inordinately directed into the winter and night, rather than the summer, and that this could be benign or even beneficial. . . . [T]he likely warming, based on the observed data [would be] between 1.0 and 1.5 degrees C. for doubling the natural carbon dioxide" (Michaels 1998) Michaels draws on research by Robert Balling, indicating "that observed changes are largely confined to winter in the very coldest continental airmasses of Siberia and northwestern North America" (Michaels N.d.). According to Michaels, atmospheric carbon dioxide is increasing at slower-than-expected levels because more of it is being captured by plants whose growth is being energized by the carbon dioxide itself.

As a result of increased evaporation from the oceans in a warmer world, Moore writes, "A warmer climate should intensify cloudiness. More cloud cover will moderate daytime temperatures while acting at night as an insulating blanket to retain heat." Moore creates a cuddly image of a warmer world, as he asserts that warmer nighttime temperatures, particularly in the spring and fall, will create longer growing seasons, and raise agricultural productivity, as "enrichment" (his word) of the atmosphere as carbon dioxide fertilizes plants and produces vigorous growth. Human-induced pollution of the atmosphere can be a good thing, Moore maintains. "Researchers have attributed a burgeoning of forests in Europe to the increased carbon dioxide and the fertilizing effect of nitrogen oxides" (Moore 1995).

Not only is pollution beneficial in Moore's view, but population growth is an index of prosperity and human well-being:

Growth in . . . population, major construction projects, a significant expansion in arts and culture—all indicate that society is prosperous. If the population is expanding, food must be plentiful, disease cannot be overwhelming, and living standards must be satisfactory. If building, art, science, and literature are vigorous, the civilization must be producing enough goods and services to provide a surplus available for such activities. (Moore 1995)

Many scientists criticize Moore's analysis as simplistic. According to George M. Woodwell, president and director of the Woods Hole (Massachusetts) Research Center, evidence shows that higher temperatures will have little effect on rates of photosynthesis, a process that removes carbon dioxide from the atmosphere. Instead, warming will increase rates of respiration among some organisms, thus releasing more carbon dioxide.

A 1 degree C. (1.8 degree F.) increase in temperature often increases rates of respiration in some organisms by 10 percent to 30 percent. Warming will thus speed the decay of organic matter in soils, peat in bogs, and organic debris in marshes. Indeed, the higher temperatures of the last few decades appear to have accelerated the decay of organic matter in the Arctic tundra. (Woodwell 1999)

Woodwell suggests, as well, that global warming will tend to erode habitat for large, long-lived plants (such as trees) in favor of small plants with short life-times and rapid reproduction rates, such as shrubs and weeds. He says that the death of some plants and their decay will release more stored carbon into the atmosphere (Woodwell 1999).

There is no evidence, for example, that the higher temperatures and rising carbon-dioxide levels of recent decades [have] increased the growth of trees worldwide. Although controversy surrounds this point, many experts believe that a rapid warming will lead to the rapid loss of carbon from plants and soils and thus to an acceleration of the warming. (Woodwell 1999)

GLOBAL-WARMING SKEPTICS AND THE SUNSPOT CYCLE

Many global-warming skeptics argue that the sunspot cycle is causing a significant part of the warming that has been measured by surface thermometers during the twentieth century's final two decades. Accurate measurements of the sun's energy output have been taken only since about 1980, however, so their archival value for comparative purposes is severely limited.

Michaels, editor of the *World Climate Report*, cites a study of sunspot-related solar brightness conducted by Judith Lean and Peter Foukal, who contend that roughly half of the 0.55 degree C. of warming observed since 1850 is a result of changes in the sun's radiative output. "That would leave," says Michaels, "at best, 0.28 degree C. [due] to the greenhouse effect" (Michaels 1996). J.J. Lean and her associates also estimate that approximately one-half of the warming of the last 130 years has resulted from variations in the sun's delivery of radiant energy to the earth (Lean, Beer, and Bradley 1995).

While solar variability has a role in climate change, Martin I. Hoffert and associates (writing in *Nature*) believe that those who make it the primary variable are overplaying their hand: "Although solar effects on this century's climate may not be negligible, quantitative considerations imply that they are small relative to the anthropogenic release of greenhouse gases, primarily carbon dioxide" (Hoffert et al. 1999, 764).

ORGANIZED GREENHOUSE SKEPTICISM: THE GLOBAL CLIMATE COALITION AND THE NATIONAL CENTER FOR PUBLIC POLICY

In *The Heat is On: The High Stakes Battle Over Earth's Threatened Climate*, Ross Gelbspan examines what he calls "The campaign of deception by big coal and big oil that is keeping the [global warming] issue off the public agenda" (Gelbspan, *The Heat is On*, 1997, 5). The coal and oil industries, which generate $2 billion in sales per day worldwide, have been key advocates of early global-warming skepticism.

Sponsors of the Global Climate Coalition (GCC), a major vehicle of the skeptics, at one time included many of the world's largest fossil-fuel corporations. Beginning in the late 1990s, a number of large companies (among them General Motors, Ford, Daimler-Chrysler, Southern Co., British Petroleum, Royal Dutch Shell, and Texaco) failed to renew their memberships in the GCC (Wadman 2000, 322).

Donald Pearlman founded the climate council to represent companies whose businesses could be impacted by global climate change. He secured nongovernmental organization (NGO) status for his lobby at the United Nations, as environmentalists crowned him "King of the Carbon Club" (Gelbspan, *The Heat is On*, 1997, 120). Pearlman specialized in coaching delegates from oil-rich countries such as Kuwait, Saudi Arabia, Syria, Iran, China, and Nigeria in how to use delaying tactics at international meetings on strategies to combat global warming. Meetings of the IPCC have been a special target for this sort of activity.

The Internet web pages of the National Center for Public Policy (NCPP), another conservative think tank (with headquarters in Washington, D.C.), proudly exhibit the "Byrd-Hagel Resolution," passed in the U.S. Senate by a margin of 95–0, July 25, 1997. The resolution puts the Senate on record as opposing United States participation in any international climate treaty, notably the Kyoto Protocol of 1997, which requires the United States and other industrialized countries to cut their emissions of greenhouse gases while poorer countries endure no such requirement. The National Center for Public Policy complains that the mandated seven percent cut in greenhouse emissions by the United States (based on consumption in 1990) would represent a 30 percent reduction in production, with population increases and economic growth factored in. According to the skeptics, this would cost the United States $100 million to $300 million in lost gross national product per year, cutting personal income by 5 to 10 percent, and reducing payrolls by 500,000 people, according to NCPP estimates (Ridenour 1998).

Navigating the National Center for Public Policy's version of climatic reality, one finds nearly a mirror image of the scenario sketched by the Intergovernmental Panel on Climate Change and most mainstream scientists. The National Center for Public Policy has its own numbers, including satellite measurements indicating that global temperatures have actually *fallen* a tenth of a degree since 1979. Skeptics have come to rely on satellite temperature readings because they do not support surface measurements' indications of warming. In theory, satellite readings record nearly the whole globe, but their accuracy has been disputed.

The National Center for Public Policy has a habit of muddling even the most elementary of climatic statistics. Writers for its web pages assert that hurricanes have become less severe during the twentieth century because the number of people killed by them have declined substantially. This argument confuses the number of deaths with the severity of storms, leaving aside the fact that improved forecasts have allowed many more people to escape the storms' paths. The NCPP's spin-doctoring sometimes even denies that the atmosphere's carbon dioxide level has anything to do with temperature. Hurricane Mitch, which killed 12,000 people in Central America late in 1998, is fobbed off as an accident of nature, while, paradoxically, a hurricane that surprised (and killed) thousands of people in Galveston, Texas, a century earlier is taken to be proof that global warming has had no effect on the intensity of tropical cyclones.

The spin masters at the NCPP will exploit any angle. Readers are told to enjoy expected mild winters for their low heat bills. The NCPP gleefully calculates the jet fuel consumed by delegates attending international climate-change conventions.

Despite the debate over the accuracy of satellite readings (to be examined below), Michaels bases his assertion that global warming is a non-event solely on these readings. This record, he writes, "leaves little doubt that a dramatic warming of the atmosphere is not occurring" (Michaels N.d.). Michaels' reliance on satellite data, like that of many skeptics, approaches article-of-faith proportions. He writes that the satellite record "faithfully reproduces global mean temperatures measured between 5,000 feet and 30,000 feet by weather balloons" (Michaels N.d.). This history, he writes, "shows a statistically significant net cooling when averaged over the 18.5-year period of record" (Michaels N.d.).

Dr. John Christy, a professor of atmospheric science at Alabama's Global Hydrology and Climate Center, tracked data from U.S. weather satellites, finding, in 1998, that changes in their orbits had created errors in their data. When readings were adjusted for their true orbits, evidence of a warming trend emerged. Dian J. Gaffen, writing in *Nature*, also found that satellite temperature readings of Earth's temperature since 1979 had been distorted by orbital decay. This problem raises others for Gaffen: "The mere fact that inclusion of the effects of orbital decay on the trend can completely change its sign makes one wonder what other factors may be influencing the apparent trends. . . . The crux of the matter is that climatologists are relying on systems that were never designed for climate monitoring" (Gaffen 1998, 616). R.A. Kerr, writing in *Science*, explained that the satellite record was erroneous not only because of orbital decay, but also because of problems with splicing together of data sets from many different satellites. Making the corrections he found necessary, Kerr asserts that satellite records show a warming trend of 0.07–0.12 degrees C. increase in global temperatures per decade between 1979 and 1995. Many climate scientists have concluded that surface thermometers provide the best records, and that these records should be used to measure global warming. This view is supported supported by F. J. Wentz and M. Schabel, writing in *Nature* (Kerr 1998, 1948; Wentz and Schabel 1998, 1661).

SULFUR AS SAVIOR: THE SKEPTICS' CASE FOR AEROSOLS

Some of the skeptics favor combating one kind of pollution, greenhouse gases, with another, sulfur particles. The skeptics assert that suspended particulates caused by emissions of sulfur dioxide and other urban air pollutants (aerosols) increase the net albedo of the Earth, thus exerting a cooling influence on planetary temperatures. This is a truism of atmospheric chemistry. James E. Hansen not only agrees with the skeptics on this subject, but has estimated that aerosols cool the climate by about 1 Watt per square meter, "which has substantially offset greenhouse warming" (Hansen et. al., Missing Climate Forcing, 1997, 231). According to Hansen, temperature increases have been cut roughly in half by aerosols' presence.

The skeptics believe that sulfur does more than negate half of humankind's contribution to infrared forcing. One group of observers estimates that "This perturbation is comparable in magnitude to current anthropomorphic greenhouse gases, but opposite in sign" (Cline 1992, 25). In layperson's English, this statement means: "Human-generated sulfur will negate the effects of human-induced greenhouse gases." This simple statement is full of assumptions that fail the reality of atmospheric chemistry. Most sulfur compounds leave the atmosphere within two weeks of their generation, while most greenhouse gases remain in the air for a century or more. To negate the effects of global warming, the "pollution solution" would require a continuous feed of sulfur into the atmosphere. As has been noted above, the resulting precipitation of sulfur-enhanced acidity would have disastrous environmental effects at ground level.

Contrary to some skeptics' assumptions, an increase in atmospheric particulate matter does not always exert a cooling influence. Researchers working with the National Oceanic and Atmospheric Administration (NOAA) have assembled data indicating that periodic increases in atmospheric dust concentrations during the glacial periods of the last 100,000 years may have resulted in significant regional warming, and that this warming may have triggered some of the abrupt climatic changes observed in paleoclimate records. Jonathan T. Overpeck, working with the Paleoclimatology Program at NOAA's National Geophysical Data Center in Boulder, Colorado, led a team of scientists who, during 1996, conducted global climate model simulations to examine the potential role of tropospheric dust in glacial climates. Comparing "modern dust" with "glacial dust" conditions, they found patterns of regional warming which increased at progressively higher latitudes. The warming was greatest (up to 4.4 degrees C.) in regions with dust over areas which were covered with snow and ice (Abrupt Climate Change 1996). Under some circumstances "aerosols can reduce cloud cover and thus significantly offset aerosol-induced radiative cooling at the top of the atmosphere on a regional scale" (Ackerman et al. 2000, 1042). Simply put, soot and other such pollution may not mitigate other warming influences on climate as much as many skeptics claim.

THE SKEPTICS AND SENATOR CHUCK HAGEL

In the United States Senate, Senator Chuck Hagel of Nebraska became the skeptics' major congressional spokesman during the late 1990s. Senator Hagel subscribes to many of the skeptics' assumptions about global warming, including assertions that warming during the late twentieth century is being caused predominantly by changes in the energy output of the sun (as reflected by sunspot activity). Hagel also often cites satellite data favored by skeptics which indicates that no warming took place from 1979 to about 1995. Hagel's rhetoric indicates no knowledge of scientific findings that impeach the satellite temperature record he is using as evidence.

For his analysis of global warming induced by sunspot activity, Senator Hagel may be relying on work by Sallie Baliunas, a senior staff physicist at the Harvard-Smithsonian Center for Astrophysics and deputy director of the Mount Wilson Observatory. She has authored more than 200 scientific articles, some of which assert that global warming is caused mainly by natural variations in the sun's thermal output, a favorite theme of the skeptics. Baliunas contends that the 30-year cycle of sunspot activity and global temperature trends form a linear relationship, proof enough for her that fossil-fuel effluent isn't the major cause of rising temperatures. While her theory seemed to follow reasonably well for records between 1900 and roughly 1980, the rapid rise in temperatures since then contradicts her beliefs.

Senator Hagel's rhetoric indicates that he believes himself to be an environmentalist who wants to preserve the Earth for future generations. He is able to characterize himself this way while opposing nearly any interference with business-as-usual regarding United States emissions of fossil fuels. To the uninitiated, such a stance may seem somewhat hypocritical, but the senator appears to be a politically sincere individual. Hagel can play both sides of this question because he denies the scientific warnings assembled by the IPCC and other groups around the world.

This is not a question for a debate about who is for cleaning up the environment and who isn't. I don't know of any senator or congressman or any American who doesn't believe that the environment is critically important. We all want to leave the environment to our children and grandchildren. That's not the debate. The debate is how we do it, what is common-sense, what . . . action we should take based on the science available. . . . We should do everything we can to control greenhouse-gas emissions and pollution, no question about that. But [if] we take the . . . drastic actions that are prescribed in this treaty [the Kyoto Protocol], that's where many of my colleagues and I essentially draw the line. (Hagel N.d.)

When Senator Hagel cites "science," his citations often come from Michaels, who has argued (as cited by Senator Hagel), "Conditions in the real world simply have not matched changes projected by some computer models. Most of the warming this century occurred in the first half of the century—before significant emissions of greenhouse gases began" (Countdown 1997). Hagel also cites Dr. Richard S. Lindzen, Alfred P. Sloan Professor of Meteorology at the Massachusetts Institute of Technology: "A decade of focus on global warming and billions of dollars of research funds have still failed to establish that global warming is a significant problem" (Countdown 1997). Hagel's scientific resources also include John Christy, an associate professor of atmospheric sciences at the University of Alabama: "The satellite and balloon data show that catastrophic warming is not now occurring. The detection of human effects on climate has not been convincingly proven because the variations we have now observed are not outside of the natural variations of the climate system" (Countdown 1997).

Given Hagel's narrow knowledge base, he can logically, by his own lights, conclude,

The scientific community has simply not yet resolved the question of whether we have a problem with global warming. They have not been able to definitively conclude if the warming that has occurred in this century is due to human action or natural variations in the earth's atmosphere. So why are we rushing to sign a treaty [the Kyoto Climate Convention] in December aimed at solving a problem the scientists cannot agree that we have or that is caused by human actions? (Countdown 1997)

Hagel's actions have provoked occasional hilarity among environmentalists. On May 1, 1998, readers of the largest newspaper in Hagel's home state, the *Omaha World-Herald*, awoke to a reporter's question: "What do Ginger Spice, Chuck Hagel, and global warming have in common?" Environmentalists had complained, according to the article, that Senator Hagel had touted a petition signed by 15,000 "scientists," a list of names which included Ginger Spice (one of the singers in British pop group the Spice Girls), to bolster his opposition to the Kyoto protocol.

Senator Hagel's notions of equity include holding Third World nation-states to the same greenhouse-gas reductions as industrialized countries, including the United States. Ignoring the high cost of technological improvements of infrastructure serving almost one billion people in India and 1.3 billion in China, Hagel believes that "developing nations" (the term itself is a metaphor for rising emissions of fossil fuels) should easily be able to quickly install new, cleaner, technology.

The developing nations and the economies of the world today are now in a position to be able to take advantage of the new technologies, the new scrubbers on the smoke stacks, and all the things that the United States and the developed nations, the industrialized nations, have developed over the last 25 years. They are cheaper, they are more effective, they are a more efficient use of energy, fossil fuel energy or any kind of energy. The developing nations are the beneficiary of that and there is no reason to ask why

China, or India, or Mexico or South Korea or any other developing nation would have to go back 50 years and use 50-year-old technology. (Hagel, N.d.)

When Hagel speaks of "clean energy," he usually means nuclear power. He is fond of pointing out that no new nuclear capacity has come on-line in the United States since 1975.

On the stump, Senator Hagel certainly does not downplay the importance of global warming to future generations. He told an audience in Canberra, Australia, "How we, the nations of the world, choose to address the global climate issue may prove to be one of the most important global economic and environmental decisions of the next century" (Countdown 1997). The ecological Senator Hagel seems to be at home on the stump, where saving the Earth is politically popular. Despite his rock-solid credentials as a skeptic of global warming, Senator Hagel can sometimes sound downright sappy about the environment, as long as no one asks him to endorse a lid on fossil-fuel emissions:

It's in the nation's self-interest to protect their environment, [and] in their economic best interests, in their future, in their children's best interests, and their environmental best interests, because pollution is inefficient. Fossil fuel use . . . [is] inefficient . . . so it's in their own interest . . . to protect their own environment. (Hagel N.d.)

Hagel's environmentalism has its limits, including any law, notably the Kyoto Climate Convention, which would inhibit the usual and accustomed economic activities of the United States:

The Byrd-Hagel resolution. . . . says very clearly that the U.S. Senate would not ratify any treaty which would submit the United States, Australia and the other Annex I [industrialized] nations to legally binding reductions in greenhouse gases without requiring any new or binding commitments from the 130 developing nations such as China, Mexico, Indonesia, and South Korea. It also says that the U.S. Senate would reject any treaty or other agreement that would cause serious economic harm to the United States.

If the Annex I nations sign a treaty in Kyoto which exempts the developing world from binding reductions in greenhouse gas emissions it will not see the light of day in the United States. The rest of the world can do as it pleases, but the United States Senate will not ratify a treaty that would place a straightjacket on our national economy while leaving many of the world's nations untouched by its provisions.

If anything has become clear to me as I have studied this issue and held hearings in the U.S. Senate, it is that the scientific community has not definitively concluded that we have a problem with global warming that is caused by human actions. The science is inconclusive and often contradictory. Predictions for the future range from no significant problem to global catastrophe. (Countdown 1997)

Hagel supports the view of Fredrick D. Palmer, chief executive officer of the Western Fuels Association, that abiding by the Kyoto Protocol would have "wrenching effects on the U.S. economy and our way of life." The U.S. economy, says Hagel (quoting Palmer), "succeeds because of fossil fuel use, not in spite of it" (Palmer 1998, A-23).

Senator Hagel stresses the cost of reductions in fossil-fuel use to the United States economy. In so doing, he sometimes cites studies by another global-warming skeptic, Yale University economist Robert Mendelsohn, who urges policy makers to avoid "costly crash abatement programs" and instead focus on longer-term policies (AEI Study 1999). In a study sponsored by the American Enterprise Institute, Mendelsohn asserts that damage from climate change will probably be "much less severe than original projections." Mendelsohn concludes that human society has time to make gradual, adaptive changes, rather than urgent, abrupt ones (AEI Study 1999).

A book by Mendelsohn and James E. Neumann (1999), titled *The Impact of Climate Change on the United States Economy*, estimates that "moderate" global warming in the United States would produce a 0.2 percent annual increase in gross domestic product, including significant positive impacts on agriculture and smaller positive impacts on forestry and recreation. All other economic sectors experience negative impacts in Mendelsohn's and Neumann's estimation, but far smaller ones than those forecast by IPCC models. Mendelsohn estimates that agriculture in the United States will gain $41 billion in production by the year 2060 (Brown 1999, 1441).

Mendelsohn believes that temperature increases will be beneficial only to approximately two degrees C. Beyond that, he notes, "The result is net damages" (AEI Study 1999). Mendelsohn also warned that we should not delay action too long, simply that we don't need to rush into costly and economically damaging policies. Overall, it is still "not clear whether the net economic effects from climate change over the next century will be harmful or helpful," Mendelsohn says (AEI Study 1999). Mendelsohn argues that improved analyses of ocean-atmosphere interaction and the cooling effects of sulfites indicate that a doubling of carbon dioxide levels in the atmosphere will result in "only" a one degree C. to 3.5 degree C. increase in air temperatures, slightly less than the IPCC projections in its *Second Assessment* (Bolin et al. 1995).

Senator Hagel bases his opposition to the Kyoto Protocol on projections that indicate that holding emissions at 1990 levels would reduce economic growth in the United States by between one and three percent a year. Hagel argues that 1,250,000 to 1.5 million jobs would be lost in the United States because of limits on consumption of fossil fuels. Hagel believes that meeting the goals in the Kyoto Climate Convention would cause energy prices to rise dramatically. "Individual Americans will pay for this treaty," he says, "either in their electric bills—at the gas pump, or by losing their jobs" (Countdown 1997).

Hagel also opposes international limits on greenhouse-gas emissions because he sees in them a germ of international political control. As a private business-man before his election to the Senate, Hagel says that he "fiercely fought any attempt by my government or some international tribunal to dictate my private business" (Countdown 1997). Hagel fears that the type of climate covenants negotiated in Kyoto and Buenos Aires would increase government control on a worldwide scale.

The true long-term impact of this treaty would impose regulations, taxation, and government command and control over the fields of transportation, industry agriculture, forestry, energy, consumption and other areas of a nation's economy. At a time when the world is increasingly embracing free markets, capitalist economies, democratically elected governments, and individual responsibility, this treaty would take the world back down the failed path of government command and control. (Countdown 1997)

Congress has become a battleground for special interests in a number of venues, including climate change. When Dana Rohrabacher, California Republican who chairs the House Science Committee's energy and environment panel, characterizes government climate-change research as "scientific nonsense" (Lawler 1995, 1208), the rhetoric is more than hot air. When Rep. Rohrabacher declares that assertions of global warming are "unproven and at worst . . . liberal claptrap" (Lawler 1995, 1208), she is taking aim at government support of climate-change research. Climate-change funding is at issue in the ongoing battle between advocates of a "no-regrets" policy (favored by President George Bush during the late 1980s and 1990s) and a policy more in tune with worldwide public opinion, as endorsed by the Clinton presidency.

GET USED TO IT: S. FRED SINGER

Like many of his fellow skeptics, S. Fred Singer believes that a "warmer climate would, overall, be good for Americans, improve the economy, and put more money in the pockets of the average family" (Singer 1999). Singer, professor emeritus of environmental sciences at the University of Virginia and president of the Science and Environmental Policy Project, advises adaption to a warmer world:

Farmers are not dumb; they will adapt to changes—as they always do. They will plant the right crops, select the best seeds, and choose the appropriate varieties to take advantage of longer growing seasons, warmer nights, and of course the higher levels of carbon dioxide that make plants and trees grow faster. (Singer 1999)

Singer's scientific world much resembles Senator Hagel's. It is a place in which warming, if it occurs at all, will enhance harvests and cause sea levels to fall, not rise. This is a world in which Frederick Seitz, who wrote the foreword to one of Singer's books, pins the whole controversy on scientists greedy for fame and fortune (Singer 1997, vii). If the Earth is warming (which Singer says is debatable), he is as apt to blame the increase in temperatures on variations in solar activity as on an increase in greenhouse gases in the atmosphere (Singer 1997, 7).

As he describes increases in agricultural yields, Singer says nothing of insect pests or heat stress on plants. Singer rejects the findings of the IPCC out of hand—climate models do not work, he says. Seas will fall if the earth warms because of increased evaporation in a warmer world. The evaporated water will end up in the polar regions as increased ice and snow, which will make sea levels fall even further, so says Professor Singer.

SKEPTICS AND URBAN HEAT ISLANDS

Greenhouse skeptics often become very particular about sources of human-induced warming, shaving away from potential greenhouse-effect warming the possible contributions of various other sources, such as the sunspot cycle. Such is the case with the urban heat-island effect. Many skeptics go to great lengths to discredit urban temperature readings because they have risen with increasing urbanization. They also seem to forget that a majority of humankind lives in urban areas, where many millions of people experience the weather being measured within heat islands. Edward C. Klug rejects surface temperature measurements taken in urban areas, notably at airports, because urban heat islands are "a micro-climatic change brought about by the transformation of the immediate environment, not national, let alone global, trends" (Klug 1997).

Cities not only generate more heat than the surrounding countryside, but their structures retain accumulated heat longer than natural surroundings. Some cities record temperatures as much as 10 degrees C. (18 degrees F.) warmer than nearby countryside on calm, clear nights (Burroughs 1999, 114). The difference decreases with the speed of the wind; with a wind speed of more than 20 miles an hour, the air usually is mixed enough to spread the city's heat downwind as quickly as it is created.

Urban areas have many buildings with vertical walls. The walls act as light traps or cavities which capture and store heat during the day and then re-emit it at night. Thus, urban areas are warmer than surrounding countryside particularly at night. (This has been known since at least 1850, when people commented on the urban heat island around the town of St. Louis.) As urban areas expand, so do their heat-island effects.

The skeptics extend the same argument to weather stations at airports, which create heat islands of their own as large, paved runways soak up sunshine, raising temperatures. The burning of jet fuel also contributes heat to airports. Today a large number of surface-temperature observations are taken at airports, so skeptics object that they overstate the rate of warming.

Many greenhouse skeptics make a clear (and largely imaginary) distinction between global warming due to greenhouses gases and "urban heat islands," as if they were separate phenomena. Actually, urban heat islands are aggregates of all the ways in which humankind is raising the temperature of the Earth, of which rising greenhouse-gas forcing is one important part. Cities not only provide emissions centers for all manner of greenhouse gases, they also replace open fields and forests with concrete and asphalt, which often intensifies heat absorption. Cities contain interior spaces that sometimes kill occupants who have no air conditioning during heat waves. Urban heat islands also are centers of industrial, office, and residential waste heat created by our machine culture. The cars and trucks that ply urban areas also add both greenhouse gases and waste heat to the atmosphere.

Expanding urban areas worldwide are contributing increasing amounts of human-induced heat which adds momentum to infrared forcing. Early in the twentieth century, for example, Mexico City was a relatively small metropolis covering 86 square kilometers bordering on Lake Texcoco, with a surface area of 120 square kilometers. At that time, according to a study in *Climatic Change* [44(2000): 515–536] by Aron Jazcilevich and colleagues, Mexico City's heat-island effect averaged about 1.5 degrees C. Mexico City at the close of the twentieth century sprawled across 1,200 square kilometers, while the lake had shrunk to about 10 square kilometers. The heat-island effect at that time (combining the effects of human heat generation with loss of the lake's cooling effect) had reached 8 to 10 degrees C. (Gillon 2000, 555).

Dale Quattrochi of NASA's Marshall Space Flight Center has studied the urban heat-island phenomenon by flying NASA aircraft over cities and measuring temperatures using equipment developed for the space program. He found that huge heat "domes" form over cities, triggering thunderstorms, increasing the production of polluting ozone, and raising local temperatures by as much as 10 degrees F. (5.5 degrees C). "Over Atlanta, the heat island is causing the city to create its own weather," he said. "At the end of July and the beginning of August, we have seen a series of thunderstorms generated in the early hours of the morning—when no thunderstorm would normally occur—as a result of heat rising from the city" (Hawkes, 2000). Between 1980 and 2000, Atlanta was one of the fastest-growing cities in North America, as it lost 380,000 acres of tree cover and gained 370,000 acres of single-family housing.

SKEPTICS AND THE LIMITS OF CLIMATE MODELING

Nearly all the prominent skeptics distrust climate modeling implicitly. A favorite article of faith among skeptics is, if they can't forecast the weather five days in advance, what about a century? Additionally, many skeptics argue that the surface-temperature record is woefully incomplete, so incomplete that indications of warming may be bogus. During the late 1990s, the most sophisticated climate models tallied surface temperature readings of 20 to 40 percent of the Earth's land surface. No readings are taken in many large regions of the oceans.

During the late 1990s, the IPCC's computer models were adjusted to account for the cooling effect of particle haze, reducing, somewhat, its forecasts of temperature increases due to global warming by the end of the twenty-first century. The reductions provoked some critics to assume that the IPCC was declaring that warming was no longer a threat. The reflection of sunlight by particles partially masks, but does not eliminate, temperature rises from the greenhouse effect. If developing countries such as China move to reduce emissions of sulfur dioxide which generate these particles (as the United States, Europe, and Japan already have done in order to reduce adverse health effects of atmospheric sulfur, as well as acid rain), the cooling effect of aerosols in the atmosphere may diminish.

Much of Michael L. Parsons' (1995) *Global Warming: The Truth Behind the Myth* is a detailed attack on climate modeling. Parsons argues that the models are not sophisticated enough to tell us much of anything about future climate trends. Parsons subscribes to the viewpoint, "Models are like sausages. You don't want to know what goes into them" (Parsons 1995, 41). Parsons accuses "warmists" of using climate models to confirm predetermined conclusions.

Climate models do have shortcomings. Restrictions on the power of computers limits the number of individual data-points which may be calculated on the face of the globe. Stephen Schneider explained problems of scale in climate models:

A typical global circulation model [GCM] grid size for a "low-resolution" model is about the size of Germany horizontally and that of a "high-resolution" GCM is about the size of Belgium. . . . In the vertical dimension there are two (low resolution) up to about twenty (high resolution) vertical layers that are typically spanning the lowest 10 (up to 40) kilometers of the atmosphere. Clouds are very important to the energy balance of the earth-atmosphere system since they reflect sunlight away and trap infrared heat. But because none of us have ever seen a single cloud the size of Belgium, let alone Germany, we have a problem of scale—how can we treat processes that occur in nature at a smaller scale than we can resolve by our approximation technique of using large grid boxes? (Schneider 1999)

William Stevens elaborates on this problem:

One big limitation of the models in predicting the climate decades ahead is that despite the growing speed of computers, they are unable to calculate climatic changes everywhere in the atmosphere. Instead, they make the calculations only at widely separated points. The points form a three-dimensional grid typically rising 10 or 12 miles above the earth. A typical spacing between grid points is about 150 miles horizontally and less than half a mile vertically. This "resolution," as scientists call it, is about twice as fine as a decade ago. But it still misses many processes that happen between grid points—cloud formation, for example. So modelers approximate these factors as best they can. Coarse resolution is also the major reason why the models are not very good at simulating climate at the regional scale. (Stevens 1997)

Despite the models' difficulty in simulating small-scale (regional) details of climate change, the IPCC's confidence in them is growing. The models have become more sophisticated and detailed, the scientists say; they also have successfully simulated actual climate changes with reasonable accuracy. The models are exhibiting "a progressive convergence toward what has happened in nature," said Dr. W. Lawrence Gates, a climatologist at the Lawrence Livermore National Laboratory in California who has long been a leader of systematic international efforts, including those of the IPCC, to evaluate and thereby improve climate models (Stevens 1997). For making long-term predictions of climatic change on a continental scale, Gates said, "The ensemble of existing modern models is reliable [and] provide a firm scientific base for policy" (Stevens 1997).

Thomas R. Karl, a senior scientist at NOAA's National Climatic Data Center in Asheville, N.C., has analyzed weather data for the United States (excluding Alaska and Hawaii) over this century to determine whether observed changes are consistent with models for greenhouse warming. Karl created a greenhouse climate response index (GCRI) to measure how well actual observations fit the models. This index is the average of four indicators: the percentage of the United States with much-above average minimum temperatures, the percentage with much-above average precipitation during the cold season, the percentage of the area in extreme drought during the warm season, and the percentage of area with a much-greater-than-average proportion of precipitation derived from extreme one-day events. Karl found that since 1976, GCRI values have been higher than the average for previous years in the century, indicating, "The late-century changes in the U.S. climate are consistent with the general trends anticipated from a greenhouse-enhanced atmosphere. . . . There is only a 5 to 10 percent chance that the increase in GCRI results stem from natural variability" (Hileman 1995). Karl continued:

Night-time temperatures have generally increased more than daytime temperatures. Climatic variability, or the frequency of extreme events, has increased in some regions, although it is not known whether these have risen on a worldwide scale. For example, heavy rainfall events in the U.S. have increased in intensity, and behavior of El Niño— the warming of the eastern equatorial Pacific that sometimes brings severe droughts to some regions and heavy rains to other regions around the world—has been unusual since 1976. These occurrences fit well with complex mathematical models of global climate change. (Hileman 1995)

Karl describes more problems inherent in climate modeling:

One of the reasons temperature extremes are so difficult to model is that they are particularly sensitive to unusual circulation patterns and air masses, which can occasionally cause them to follow a trend in the direction opposite that of the mean temperature. For example, in the former Soviet Union, the annual extreme minimum temperature has increased by a degree and a half, whereas the annual extreme maximum showed no change. (Karl et al. 1997)

In other evidence that global climate models have accurately predicted some of the nuances of a warmer world, the gap between daytime highs and nighttime lows has narrowed over land since 1950. Nighttime temperatures have increased at almost twice the rate of daytime temperatures since 1950 (roughly 0.9 degrees

C. versus 0.5 degrees C.). Rising nighttime temperatures exacerbate heat waves and reduce the beneficial effects of frost in killing pests.

Increasingly sophisticated climate models strictly conserve energy, mass, momentum, and water substance as they calculate atmospheric motions, temperatures, clouds, precipitation, and evaporation at thousands of points around the globe in response to changing solar insulation, greenhouse gases, and aerosols (Hansen et al. 1983). These models are "coupled" (integrated) with similar ones for the oceans to simulate real-world influences of one upon the other. By comparison, computing the spacecraft trajectory to the outer reaches of the solar system is a rather simple task. Climate models are complicated by the multitude of interactions between different components of the climate system (Hansen et al. 1983). Cloud feedbacks also introduce a degree of uncertainty into climate models. An increase in cloud cover or optical depth of low clouds will cool the climate, while an increase in cloud cover, optical depth, or height of high clouds would act to warm it.

Many of these interactions cannot be represented (or resolved) directly in the models, but must be "parameterized" so that computations may be performed within the restraints of available computing facilities. As computing facilities improve, so does the ability of climate models to represent atmospheric processes. With the emergence of global warming as a worldwide concern, climate modeling has become a "big science" field which is being developed by teams of scientists in many places, such as the Goddard Institute for Space Studies and National Center for Atmospheric Research (NCAR) in the United States, the Hadley Center in the United Kingdom, and the Max Planck Institute in Germany. Scientists are challenged by such things as wind eddies, convection, and turbulence which take place on a scale that is smaller than any given climate model's grid size. It is probably fair to say that atmospheric dynamics are sufficiently well understood to develop the physical parameterizations (characteristics) and approximations that permit climate models to reproduce the basic structure of atmospheric wind patterns.

The physical interactions that produce biotic feedbacks are difficult to model in climate simulations. This is what produces the uncertainty in the expected temperature increase for a given carbon dioxide increase. Into this formulation must be factored the heat capacity of the ocean and the length of time required for a given amount of heat to diffuse into the deep ocean.

The radiative forcings of climate can be accurately determined if we know the greenhouse gas and anthopogenic aerosol changes that have taken place. Measurements of greenhouse-gas changes are known with good accuracy. Knowledge of aerosol changes and radiative properties, and hence their effect on climate, are much more problematic. With climate feedbacks reasonably well understood, most of the uncertainty in predicting future trends in global temperature lies in providing reliable estimates for future increases in greenhouse gases and anthropogenic aerosols. Global temperature changes are easier to predict than changes in regional temperatures and precipitation, which are of prime interest to local residents.

The June 1991 eruption of Mount Pinatubo provided a test for climate modelers. Accurate measurements were made of the extent, duration, and radiative properties of the eruption's aerosol cloud. With this information in hand, accurate radiative forcing changes could be computed. Hansen and associates, writing in *Geophysical Research Letters*, used this information to predict a global cooling of about 0.5 degrees C. for the winter of 1992, with temperatures slowly returning to average by 1995 (Hansen et al. 1992). The global temperature curve very closely followed this prediction, providing evidence that the climate models being developed by the Goddard Institute for Space Sciences correspond closely with climatic reality.

QUESTIONING THE MOTIVES OF "GLOBAL WARMISTS"

Some of the skeptics do not restrict themselves to debating points of scientific veracity. Occasionally, the debate descends to questions of personal or political motivation. The IPCC, as a deputy body of the United Nations, sometimes becomes the target of a general fear, prominent in very conservative U.S. political circles since the days of the John Birch Society, that climate diplomacy, and especially the Kyoto Protocol, is a cover for an international plot to strip the United States of its national sovereignty. Other critics argue that the members of the IPCC and other "global warmists" are ringing imaginary alarm bells to make themselves famous. Michael L. Parsons (1995), in *Global Warming: The Truth Behind the Myth*, maintains that environmental activists have overplayed global warming out of self-interest in maintaining their own influence.

In the Foreword to Robert C. Balling's (1992) book, *The Heated Debate: Greenhouse Predictions Versus Climate Reality*, Singer writes that "These [forecast] disasters are not grounded in fact, but spring from the feverish imagination of activists and their ideological desire to impose controls on energy use and to stop—or at least micromanage—economic growth" (Singer, Foreword, 1992, vii). Singer's point of view has been borrowed, with his usual rhetorical flourish, by Rush Limbaugh, who insists that the whole global warming case is fiction invented by "environmental whackos" (Limbaugh 1992, 168). Since this point of view conceives of global warming as a problem of (at most) a few tenths of a degree's worth of temperature and a couple of inches of sea level, Singer and Limbaugh both believe that warming will be good for the economy. It will, they say, increase agricultural production. The *real* problem, writes Singer, is that the Earth is overdue for a new ice age. Burning more carbon-based fuels may help human civilization escape this new ice age, Singer asserts.

CARBON AS COOLING AGENT?

Edward C. Klug adds an extra twist to the debate over greenhouse warming. He believes that increasing atmospheric carbon dioxide levels actually may *cool* the lower atmosphere:

[C]ontrary to popular perception, CO2 plays only a minor role in regulating the temperature of planet Earth, and an increase in atmospheric concentrations will have negligible influence on global temperatures. Indeed, beyond a point, increasing CO2 levels may even cool the Earth a little. (Klug 1997)

The "warmists," of course, argue that Klug and Parsons have invented an egg that lays a chicken. According to Parsons' line of reasoning, "[T]he Earth's temperature determines the amount of carbon dioxide in the atmosphere, and not the other way around" (Parsons 1995, 149).

Sherwood Idso is one of a number of greenhouse skeptics who believe that life on earth should be genetically modified to adapt to conditions imposed by humankind's pollution of the atmosphere. Such genetic modification will probably not be necessary, according to Idso, because his version of climate science indicates that any warming of the atmosphere produced by human combustion of fossil fuels will be slight—just enough to increase the yields of carbon-hungry food crops and ameliorate the rough edges of continental winters in the temperate zones. To Idso, a rising level of carbon dioxide is a "friend," not a "foe."

Idso cites approvingly a study indicating that a doubling of carbon dioxide levels will raise global average temperatures by only 0.25 degrees C. (Newell and Dopplick N.d.) Idso next takes his argument a step further by arguing that increasing levels of carbon dioxide may actually cause temperatures to *drop*, not rise. Idso's theory appears to stand the physics of the atmosphere on its head:

CO2 significantly reduces the short-wave energy absorbed by the surface of snow and water, and . . . this energy deficit—when not augmented by downward atmospheric thermal radiation—may delay the recrystalization of snow and the dissipation of pack ice, resulting in a cooling rather than a warming effect. (Idso 1982, 44)

Idso supports his theory with a report by K.F. Dewey and R. Heim of the University of Nebraska, who studied snow packs across North America between 1965 and 1980 for the Earth Satellite Service of the National Oceanographic and Atmospheric Administration. The two scientists found that snow cover in North America increased three million square kilometers during that decade and a half (Dewey and Heim 1981). Consistent with this increase, Idso writes, "There has also been a trend toward earlier, more extensive snow cover in the fall and slower ablation [melting] in the spring" (Idso 1982, 44). Idso also calls on the work of B. Choudhury and B. Kukla who contend that rising carbon dioxide levels cause snow and water-covered surfaces to cool, not warm (Choudhury and Kukla, 1979).

Idso then argues that after global temperatures rose rapidly from roughly 1880 to 1945, the averages reversed trend and declined slightly from 1945 to 1975. He then leaps to the conclusion that increasing carbon dioxide levels in the atmosphere are causing temperatures to fall. In 1982, when this case was published, Idso had an arguing point (if one disregards his sloppy handling of cause-and-effect relationships). Global temperatures at that time had been falling (slightly) for three decades. In the year 2000, however, after two decades of sharply rising temperatures on a global scale, the foundations of Idso's original case have melted from under him. Idso also seems to assume that warming or cooling manifests itself in the atmosphere instantly after a given amount of carbon has entered the atmosphere, ignoring the copious literature on thermal equilibrium which delays heating or cooling several decades after the circumstances which initiate it.

Richard Lindzen theorizes that the movement of carbon dioxide in the atmosphere will nullify the effects of global warming. This movement will dry the air in the upper atmosphere and put a cap on heat retention, Lindzen theorizes (Lindzen 1990, 288). In support of Lindzen's theory, NASA's Roy Spencer asserts that the atmosphere from 25,000 to 50,000 feet is much drier than most climate modelers have thought. In theory, according to the skeptics, lower humidity in this layer of the atmosphere will limit global warming at the surface (Spencer and Braswell 1997).

Lindzen's theory has been contradicted by satellite and balloon observations indicating that a lower level temperature rise increases humidity at the lower levels of the stratosphere, not vice versa. In March, 1995, NOAA researchers found evidence of increasing moisture levels in the lower stratosphere. These readings, taken over Boulder, Colorado, contradict Lindzen's theory. Lindzen's idea that drying in the stratosphere will counteract warming in the lower atmosphere and at the surface was dismissed by the IPCC as "small and greatly outweighed by other effects of greenhouse warming" (Cline 1992, 33).

Several studies (Rind et al. 1991; Inamdar and Ramanathan 1994; Soden and Fu 1995; Sun and Held 1996; Soden 1997) contradict Lindzen's assertion that increased convection in a warmer climate would dry the upper troposphere and retard warming at the surface. These studies maintain, contrary to Lindzen, that increased convection will result in increased upper tropospheric moisture, producing a positive feedback which enhances potential global warming.

By the late 1990s, new research was indicating that surface warming and cooling of the stratosphere are related and that warming near the surface may be accelerating ozone depletion in the stratosphere. (See in Chapter 2, Ozone Depletion's Relationship to the Greenhouse Effect.)

WARMER PLANET = MORE SNOW?

A minority of scientists reject the notion that global warming will reduce polar and glacial snowpack leading to a rise of sea levels. The possibility that increased snowfall in the high latitudes could actually *reduce* net sea level has been addressed by Hengchun Ye and John Mather. They estimated the change in the amount of water that would be stored annually as snow over the polar regions, using general-circulation model estimates of temperature and precipitation assuming doubling of the carbon dioxide level in the atmosphere. The study indicates that snowfall at many high-altitude locations will rise with temperatures, because warmer (but still freezing) air would carry more moisture than colder air.

Based on their predictions, Ye and Mather suggest that the Northern Hemisphere will experience a small net snowfall increase as Antarctica's snowfall increases on the order of 100 to 300 millimeters of water equivalent (about 6 to 10 inches) of snow per year. The authors assert that "This accumulation would be most noticeable in the south polar region and total accumulation of water in the snow cover . . . [could] result in thicker ice caps, especially in Antarctica and central to northern Greenland" (Ye and Mather 1997, 155).

Ignoring copious reports of significantly rising temperatures in the Arctic and Antarctic, Edward C. Klug argues, "Far from melting in preparation for the inundation of coastal areas worldwide the polar ice sheets are growing and polar temperatures are dropping. . . . More than half of the world's glaciers are now growing" (Klug 1997). Klug asserts, without documentation, that frost frequency in Florida has increased notably, and that the citrus industry has been forced southward to avoid devastating freezes.

Contrary to eyewitness reports from the ice ship Sheba (described in Chapter 4), Klug asserts that ice cover in the Arctic has increased by 600,000 square kilometers during the summer. While Inuit on the North Slope of Alaska report that polar bears are pacing the beaches because the ice is late, Klug asserts, "[T]he summer season is now considered shortened by one month" (Klug 1997). Klug also endorses "previously inaccessible Soviet data [which] shows that over the past four decades the Arctic Ocean has cooled 4.4 degrees F." (Klug 1997). Increasing Arctic cold has caused forests in Russia to retreat south and the tundra to advance, according to Klug, as he continues to contradict the weight of contrary evidence.

Klug has good news for the "effluent society." According to his analysis, garbage and trash in landfills may be an unrecognized carbon sink. The accumulation of garbage in landfills is good for the future, Klug states. "[W]aste paper, sewage sludge and yard wastes are largely carbon compounds and insofar as they are landfilled they represent a massive human-caused withdrawal of carbon from the atmosphere that grew their raw materials" (Klug 1997).

Klug refuses even to consider the idea that industrial activity is causing the level of atmospheric carbon dioxide to rise. Klug thus concludes that the Keeling Curve is a fantasy. He faults Keeling's readings at the Mauna Loa observatory, Hawaii, because it is 17 miles from the world's largest continuously active volcano, Kilauea. He argues that gas venting from the volcano is skewing the readings. Klug ignores the fact that the Mauna Loa readings have been substantiated by many other observations around the world. Having deconstructed the case for warming, Klug concludes, "Global warming scaremongering is utterly without empirical foundation. It is a set of implausible speculations based on rigged data" (Klug 1997).

CARBON DIOXIDE "ENRICHMENT"

The case for atmospheric carbon dioxide as agricultural bromide is supported by some greenhouse skeptics with some very absolute and simplistic statements about its assumed benefits. One such statement comes from Sylvan Wittner, director of the Michigan Agricultural Experiment Station and chairman of the agricultural board of the National Research Council:

Flowers, trees and food crops love carbon dioxide, and the more they get of it, the more they love it. Carbon dioxide is the basic raw material that plants use in photosynthesis to convert solar energy into food, fiber and other forms of biomass. Voluminous scientific evidence shows that if CO_2 were to rise above its current ambient level of 360 parts per million, most plants would grow faster and larger because of more efficient photosynthesis and a reduction of water loss. (Klug 1997)

Like Wittner, Idso asserts that the Earth's ecosystem is starved for carbon dioxide. Idso paints a wondrous picture of giant plants nourished by carbon dioxide levels "in the thousands of parts per million" (Klug 1997). Idso calls such levels "a breath of fresh air for the planet's close-to-suffocating vegetation" (Klug 1997). The word "rejuvenation" is prominent in Idso's case, describing his vision of a carbon-enriched world atmosphere. Idso does not tell his audience that such carbon dioxide levels were experienced at a time when global temperatures were at a level that would make climate over much of the Earth's surface virtually insufferable during summers for warm-blooded animals.

Idso has written,

The whole face of the planet will likely be radically transformed—rejuvenated as it were—as the atmospheric CO_2 content reverses its long history of decline and returns, in significant measure, to conditions much closer to those characteristic of the Earth at the time when the basic properties of plant processes were originally established. (Idso 1982, 9)

A slight rise in temperatures and carbon dioxide levels would stimulate the growth of many plants, but any gardener knows that each plant has its heat limit. These levels vary considerably between species. Every plant has a level of temperature at which the rate of respiration surpasses the rate of photosynthesis, causing the organism to lose carbon and eventually die. Every plant also requires moisture in certain amounts. Too little (or, conversely, too much) moisture will kill many plants. Nighttime temperatures also play a crucial role in relieving moisture stress for some species. If nighttime temperatures are too high, some trees will die.

Levels of carbon dioxide above a certain level may actually stunt the growth of many plants. Many insects are very sensitive to temperature, as well as humidity, and even a small rise can cause massive outbreaks, as witnessed by the insect population explosion in New Orleans during 1995, following five years without a killing frost.

Researchers Roseanne D'Arrigo and Gordon Jacoby of the Lamont-Doherty Earth Observatory studied tree rings near the timberline in northern and central Alaska's boreal forests. They found that growth increased with modest increases in warming during the 1930s and 1940s. After that, increased warming produced heat and drying that, along with increasing activity by insects, stressed the trees and caused tree growth to slow most years. Growth was higher than usual during years when moisture was plentiful, but below average during relatively warm, dry years. In the meantime, insects usually flourish in warmer weather. The bark beetle, a major pest in Alaska's forests, has shortened its reproductive cycle to half due to warming, according to Jacoby and D'Arrigo (Gelbspan, *The Heat is On*, 1997, 141). In the meantime, David Deming has measured soil temperatures in northern Alaska. His records indicate that the soil there has warmed by two to five degrees C. (3.6 to 9 degrees F.) during the twentieth century.

CORPORATE SPONSORSHIP OF THE SKEPTICS

The fossil-fuel industry had its own response to the notably hot summer of 1988. In 1989, industrial interests formed the Global Climate Coalition (GCC), to act as the "leading voice for business and industry in the climate change debate" (Ewen 1999, 27).

At its inception the GCC included corporate members as well as industry lobbying groups such as the American Petroleum Institute, the American Iron and Steel Institute, the American Automobile Manufacturers Association, the American Forest and Paper Association, the Air Transport Association, the Association of American Railroads, the Chemical Manufacturers Association, the National Association of Manufacturers, the National Mining Association, the Society of the Plastics Industry and the U.S. Chamber of Commerce.

The GCC's agenda, described by Alexander Ewen, is

to prevent any attempt to regulate its members or hold them responsible for global warming and its effects. Although the GCC purportedly believes that science should rule the debate on climate change, like all of the industry-sponsored media initiatives, it has vigorously attacked any science that links global warming to greenhouse gases, and promotes any view that casts doubt upon global warming. (Ewen 1999, 27–28)

The GCC and other industry lobbies sought to mount an attack on the scientific basis supporting the assertion that the Earth is warming because of fossil-fuel combustion. To do this, the GCC hired its own scientists. Thus was formed the Information Council on the Environment (ICE) which was initially funded by the Edison Electric Institute, the Southern Company, and the Western Fuels Association.

The ICE, founded in 1991, was designed to raise doubts by placing scientists who were skeptical about global warming in a position to address large audiences via various news media. Scientists emphasized that global warming was theory, not fact, and that the threat, if any existed, was not as serious as other scientists have asserted. Three major skeptics whose views were promoted by ICE included Michaels, Balling, and Idso.

Michaels and the ICE cemented their ideological marriage in 1989, after he wrote in the *Washington Post* that the main cause of concern regarding global warming was "apocalyptic environmentalism," which he described as "the most popular new religion to come along since Marxism" (Ewen 1999, 28). Michaels moaned that special interests and "consensus science" were exploiting fear on the subject. Michaels and other scientists attacked the credibility of climate models used to forecast global warming, and asserted that no action should be taken until all scientists were certain that global warming is really a problem.

Stephen H. Schneider engaged Michaels in debate regarding the nature of "consensus science."

In a debate in London in September of 1994 sponsored by the Royal Society and the Ciba Scientific Foundation . . . contrarian climatologist Patrick Michaels . . . accused me of preaching "science by consensus," rather than science by the scientific method. "No, Pat," I recall saying, "I don't believe in doing *science* by consensus. But I do believe consensus is the best way to approach *science policy*." I went on to suggest that while it may be so that for every Ptolomey there eventually comes a Copernicus, I'll also bet that for every true Copernicus, there are at least 1,000 pretenders. Where I think Michaels fits in that analogy, I'll leave to your imagination. But I will state directly that the reason people who overthrow dogma are so celebrated is precisely because they are so rare. (Schneider, 2001)

The Western Fuels Association, representing the coal-mining industry in the United States, funded a magazine, *World Climate Review* (later *World Climate Report*). In 1991, the ICE also spent about $250,000 on a video, "The Greening of Planet Earth," in which "scientific" testimony was used to support an assertion that global warming will increase agricultural yields (including corn, wheat, and cotton) by 30 to 60 percent.

"The Greening of Planet Earth," which was screened in President George Bush's White House, asserts that vast areas of desert will be reclaimed by new grasslands brought to life by rising carbon dioxide levels, as Earth's forests spread under a carbon-enriched atmosphere. The video "overlooks the bugs" (Gelbspan, *The Heat is On*, 1997, 37).

During 1992, Michaels assumed the editorship of the *World Climate Review*. Michaels also received funding from the German Coal Mining Association, the Edison Electric Institute, and Cyprus Minerals, one of the leading backers of the anti-environmental Wise Use Movement. By 1999, Michaels had received more than $165,000 in research grants from the energy industry. Balling, who is Director of the Office of Climatology at Arizona State University, has taken in more than $700,000 from energy interests (Ewen 1999, 28).

The ICE funded a public-relations campaign, beginning shortly after 1990, to "reposition global warming as theory rather than fact" (Gelbspan, *The Heat is On*, 1997, 34). The strategy was to utilize the words of a few scientists to create general skepticism among the public and thereby forestall pressure from government. The anti-greenhouse "experts" heard most often included Michaels, Balling, Idso, Singer, and Lindzen, who, wrote Ross Gelbspan, "have proven extraordinarily adept at draining the [global warming] issue of all sense of crisis . . . [to] . . . create a broad public belief that the question of climate change is hopelessly mired in unknowns" (Gelbspan, *The Heat is On*, 1997, 35).

Gelbspan (1997, *The Heat is On*, 38–48) provides a detailed catalogue of grants by various fossil-fuel-supported groups to the most prominent skeptics of the greenhouse effect. According to Gelbspan, Lindzen charges oil and coal interests $2,500 a day for consulting services. Balling received more than $200,000 from coal and oil interests in Great Britain, Germany, and elsewhere. Michaels has received more than $115,000 from coal and energy interests (Gelbspan 1995).

During the early and middle 1990s, fossil-fuel industrial interests bankrolled several public-relations efforts. The Global Climate Coalition spent more than a million dollars a year to downplay the threat of climate change. The National Coal Association spent more than $700,000 on the global climate issue in 1992 and 1993. During the same year, the American Petroleum Institute, which also sits on the GCC, paid $1.8 million to the public relations firm of Burson-Marsteller, mainly to lobby against a proposed tax on fossil fuels. These expenditures roughly matched the $2.1 million being spent annually on the other side of the issue by the Environmental Defense Fund, the Natural Resources Defense Council, the Sierra Club, the Union of Concerned Scientists, and the World Wildlife Fund (Gelbspan 1995).

On Earth Day, 1998, Fredrick Palmer, chief executive officer of the Western Fuels Association, launched the Greening Earth Society, which preaches the benefits of "moderate" global warming. The Greening Earth Society, with a head office in Arlington, Virginia, asserts "that humankind's industrial revolution is good and not bad and that humans utilizing fossil fuels to enable our economic activity is as natural as breathing" (Johnson 1999, 20). The Greening Earth Society believes that higher carbon-dioxide levels will "produce a tremendous greening of planet Earth" (Johnson 1999, 20).

Following the announcement of a general scientific consensus regarding global warming by the IPCC in its *Second Assessment*, coal and oil companies in the United States mobilized a multi-million dollar public relations plan (Cushman 1998, A-1). The same coalition also lobbied against the Kyoto protocol. To undermine planning for the Kyoto agreement, during 1997, several energy corporations planned, according to documents obtained by the *New York Times*, "to recruit a cadre of scientists who share the industry's views of climate science and to train them in public relations so they can help convince journalists, politicians, and the public that the risk of global warming is too uncertain to justify controls on greenhouse gases like carbon dioxide that trap the sun's heat near Earth" (Cushman 1998, A-1). This plan was spearheaded by Joe Walker, a public relations representative of the American Petroleum Institute (API). Scientific support for this public relations campaign was recruited by Frederick Seitz, a physicist who has served as president of the American Physical Society, president of the National Academy of Sciences, and president emeritus of Rockefeller University. Seitz is also a severe critic of the idea that human activity is fundamentally altering the Earth's climate. He has consulted for the George C. Marshall Institute and the Advancement of Sound Science Coalition (Montague N.d.). Seitz calls rising carbon dioxide levels in the atmosphere "a wonderful and unexpected gift from the Industrial Revolution" (Stevens, *The Change*, 1999, 244).

During 1998, the American Petroleum Institute assembled plans to spend more than $5 million on global-warming spin-control. The goal of this effort is to convince the American public that no scientific basis exists for measures, especially restrictions on use of fossil fuels, to counter global warming. Spin-control was to be practiced by the API with the aid of willing scientists trained in public relations. The API sought, according to Alexander Ewen, to

Inject credible science and scientific accountability into the global climate debate, thereby raising questions about and undercutting the "prevailing scientific wisdom." A team of five "independent scientists" would be identified, recruited and trained to participate in media outreach—that is, radio and television talk shows—and $600,000 was earmarked for science writers, editors and journalists. All, in the words of the institute, geared to "maximize the impact of scientific views consistent with ours on Congress, the media and other key audiences." Among those corporations actively involved in drafting this plan were Exxon, Chevron and the Southern Co. (Ewen 1999, 27)

As the 1990s closed, Idso continued to promote global warming as a boon to agriculture. Idso, an agriculturist with the U.S. Department of Agriculture Water Conservation Laboratory, authored *Carbon Dioxide and Global Change: Earth in Transition*. After watching two decades of accelerating global temperatures, Idso had stopped arguing that rising atmospheric levels of carbon dioxide might cool the Earth, as he did until 1982. Instead, Idso believes that a doubling of greenhouse-gas levels will raise global temperatures only 0.4 degrees C. A Western Fuels Association video, which featured Idso, promoted the view that a little warming will produce a planet clothed in verdant green, nourished by perpetual springtime. Idso's two sons, both formerly under contract with Western Fuels Association, by the late 1990s headed the Center for the Study of Carbon Dioxide and Global Change, another group which promotes the benefits of rising carbon dioxide levels (Ewen 1999, 28).

Ewen, writing in *Native Americas*, described the public relations full-court press which the industry groups assembled:

Major damage-control public relations specialists, such as Burson-Marsteller, who spun the Bhopal, Exxon Valdez and James Bay campaigns, were employed. Industry associations, such as the National Coal Association, the American Petroleum Institute, and others, made opposition to greenhouse-gas emission controls their number-one agenda item. For years, Mobil placed advertising on the op-ed page the *New York Times* prominently featuring the work of Balling and Singer as part of the company's assertions that global warming was nothing to worry about. (Ewen 1999, 31)

The industry groups networked with far-right political groups throughout the United States and adopted parts of their ideologies. They then began a campaign to portray global-warming diplomacy accords as a conspiracy by big government to aid a United Nations takeover of the United States which would deprive Americans of their land, sovereignty, and way of life.

A group calling itself the Global Climate Control Project spent more than $13 million in the months preceding the 1997 Kyoto Conference. The GCCP's biggest budget item was advertising aimed mainly at the U.S. Senate, which passed, 95 to 0, a resolution stating flatly that it would not ratify any global-warming treaty which exempted developing nations from worldwide emission cuts.

SOME CORPORATIONS BREAK RANK WITH THE SKEPTICS

By the late 1990s, some notable cracks were forming in corporate diffidence toward global warming. The chief executive officer of British Petroleum (BP) Co., the world's third-largest publicly owned oil-and-gas company, pledged on September 18, 1998, to reduce the company's greenhouse-gas emissions to 10 percent below 1990 levels by the year 2010. In May, 1997, Sir E. John P. Browne, British Petroleum's chief, became the first oil-industry official to publicly accept the role which fossil fuels play in global warming. During 1999, Royal Dutch Shell joined British Petroleum with a pledge to cut its contributions to global warming. The company also was investing in alternative fuels, including solar and wind power.

British Petroleum (which has merged with Amoco) positioned itself as "the world's largest solar-energy company" (Sudetic 1999, 129), as it seeks ways to reduce greenhouse-gas emissions from its refineries and increase use of cleaner natural gas (as opposed to oil). Company officials believe they can reduce BP/Amoco's greenhouse-gas emissions "without compromising either growth or profits" (Hamilton 1998, A-6). In 1990, the company's operations produced roughly 40 million tons of carbon dioxide and other greenhouse gases. Browne, who made his pledge during a speech at Yale University's School of Management, said that British Petroleum's 90 business units plan to reduce greenhouse-gas emissions mainly by improving energy efficiency and using new technology. For example, Browne said that BP has installed new technology to enhance oil flow, and this will eliminate a number of pumping stations on the Trans-Alaska Pipeline, a move that reduced carbon dioxide emissions on that site by 236,000 tons a year. The oil company plans to meet its goals with advice from groups such as Environmental Defense, as well as its own engineers and other employees.

Dissent to the skeptics' beliefs was being heard during the late 1990s even within the boardrooms of some coal-fired electric utilities. Dale Heydlauff, American Electric Power's (AEP) vice president for environmental affairs, said, "You can't help but be alarmed at the temperature record of the last two decades. There are enough warning signs out there that we [have] recognized [that] we need to do our part to begin looking at options to solve this problem" (Sudetic 1999, 129). AEP's environmental course is not an easy one politically. As a primary customer for coal mined in the eastern United States, the utility is constantly under pressure to help maintain payrolls in the mining industry.

Robert N. Stavins, an environmental economist at the John F. Kennedy School of Government at Harvard University, cautioned that reducing greenhouse-gas emissions in oil and gas production does not affect greenhouse-gas pollution from the burning of gasoline and other fuels manufactured by energy companies. "They're of much greater magnitude than the emissions that come from production," he said (Hamilton 1998, A-6). "But I hope we can show—in a small way—what can be done. And I hope we can help the process forward" (Hamilton 1998, A-6).

An executive of the American Petroleum Institute said that the move by BP may be having an effect on other oil companies. "A number of companies are already doing things to control their carbon-dioxide emissions, but I don't know if any companies will set their own targets and announce it," said William F. O'Keefe, executive vice president at API (Hamilton 1998, A-6). In the meantime, Texaco was announcing the acquisition of a fuel-cell company, and investment in a Mississippi Valley reforestation project. Texaco president Peter Bijur gave a little-noticed speech that "posited a world where we will see multiple ways to power cars—hybrids, advanced batteries, fuel cells—even cars that run on pure hydrogen" (Liesmen 2000, B-4).

By the late 1990s, a number of corporations in industries other than energy also were breaking ranks with anti-warming lobbying efforts. The Chicago-based Boeing Company was working to become a leader in the corporate battle against global warming. During May of 1998, Boeing joined 12 other Fortune 500 corporations in a media campaign coordinated by the Pew Center for Global Climate Change. The media campaign supported the scientific basis supporting global warming. Boeing is planning for a time when aircraft emissions will be controlled more strictly. "Moreover," wrote Chris Carrel in the *Seattle Weekly*, "lessening its airplanes' impacts on the atmosphere may give Boeing a competitive advantage over its rival, Airbus" (Carrel 1998). Ford Motor Co. failed to renew its membership in the Global Climate Coalition during 1999. Daimler Chrysler and General Motors also withdrew in 1999. At about the same time, Kelly Sims, science-policy director with Ozone Action, said,

With technology-oriented corporations such as Boeing, Maytag, Whirlpool, United Technologies, Minnesota Mining, Milling, and Manufacturing (3M), and American Electric Power worrying out loud about the weather, the climate-change contrarians are becoming increasingly marginalized. (Carrel 1998)

Among some corporate stockholders, accountability for greenhouse-gas emissions became an issue during the late 1990s. Twenty-eight percent of Chevron Oil shareholders supported a shareholder proposal demanding that the company document its emissions of pollution that cause global warming and assess the resulting financial liabilities. Such resolutions are one of the few democratic measures by which small shareholders may raise politically sensitive subjects before the owners and managers of a company. The resolutions are not legally binding, but they are intended to focus management's attention on a given issue.

The Chevron proposal was sponsored by investors representing major religious groups. It asked Chevron to catalog actions by the company or its trade associations "promoting the view that the issue of climate change is exaggerated, not real, or that global warming may be beneficial" (Chevron 1999). The Chevron resolution was initiated by the Ancilla Domini Sisters and the Benedictine Sisters of Boerne, Texas, along with the Congregation of the Sisters of Charity of the Incarnate Word in Houston, as well as the Missionary Oblates of Mary Immaculate, Franciscan Fathers, and the School Sisters of Notre Dame Cooperative Investment Fund.

"This vote should make companies like Exxon and others very nervous," said Sister Patricia Daly, of the Sisters of St. Dominic of Caldwell, New Jersey. "Like Chevron, they must start factoring global warming into their long-term plans or risk being outpaced by more forwarding-looking competitors" (Chevron 1999). By the summer of 1999, other companies, including Ford, General Motors, Allegheny Power, and Reynolds Aluminum, faced comparable shareholder actions.

By the middle 1990s, London's Delphi Group, which advises large institutions on their investment strategies, recommended that banks, insurers, and other large institutional investors begin to withdraw their investments from oil and coal companies, with their traditionally lucrative returns. Gelbspan wrote,

The report noted that continuing disturbances in the global climate could easily lead to high carbon taxes and enforced reductions on oil and coal use. "As a result," noted the report's author, Mark Mansley, a former financial analyst for Chase Manhattan Bank, "Climate change presents major long-term risks to the carbon fuel industry [that have] not been adequately discounted by the financial markets." (Gelbspan, A Global Warming, 1997)

At a 1995 conference on climate change for leaders of financial institutions, Sven Hansen, vice president of the Union Bank of Switzerland, called climate change the "single most important environmental problem for the world today" (Gelbspan, A Global Warming, 1997).

An influential business group, the International Climate Change Partnership (ICCP) includes General Electric, AT&T, Allied Signal, Dow Chemical, 3M, Dupont, Enron, and Electrolux. ICCP members agree on a need for an enforceable set of carbon dioxide reductions. "The fundamental science of global warming is pretty basic," ICCP Executive Director Kevin Fay has been quoted as saying. "There is some uncertainty about specific effects and impacts, but we understand that there is a long lag time for atmospheric greenhouse gases—and that it also takes a long time to develop remedies for the problem" (Gelbspan, A Global Warming, 1997). Many ICCP members believe they can profit from the development of renewable energy technologies. These corporations, like others, are beginning to realize that a paradigm change in energy-generation technology could be good for businesses in which they are positioned to participate.

In a reflection of rising corporate concern regarding global warming, in the middle of 2000, the *Harvard Business Review* published an article titled "What Every Executive Needs to Know About Global Warming," which said, "While a few skeptics still question the evidence, most experts believe that human activity is contributing, at least in part, to an increase in the Earth's average surface temperature" (Packard and Reinhardt 2000, 130). "It's no wonder," Kimberly O'Neill Packard and Forest Reinhardt wrote, "that business leaders at the most recent World Economic Forum in Davos, Switzerland, voted global climate change as the most pressing issue confronting the world's business community" (Packard and Reinhardt 2000, 130). The reinsurer Swiss Re by the late 1990s employed in-house specialists to track the emerging risks of climate change.

JAMES HANSEN AS A SKEPTIC?

By the turn of the millennium, global-warming skeptics seemed to be grasping at straws to keep their case alive, as world temperatures continued to escalate rapidly. When James E. Hansen and several colleagues published two introspective pieces in scientific journals on the role of chaos (among other non-greenhouse factors) in climate change, many skeptics rushed to their Internet Web pages with declarations that their long-time adversary was backing off his belief that infrared forcing is a major cause of contemporary global warming.

Hansen and colleagues published "Climate Forcings in the Industrial Era" in the *Proceedings of the National Academy of Sciences in the United States.* What caught the skeptics' eyes in this article was the first sentence of the abstract: "The forcings that drive long-term climate change are not known with accuracy sufficient to define future climate change" (Hansen et al., Climate Forcings, 1998, 12753). What the skeptics took to be a confirmation of their biases actually develops into a considered explication of the problems facing scientists who attempt to forecast how much the Earth may warm during coming decades.

The major part of this paper estimates the relative values of various "forcings"—"imposed perturbation[s] of the Earth's energy balance with space" (Hansen et al., Climate Forcings, 1998, 12753), including greenhouse gases, aerosols, changes in cloud behavior, land-use changes, solar irradiance, and volcanic eruptions. Some of these factors (especially the behavior of clouds) are so uncertain that quantifying them does little but accentuate the risky nature of the enterprise.

Hansen and 42 co-authors elaborated their attempts to balance the relative strengths of various climatic forcings in the November 27, 1997 issue of the *Journal of Geophysical Research*. Two sentences in the abstract of this article, taken out of context, compelled some skeptics to claim that Hansen was going soft on the greenhouse effect: "The experiments [summarized in the article] suggest that most interannual climate variability in the period 1979–1996 at middle and high latitudes is chaotic. . . . The net calculated effect of all measured radiative forcings is approximately zero surface temperature trend and zero heat storage in the ocean for the period 1979–1996" (Hansen et al., Forcings and Chaos, 1997, 25,679).

At this point, the skeptics seem to have thrown up their hands in joy, and quit reading. Relishing the irony of what he took to be Hansen's newfound climate skepticism, Michaels wrote:

Hansen et al. . . . recently calculated that the concentrations of carbon dioxide in the atmosphere are increasing at approximately 60 percent of the rate that is normally projected [*sic*]. Notably, he argues that the biosphere is absorbing CO2 at a rate much faster than anticipated, as he wrote that "Apparently the rate of uptake by CO2 sinks, either the ocean, or, more likely the forests and soils . . . has increased." (Michaels 1998)

The Global Climate Coalition quickly embraced Hansen's statements out of context, without so much as asking him what they meant. Suddenly, the Internet was popping with "anti-warmists" exuding joy over the supposed abdication of the "global warmist" whom they had most loved to hate. The antiwarming lobby was elated by out-of-context citations of comments by Hansen and his coauthors.

Hansen's abstract goes on to say that most of the heat the research team is measuring has not yet been expressed in the temperature of the ocean. Expression of warming in the oceans lags emission of greenhouse gases in the atmosphere by several decades. The abstract explains that the real news is yet to come: "One implication of the disequilibrium forcing is an expectation of new record global temperatures in the next few years" (Hansen et al., Forcings and Chaos, 1997, 25,679). The text of the article, which was prepared in 1997, warned that worldwide temperatures would rise to records during the next three years. The next year, 1998, they did just that. Hansen said, "In my opinion, the rate of ocean heat storage is the most fundamental number for our understanding of long-term climate change" (Stevens, Oceans Absorb, 2000, A-16).

Hansen and colleagues closed their article in the *Journal of Geophysical Research* with a warning: "Quantification of the radiation imbalance has practical importance, because it is the possible existence of such 'unrealized warming' that makes a global climate policy of 'wait and see' problematical and perhaps dangerous" (Hansen et al., Forcings and Chaos, 1997, 25,714). In both papers, Hansen and colleagues also noted a declining trend in the rate of atmospheric carbon dioxide accumulation since 1975, as well as a decline in methane levels during the 1990s, and asked whether these changes are long-term, or "a temporary biospheric uptake presaging a later burst of CO2 growth" (Hansen et al., Climate Forcings, 1998, 12,757). In their final analysis, Hansen and colleagues still expect a doubling of carbon dioxide levels within a century, along with a three degree C. rise in temperature, plus or minus one degree (Hansen et al., Climate Forcings, 1998, 12,757).

A close look at Hansen's recent writings does not indicate that he has backed off any of his warnings about the dangers of global warming. Hansen has been attempting to estimate how much global warming must manifest itself in the lives of the world's residents to create the political consensus necessary for fundamental movement away from the fossil-fueled economy which now dominates worldwide commerce. In certain areas of Asia and Alaska, the degree of warming is already perceptible to most people in their daily experiences. "If global warming proceeds according to our climate-model projections," wrote Hansen and colleagues, "There should be a large increase of the area with obvious climate change during the next several years" (Howler 1999). Hansen commented on an Internet page under the aegis of a Boston television station,

My colleagues at the Goddard Institute and I have proposed a simple index, the Common Sense Climate Index (CSCI), that measures the degree, if any, to which practical climate change is occurring. The Index is a composite of climate quantities that are noticeable to people such as the frequency of extreme temperatures and precipitation. . . . [T]he record does not show convincing evidence of a long-term change in climate during the past 50 years. A global picture of the Climate Index for the most recent three years of analyzed data (1995–1997) indicates that most of the world is positive, consistent with global warming. . . . The only areas where warming does not seem to have set in is the eastern half of the United States and a large swath running southeastward from Eastern Europe to China. (Hansen 1999; Hansen et al., Common-sense, 1998)

Far from dismissing the dangers of global warming due to human activity, Hansen and associates are concerned with "the practical detection issue. . . . [W]hen will global warming be large enough to be obvious to most people? Until then, it may be difficult to achieve consensus on actions to limit climate change" (Howler 1999). In an October 1998 forum on "Climate Change and the Oil Industry," Hansen said,

So, the question is, what is the bottom line? What is the implication for fossil fuels and the fossil-fuel industry? I think that this summary paragraph from our paper two decades ago is still valid. But the evidence of global warming has become a lot stronger, and I think it's going to be more obvious to more people soon. . . . I think that actions are, in fact, called for now. (Howler 1999)

Hansen did not sound like much of a skeptic when he observed, on a website maintained by NASA,

The rapid warming of the past 25 years undercuts the argument of 'greenhouse skeptics,' who have maintained that most of the global warming occurred early this century while greenhouse gases were increasing more slowly—in fact, the fastest warming is occurring just when it is expected. . . . There should no longer be an issue about whether global warming is occurring, but what is the rate of warming, what is its practical significance, and what should be done about it. (Global Temperature N.d.)

In the August 29, 2000 issue of the *Proceedings of the National Academy of Sciences*, Hansen and colleagues wrote that the quickest way to slow global warming could be first to cut emissions of gases other than carbon dioxide. Their study concludes that roughly half of the greenhouse-gas-induced warming during the twentieth century was caused by gases other than carbon dioxide, most notably methane, chlorofluorocarbons, black particles of diesel and coal soot, and compounds which create ozone.

"We argue that rapid warming in recent decades has been driven mainly by non-CO2 greenhouse gases," Hansen and colleagues wrote. "The growth rate of non-CO2 greenhouse gases has declined in the past decade" (Hansen et al. 2000, 9875). In their paper, Hansen and associates suggest that existing technology provides ways to reduce the net change in atmospheric methane levels to nearly zero by the year 2050. Methane levels, which flattened during the 1990s, seem to be easiest to influence, according to Hansen, by such measures as controlling leakage from energy plants and pipelines, changing the diets of ruminant animals, and changing some of the ways in which rice is cultivated. Methane also has a relatively short lifetime in the atmosphere (about eight years) so reductions in its production will have an impact in the atmosphere more quickly than reductions in carbon dioxide, with an atmospheric lifetime of a century or more.

Because many of the greenhouse gases (other than carbon dioxide) cause serious, costly health problems or can harm agriculture, there are "strong economic reasons for wanting to eliminate them," Hansen and colleagues wrote (Revkin 2000, A-13). "A key feature of this strategy is its focus on air pollution, especially aerosols and tropospheric ozone," Hansen and associates wrote (Hansen et al. 2000, 9879).

Lest the skeptics attempt once again to enlist Hansen's measured optimism in support of their belief that global warming is not a threat, Hansen and colleagues wrote that "improved energy efficiency and a continued trend toward decarbonization of energy sources" will be required to stem global warming. They suggest, "There are opportunities to achieve reduced emissions consistent with strong economic growth" (Hansen et al. 2000, 9878). "The prospects for having a modest climate impact instead of a disastrous one are quite good, I think," Hansen said (Revkin 2000, A-13). Hansen stressed that he is still convinced global warming is underway and that human activity is a significant cause. In fact, the National Academy of Sciences paper prepared by Hansen and associates suggests that in coming years the role of carbon dioxide in climate change could increase: "CO2 will become the dominant climate forcing, if its emissions continue to increase and aerosol [cooling] effects level off [from pollution controls]" (Hansen et al. 2000, 9880).

Hansen's most recent analysis faulted some climate models for failing to consider improved energy efficiencies which seem to be keeping the increase in atmospheric carbon dioxide below the rate of economic growth worldwide. At the same time, the analysis of Hansen and his co-authors acknowledges considerable uncertainty about the future course of climate due to human-induced "forcings" of the atmosphere. Reductions in carbon dioxide accumulation rates during the late twentieth century may have resulted in part from thus-far unquantified natural carbon sinks "which may be temporary" (Hansen et al. 2000, 9877).

Robert T. Watson, chairman of the United Nations Intergovernmental Panel on Climate Change, said he had received a flurry of phone calls from other scientists expressing worries that Hansen's study could be misportrayed. "They said this could easily be interpreted as the guy who got Vice President Gore all excited about global warming now saying everything's fine," Dr. Watson said, adding, "If this paper is viewed that way, it'd be a horrible distortion. If anything, our projections for warming are higher than they used to be, and we're seeing discernible changes in ecosystems that we can link back to climate" (Revkin 2000, A-13).

REFERENCES

Abrupt Climate Change During Last Glacial Period Could be Tied to Dust-Induced Global Warming. 1996. Press Release NOAA 96–78, 4 December. [http://www.noaa.gov/public-affairs/pr96/dec96/noaa96–78.html]

Ackerman, A.S., O.B. Toon, D.E. Stevens, A.J. Heymsfield, V. Ramanathan, and E.J. Welton. 2000. "Reduction of Tropical Cloudiness by Soot." *Science* 288 (12 May): 1042–1047.

AEI Study Calls for Delayed Action on Global Warming. 1999. 28 July. [http://www.weathervane.rff.org/negtable/AEIStudy.html]

Baliunas, Sallie, and Willie Soon. N.d. "A Scientific Discussion of Climate Change: Comments on 'The Truth About 10 Leading Myths.' " Marshall Institute. [http://www.marshall.org/response.html]

Balling, Robert C., Jr. 1992. *The Heated Debate: Greenhouse Predictions Versus Climate Reality*. San Francisco: Pacific Research Institute for Public Policy.

Bolin, Bert, et al. 1995. *Intergovernmental Panel on Climate Change. Second Assessment Synthesis of Scientific-Technical Information Relevant to Interpreting Article 2 of the United Nations Framework Convention on Climate Change*. Approved by the IPCC at its eleventh session, 11–15 December, Rome. [http://www.unep.ch/ipcc/pub/sarsyn.htm.]

Brown, Kathryn. 1999. "Climate Anthropology: Taking Global Warming to the People." *Science* 283 (5 March): 1440–1441.

Burroughs, William J. 1999. *The Climate Revealed*. Cambridge, U.K.: Cambridge University Press.

Carrel, Chris. 1998. "Boeing Joins Fight Against Global Warming." *Seattle Weekly*, 17 September. [http://climatechangedebate.com/archive/09–18_10–27_1998.txt]

Charlson, R.J., S.E. Schwartz, J.M. Hales, R.D. Cess, J.A. Coakley, Jr., J.E. Hansen, D.J. Hofmann. 1992. "Climate Forcing by Anthropogenic Aerosols." *Science* 255 (24 January): 423.

Chevron Stockholder Vote on Global Warming Resolution Surprises Annual Meeting. 1999. Environmental Media Services/Ozone Action, 29 April. [http://www.corpwatch.org/trac/corner/worldnews/other/355.html]

Choudhury, B., and G. Kukla. 1979. "Impact of CO_2 on Cooling of Snow and Water Surfaces." *Nature* 280:668–671.

Christianson, Gale E. 1999. *Greenhouse: The 200-Year Story of Global Warming*. New York: Walker and Company.

Cline, William R. 1992. *The Economics of Global Warming*. Washington, D.C.: Institute for International Economics.

Countdown to Kyoto: The Consequences of Mandatory Global CO_2 Emission Reductions. 1997. Remarks by United States Senator Chuck Hagel, 21 August, Canberra, Australia. Office of Senator Chuck Hagel. [http://www.altgreen.com.au/gge/CtK_Hagel.html]

Cushman, John H., Jr. 1998. "Industrial Group Plans to Battle Climate Treaty." *New York Times*, 26 April, A-1, A-24.

Dewey, K.F., and R. Heim, Jr. 1981. "Satellite Observations of Variations in Northern Hemisphere Seasonal Snow Cover." NOAA Technical Report NESS 87. Washington, D.C.: U.S. Department of Commerce.

Ewen, Alexander. 1999. "Consensus Denied: Holy War Over Global Warming." *Native Americas* 16, no. 3/4(Fall/Winter): 26–33. [http://nativeamericas.aip.cornell.edu]

Friis-Christensen, E., and K. Lassen. 1991. "Length of the Solar Cycle: An Indicator of Solar Activity Closely Associated with Climate." *Science* 254:698–700.

Gaffen, Dian J. 1998. "Falling Satellites, Rising Temperature?" *Nature* 394 (13 August): 615–616.

Gelbspan, Ross. 1995. "The Heat is On: The Warming of the World's Climate Sparks a Blaze of Denial." *Harper's Magazine*, December. [http://www.dieoff.com/page82.htm]

———. 1997. *The Heat is On: The High Stakes Battle Over Earth's Threatened Climate*. Reading, Mass.: Addison-Wesley Publishing Co.

————. 1997. "A Global Warming." *American Prospect* 31 (March/April). [http://www.prospect.org/archives/31/31gelbfs.html]

Gillon, Jim. 2000. "The Water Cooler." *Nature* 404 (6 April): 555.

Global Temperature Trends: 1998 Global Surface Temperature Smashes Record. N.d. [http://www.giss.nasa.gov/research/observe/surftemp]

Global Warming Pests and Pestilence. 1995–2000. In *What's Hot: World Climate Report Archives: 1995–2000.* Vol. 2, no. 2. [http://www.nhes.com/back_issues/Vol1and2/WH/hot.html]

Hagel, Chuck, Senator. N.d. Interview on "Global View." [http://www.oche.de/~norb/chuck.html]

Hamilton, Martha M. 1998. "British Petroleum Sets Goal of 10 per cent Cut in 'Greenhouse Gases.' " *Washington Post*, 18 September, A-6. [http://climatechangedebate.com/archive/09–18_10–27_1998.txt]

Hansen, James E. 1999. "Global Warming, a Time for Action." WHOH-TV, Boston, Mass. [http://www.7almanac.com/1999/pg10.htm]

————. 1997. "Radiative Forcing and Climate Response."*Journal of Geophysical Research* 102:6831–6864.

Hansen, J.E., and A.A. Lacis. 1990. "Sun and Dust Versus Greenhouse Gases: An Assessment of Their Relative Roles in Global Climate Change." *Nature* 346:713–718.

Hansen, James, Andrew Lacis, Reto Ruedy, and Makiko Sato. 1992. "Potential Climate Impact of the Mount Pinatubo Eruption." *Geophysical Research Letters* 19:2 (24 January): 215–218.

Hansen, James, Makiko Sato, Jay Glascoe, and Reto Ruedy. 1998. "A Common-sense Climate Index: Is Climate Changing Noticeably?" *Proceedings of the National Academy of Sciences of the United States of America* 95 (April): 4113–4120.

Hansen, James E., Makiko Sato, Andrew Lacis, Reto Ruedy, Ina Tegen, and Elaine Matthews. 1998. "Climate Forcings in the Industrial Era." *Proceedings of the National Academy of Sciences of the United States of America* 95 (October): 12,753–12,758.

Hansen, James, Makiko Sato, Reto Rueddy, Andrew Lacis, and Valdar Oinas. 2000. "Global Warming in the Twenty-first Century: An Alternative Scenario." *Proceedings of the National Academy of Sciences of the United States of America* 97, no. 18 (29 August): 9875–9880.

Hansen, James E., W. Rossow, B. Carlson, A. Lacis, L. Travis, A. Del Genio, I. Fung, B. Cairns, M. Mishchenko, and M. Sato. 1995. "Low-cost, Long-term Monitoring of Global Climate Forcings and Feedbacks." *Climatic Change* 31:117–141.

Hansen, James E., G. Russell, D. Rind, P. Stone, A. Lacis, S. Lebedeff, R. Ruedy, and L. Travis. 1983. "Efficient Three-dimensional Global Models for Climate Studies: Models I and II." *Monthly Weather Review* 111:609–662.

Hansen, J.E., M. Sato, and R. Ruedy. 1995. "Long-term Changes of Diurnal Temperature Cycle: Implications About Mechanisms of Global Climate Change." *Atmospheric Research* 37:175–209.

Hansen, J., M. Sato, and R. Ruedy. 1997. "The Missing Climate Forcing." *Philosophical Transactions of the Royal Society of London* B352:231–240.

Hansen, James E., M. Sato, R. Ruedy, A. Lacis, K. Asamoah, K. Beckford, S. Borenstein, E. Brown, B. Cairns, B. Carlson, B. Curran, S. de Castro, L. Druyan, P. Etwarrow, T. Ferede, M. Fox, D. Gaffen, J. Glascoe, H. Gordon, S. Hollandsworth, X. Jiang,

C. Johnson, N. Lawrence, J. Lean, J. Lerner, K. Lo, J. Logan, A. Luckett, M.P. McCormick, R. McPeters, R. Miller, P. Minnis, I. Ramberran, G. Russell, P. Russell, P. Stone, I. Tegen, S. Thomas, L. Thomason, A. Thompson, J. Wilder, R. Wilson, and J. Zawodny. 1997. "Forcings and Chaos in Interannual and Decadal Climate Change." *Journal of Geophysical Research* 102, no. D22 (27 November): 25,679–25,720.

Hawkes, Nigel. 2000. "Giant Cities are Creating Their Own Weather." *Times of London*, 23 February, in LEXIS.

Hileman, Bette. 1995. "Climate Observations Substantiate Global Warming Models." *Chemical & Engineering News*, 27 November. [http://pubs.acs.org/hotartcl/cenear/951127/pg1.html]

Hobbs, P.V., J.S. Reid, and R.A. Kotchenruther. 1997. "Direct Radiative Forcing by Smoke from Biomass Burning." *Science* 275:1776–1778.

Hoffert, Martin, Ken Caldeira, Curt Covey, Philip P. Duffy, and Benjamin D. Santer. 1999. "Solar Variability and the Earth's Climate." *Nature* 401 (21 October): 764–765.

"The Howler Epilogue: A Global Warning." 1999. *The Daily Howler*, 3 May. [http://www.dailyhowler.com/h050399_html]

Hoyt, Douglas V. 2000. Greenhouse Warming: Fact, Hypothesis, or Myth? January. [http://users.erols.com/dhoyt1]

Idso, Sherwood B. 1982. *Carbon Dioxide: Friend or Foe?* Tempe, Ariz.: IBR Press.

———. 1989. *Carbon Dioxide and Global Change: Earth in Transition.* Tempe, Ariz.: IBR Press.

Inamdar, A.K., and V. Ramanathan. 1994. "Physics of the Greenhouse Effect and Convection in Warm Oceans." *Journal of Climate* 7:715–731.

Jazcilevich, Aron, Vicente Fuentes, Ernesto Jauregui, and Estaben Luna. 2000. "Simulated Urban Climate Response to Historical Land-use Modification in the Basin of Mexico." *Climatic Change* 44:515–536.

Johnson, Tim. 1999. "World out of Balance: In a Prescient Time Native Prophecy Meets Scientific Prediction." *Native Americas* 16, no. 3/4(Fall/Winter): 8–25. [http://nativeamericas.aip.cornell.edu]

Karl, Thomas R., Neville Nicholls, and Jonathan Gregory. 1997. "The Coming Climate: Meteorological Records and Computer Models Permit Insights Into Some of the Broad Weather Patterns of a Warmer World." *Scientific American* 276:79–83. [http://www.scientificamerican.com/0597issue/0597karl.htm]

Kerr, R.A. 1998. "Among Global Thermometers, Warming Still Wins Out." *Science* 281 (25 September): 1948–1949.

Klug, Edward C. 1997. "Global Warming: Melting Down the Facts About This Overheated Myth." CFACT Briefing Paper #105, November. [http://www.cfact.org/IssueArchive/greenhouse.bp.n97.txt]

Lawler, Andrew. 1995. "NASA Mission Gets Down to Earth." *Science* 269:1208–1210.

Lean, J., J. Beer, and R. Bradley. 1995. "Reconstruction of Solar Irradiance Since 1610: Implications for Climate Change." *Geophysical Research Letters* 22:3195–3198.

Liesman, Steve. 2000. "Texaco Appears to Moderate Stance on Global Warming: New Hires and Investments Move the Oil Giant in a Greener Direction." *Wall Street Journal*, 15 May, B-4.

Limbaugh, Rush. 1992. *The Way Things Ought to Be.* New York: Pocket Books.

Lindzen, Richard S. 1990. "Some Coolness Concerning Global Warming." *Bulletin of the American Meteorological Society* 71:288–299.

McKibben, Bill. 1993. "James Hansen: Getting Warmer." *Outside*, May, 116–189.

Mendelsohn, Robert. 1999. *The Greening of Global Warming*. Washington, D.C.: American Enterprise Institute.

Mendelsohn, Robert, and James E. Neumann. 1999. *The Impact of Climate Change on the United States Economy*. Cambridge: Cambridge University Press.

Michaels, Patrick J. 1998. Kyoto Protocol: A Useless Appendage to an Irrelevant Treaty. Testimony of Patrick J. Michaels, Professor of Environmental Sciences, University of Virginia, and Senior Fellow in Environmental Studies at Cato Institute before the Committee on Small Business, 29 July, United States House of Representatives, Washington, D.C. [http://climatechangedebate.com/archive/09–18_10–27_1998.txt]

———. 1996. "Solar Energy." In *What's Hot: World Climate Report Archives: 1995–2000*. Vol. 1, no. 15. [http://www.nhes.com/back_issues/Vol1and2/WH/hot.html]

———. N.d. "The Decline and Fall of Global Warming." Fraser Institute. [http://www.fraserinstitute.ca/publications/books/g_warming/decline.html]

Montague, Peter. N.d. "A New Disinfomation Campaign." Environmental Research Foundation. [http://www.eieio.org/airquality/news/rachelsglobalwarming.html]

Montague, Peter, ed. 1992. "Global Warming Part I: How Global Warming is Sneaking Up on Us." *Rachel's Environment and Health Weekly*, no. 300, 26 August. Annapolis, Md.: Environmental Research Foundation. [http://www.monitor.net/rachel/r300.html]

Moore, Thomas Gale. 1995. "Why Global Warming Would be Good for You." *Public Interest*, 1 January, 83. [http://www.cycad.com/cgi-bin/Upstream/Issues/science/WARMIN.html]

———. 1996. "Health and Amenity Effects of Global Warming." Working Papers in Economics, E-96–1. The Hoover Institution.

———. 1997. "Happiness Is A Warm Planet." *Wall Street Journal*, 7 October. [http://www.freerepublic.com/forum/a182.htm]

———. 1998. *A Politically Incorrect View of Global Warming: Foreign Aid Masquerading as Climate Policy*. Washington, D.C.: Cato Institute.

Newell, R.E. and T.G. Dopplick. 1979. "Questions Concerning the Possible Influence of Anthropogenic CO2 in Atmospheric Temperature." *Journal of Applied Meteorology* 18:822–825.

North, Gerald R., Jurgen Schmandt, and Judith Clarkson. 1995. *The Impact of Global Warming on Texas*. Austin: University of Texas Press.

Packard, Kimberly O'Neill, and Forest Reinhardt. 2000. "What Every Executive Needs to Know About Global Warming." *Harvard Business Review* 78:4 (July/August): 128–135.

Palmer, Frederick. 1998. "Not So Hot." *Washington Post*, 12 December, A-23.

Parsons, Michael L. 1995. *Global Warming: The Truth Behind the Myth*. New York: Plenum Press/Insight.

Revkin, Andrew C. 2000. "Study Proposes New Strategy to Stem Global Warming." *New York Times*, 19 August, A-13.

Ridenour, David. 1998. "Hyprocrisy in Buenos Aires: Millions of Gallons of Fuel to be Burned by Those Seeking Curbs on Fuel Use." *National Policy Analysis*, no. 217

(October). National Center for Public Policy Research. [http://nationalcenter.org/NPA217.html]

Rind, D., E.W. Chiou, W. Chu, J. Larsen, S. Oltmans, J. Lerner, M.P. McCormick, and L. McMaster. 1991. "Positive Water Vapour Feedback in Climate Models Confirmed by Satellite Data." *Nature* 349:500–503.

Schneider, Stephen H. 2001. "No Therapy for the Earth: When Personal Denial Goes Global." In Michael Aleksiuk and Thomas Nelson, eds., *Nature, Environment & Me: Explorations of Self In A Deteriorating World*. Montreal: McGill-Queens University Press.

———. 1999. Modeling Climate Change Impacts and Their Related Uncertainties. Paper for Conference on Social Science and the Future, Somerville College, Oxford, July 7–8, 1999. Ms. copy provided by the author.

Singer, S. Fred. 1999. "Global Warming Whining." *Washington Times*, 16 April. [http://www.cop5.org/apr99/singer.htm]

———. 1997. *Hot Talk/Cold Science: Global Warming's Unfinished Debate*. Oakland, Calif.: Independent Institute.

———. 1992. "Foreword." In Robert C. Balling, Jr., *The Heated Debate: Greenhouse Predictions Versus Climate Reality*. San Francisco: Pacific Research Institute for Public Policy.

Soden, Brian J. 1997. "Variations in the Tropical Greenhouse Effect during El Nino." *Journal of Climate* 10:1050–1055.

Soden, B.J., and R. Fu. 1995. "A Satellite Analysis of Deep Convection, Upper Tropospheric Humidity, and the Greenhouse Effect." *Journal of Climate* 8:2333–2351.

Spencer, R.W., and W.D. Braswell. 1997. "How Dry is the Tropical Free Tropical? Implications for Global Warming Theory." *Bulletin of the American Meteorological Society* 78:1097–1106.

Stevens, William K. 1997. "Computers Model World's Climate, But How Well?" *New York Times*, 11 April. [http://benetton.dkrz.de:3688/homepages/georg/kimo/0254.html]

———. 1999. *The Change in the Weather: People, Weather, and the Science of Climate*. New York: Delacorte Press.

———. 2000. "The Oceans Absorb Much of Global Warming, Study Confirms." *New York Times*, 24 March, A-16.

Sudetic, Chuck. 1999. "As the World Burns." *Rolling Stone*, 2 September, 97–106, 129.

Sun, D.Z., and I.M. Held. 1996. "A Comparison of Modelled and Observed Relationships Between Interannual Variations of Water Vapor and Temperature." *Journal of Climate* 9:665–675.

Wadman, Meredith. 2000. "Car Maker Joins Exodus from Anti-Kyoto Coalition." *Nature* 404:322.

Wentz, F.J., and M. Schabel. 1998. "Effects of Orbital Decay on Satellite-Derived Lower Tropospheric Temperature Trends." *Nature* 394 (13 August): 661–664.

Woodwell, George M. 1999. The Global Warming Issue. [http://www.gibbons.freeonline.co.uk/Articles/The_Global_warming_issue.htm]

Ye, H., and J.R. Mather. 1997. "Polar Snow Cover Changes and Global Warming." *International Journal of Climatology* 17:155–162.

CHAPTER 4

Icemelt: Glacial, Arctic, and Antarctic

What a show Hollywood could make of global warming, if only the West Antarctic Ice Sheet (which is roughly the size of Mexico) would disintegrate all at once, provoking an overnight rise in sea levels of perhaps 15 to 20 feet. Such an event would be catastrophic for many millions of people around the world who live in large coastal urban areas, from New York City to Shanghai. The rising oceans would inundate intensely farmed river deltas, such as the Nile and Ganges. In the real world, however, the melting of the West Antarctic Ice Sheet may take several centuries to unfold, and thus will not reach its climatic end until long after the last members of any present day audience will have left the theater. Ice turns to water one prosaic drop at a time.

Even though the melting of the world's ice usually lacks Hollywood-style flair, the eventual result may be climate change on a scale that the Earth has not experienced since the days of the dinosaurs. Human-induced climate change may be ending the cycle of glaciation that has been the norm on Earth since the end of the age of the dinosaurs, about 65 million years ago. Viewed on a timescale of the last 540 million years, however, glacial cycles have been relatively rare events. For three-quarters of the last 540 million years, the Earth's ice caps have been negligible or non-existent (Huber et al. 2000, xi).

As the twenty-first century opened, ice was melting around the world. The Arctic ice cap was thinning, and scientists were speculating over how many years will pass before a summer during which most of it melts. The Intergovernmental Panel on Climate Change's models project that between one-third and one-half of existing mountain glacial mass could disappear over the next 100 years. Sometime during the present century, the last glacier may melt in Glacier National Park.

Many lakes and rivers in the Northern Hemisphere usually freeze about a week later and thaw out 10 days sooner than a century and a half ago, according to John J. Magnuson and colleagues. In some areas (such as the harbor of

Toronto, Ontario), heat contributed by urbanization, as well as general warming, may have added a month or more to the ice-free season. The warming trend appears to have begun at least half a century before the significant buildup of greenhouse gases caused by the burning of fossil fuels, leading to speculation that a long-term climate cycle is accelerating changes induced by infrared forcing.

An international team of scientists analyzed written accounts, some of them centuries old, including newspaper reports, fur traders' records, ships' navigation logs, and descriptions of religious events. These results illustrate "a very clear record of the response of aquatic systems to [global] warming," said Magnuson (Suplee 2000, A-2). The authors of this study estimate that the changes in freezing and thawing dates that they found required an average temperature increase of about three degrees F. during the last 150 years (Magnuson et al. 2000). "We know that the regions of Eurasia and North America, where most of these [lake and river sites] are, has indeed warmed according to thermometers at a rate of about double that for the globe—both recently and for the past century," said Kevin Trenberth, head of climate analysis at the National Center for Atmospheric Research, who did not participate in the research (Suplee 2000, A-2).

Observed icemelt probably accounts for only a fraction of the impact to come, as the world's oceans and atmosphere factor increased levels of "greenhouse forcing" into their dynamics. Jonathan Overpeck, a paleoclimate specialist with the National Oceanic and Atmospheric Administration (NOAA) in Boulder, Colorado, speculates that human activity may be short circuiting the glaciation cycle which has dominated the Earth's climate for millions of years.

If we left Mother Nature to do what she wanted to do, we would be going back into another ice age in the next 10,000 years. Now, because of what humans are doing, it's unlikely that we'll be going back into another ice age. Instead, glaciers around the world are receding. The Earth is warming up, and it will likely continue to warm up. (Sudetic 1999, 102)

As the Earth warms, increased advection transports surplus heat from equatorial to polar areas (heat flows from hotter to colder regions), so temperature changes tend to be larger at the poles than at mid-latitudes. In the tropics, warming generally has been more pronounced at higher altitudes than closer to sea level.

Harvard epidemiologist Paul Epstein comments,

Some of the strongest evidence of a warming trend comes from mountainous regions, where summit glaciers are melting on six continents. Many may soon disappear. In tandem with the retreat of peak ice sheets, plants are migrating up mountain slopes on 26 Swiss Alpine crests, and in the U[nited] S[tates] Sierra Nevada, Alaska and New Zealand. (Epstein et. al. 1996)

The melting of Arctic and Antarctic ice will do more than inconvenience coastal urban dwellers. Melting is already destroying an ecosystem built around sea ice. A report by the World Wildlife Fund and the Marine Conservation Biology Institute sketches the vital role of sea ice in polar ecosystems:

> Sea ice is fundamental to polar ecosystems: it provides a platform for many marine mammals and penguins to hunt, escape predators, and breed. . . . Its edges and undersides provide vital surfaces for the growth of algae that forms the base of the polar food web. In areas with seasonal ice cover, spring blooms of phytoplankton occur at ice edges as the ice cover melts, boosting productivity early in the season. But sea ice is diminishing in both the Arctic and the Antarctic. As this area diminishes, so does the food available to each higher level on the web, from zooplankton to seabirds. Higher temperatures predicted under climate change will further diminish ice cover, with open water occurring in areas previously covered by ice, thereby diminishing the very basis of the polar food web. (Mathews-Amos and Berntson 1999)

Reduced ice cover changes the Earth's albedo (reflectivity), which could become a factor in global warming, causing the earth to absorb more solar energy, as ice and snow (which reflect 75 percent or more of incoming sunlight) is replaced by bare soil, which reflects 10 to 25 percent. Ice and snow, in some cases, may be replaced by liquid seawater, which reflects 10 to 70 percent of incoming sunlight, depending on the sun's angle.

WHITHER THE WEST ANTARCTIC ICE SHEET?

The idea that global warming could provoke the disintegration of the West Antarctic Ice Sheet was aired as theory by glaciologists as early as 1979 (Hughes et al. 1979; Bentley 1980). J.H. Mercer has suggested that the West Antarctic Ice Sheet fell apart during an interglacial period about 125,000 years ago without an added boost from the burning of carbon-based fuels. T.J. Hughes has examined the geophysical mechanisms which may cause the West Antarctic Ice Sheet to collapse, and J. T. Hollin has examined evidence of major ice-sheet "surges" in the past which led to 10 to 30 meter rises of sea level in less than 100 years (Barry 1978).

During the 1990s, the stability of the West Antarctic Ice Sheet (which comprises about a quarter of the Earth's largest mass of frozen water) became a subject of intense scientific inquiry. A vibrant debate has grown up regarding the future of the ice sheet, with assurances of stability on one side, and speculation of future collapse on the other.

A report issued by the National Academy of Sciences (NAS) during 1991 asserted that melting of the West Antarctic Ice Sheet is unlikely, "and virtually impossible before the end of the next century" (National Academy 1991, 23). According to climate models used in this report, several centuries of rising temperatures will be required before the ice sheet disintegrates.

Reports authored by committees sometimes illustrate disagreements between the lines. One page after it dismisses the prospect that the West Antarctic Ice Sheet may melt, the same NAS report says that existing climate models may not allow for unforeseen "surprises." The literature of global warming is studded with admissions of this kind, which lend a startling degree of uncertainty to any attempt to forecast future climatic conditions and their impact around the world. For example, the same NAS report says, "CH4 [methane] could be released as high-latitude tundra melts, providing a sudden increase in CH4, which would add to the greenhouse warming" (National Academy 1991, 24), and "There could be a significant melting of the West Antarctic Ice Sheet resulting in a sea-level several meters higher than it is today" (National Academy 1991, 24).

During 1998, an international team of scientists analyzed five years of satellite radar measurements covering a large part of the West Antarctic Ice Sheet to determine whether it is becoming more unstable. The team concluded that the ice sheet is not melting rapidly and remains reasonably stable, as it has been for more than a century. "Based on our short, five-year period of observation of the interior of Antarctica, we do not seem to detect that the ice is melting more than one centimeter per year," explained C.K. Shum, an associate professor of civil and environmental engineering at Ohio State University. "That would mean that the interior Antarctic ice sheet does not seem to be contributing to sea-level rise more than 1 millimeter per year. We assume that global warming is under way now, and it may be enhanced by human activities but, until now, its effect on ice loss in Greenland and the Antarctic has been mostly speculation" (West Antarctic 1998).

The rate of ice decay on the Antarctic Peninsula accelerated markedly during the record global warmth of the 1990s. A 48-by-22-mile chunk of the Larsen Ice Shelf broke off during March, 1994, exposing rocks that had been buried for 20,000 years, prompting Rodolfo del Valle, director of geoscience at the Argentine Antarctic Institute in Buenos Aires, to tell the Associated Press, "Last November we predicted the [ice shelf] would crack in ten years, but it has happened in barely two months" (Gelbspan 1995). Del Valle was one of a team of scientists who witnessed the collapse of the Larsen A Ice Shelf, a floating ice sheet as large as Rhode Island which had averaged 500 feet in thickness. One day, according to *Christian Science Monitor* correspondent Colin Woodard, the shelf collapsed with a thunderous roar. In two hours, the Argentine team found itself standing on an island surrounded by open water and enormous icebergs. "I felt a sadness, a pain in my heart for the loss of a place that had become like a home to me," del Valle recalls. "I've experienced strong earthquakes on land, but this was different. After an earthquake something remains. But not with the ice shelf—it was completely destroyed" (Woodard 1998).

In early 1995, the Larsen A Ice Shelf completely disintegrated during a single storm after years of shrinking gradually. "The speed of the final breakup was unprecedented, and followed several of the warmest summers on record for this portion of the Antarctic," said Ted Scambos of the National Snow and Ice Data

Center based at the University of Colorado at Boulder (Satellite Images 1998). "Ice shelves appear to be good bellwethers for climate change, since they respond to change within decades, rather than the years or centuries sometimes typical of other climate systems," said Scambos (Satellite Images 1998).

During 1998 and 1999, several reports described the retreat of ice along the shores of the Antarctic Peninsula. Reports indicated that an additional 1,100 square miles of ice had melted during 1998. At about the same time, David Vaughan, a researcher with British Antarctic Survey, and Scambos reported that the Larsen B and Wilkins shelves on the Antarctic Peninsula were in "full retreat" (Britt 1999). These ice sheets had been retreating slowly for about 50 years, losing about 2,700 square miles during that period. A loss of 1,100 square miles in one year (1998) thus represented a major acceleration of the ice sheets' erosion. During that year as much ice was lost as had melted during the preceding four decades.

"This may be the beginning of the end for the Larsen Ice Shelf," said Scambos as the ice sheet crumbled in 1999. "This is the biggest ice shelf yet to be threatened," Scambos said. "The total size of the Larsen B Ice Shelf is more than all the previous ice that has been lost from Antarctic ice sheets in the past two decades" (Satellite Images 1998). Until its dissolution, the Larsen B was the northernmost ice shelf in Antarctica, and therefore "on the front line of the warming trend," said Scambos (Satellite Images 1998).

Radar images from satellite observations of the Pine Island glacier in Antarctica taken during the 1990s indicate that it has been shrinking rapidly. The shrinking of this glacier is important "because it could lead to a collapse of the West Antarctic Ice Sheet," said Eric Rignot, a computer-radar scientist at the Jet Propulsion Laboratory in California, who led a study of the glacier. "We are seeing a . . . glacier melt in the heart of Antarctica" (Melting 1998). "The continuing retreat of Pine Island glacier could be a symptom of the WAIS [West Antarctic Ice Sheet] disintegration," said Craig Lingle, a glaciologist at the University of Alaska in Fairbanks, who is familiar with the study (Melting 1998).

The Pine Island glacier is important because it is part of a stream of ice that moves more rapidly than the ice cap surrounding it. This glacier is part of an ice stream which runs from the interior of the West Antarctic Ice Sheet into the surrounding ocean waters. If a glacier in this ice stream melts more quickly from the bottom than snow accumulates on its top, the net icemelt goes into the ocean, raising sea levels.

A "disaster scenario," as described by Richard Alley, a glaciologist at Pennsylvania State University, has the Pine Island glacier retreating enough to "make a hole in the side of the ice sheet. . . . The remaining ice would drain through that hole" (Melting 1998). Once enough ice had drained through the hole, the West Antarctic Ice Sheet might eventually collapse, raising average sea levels around the world 15 to 20 feet in a few years. Such an increase in mean sea level would flood roughly 30 percent of Florida and Louisiana, 15 to 20 percent of the District of Columbia, Maryland, and Delaware, and 8 to 10 percent of the Carolinas and New Jersey. Inundation of coastal areas would have a similar impact around the world (Schneider and Chen 1980). Among the flooded areas would be the centers of some of the world's great urban and commercial centers, from New York City to Bombay, Calcutta, and Manila.

During 1998, Eric Rignot wrote in *Science* that West Antarctica's Pine Island Glacier was retreating at 1.2 kilometers a year (plus or minus 0.3 kilometers) and that its ice was thinning 3.5 (plus or minus 0.9) meters per year. "The fast recession of the Pine Island Glacier, predicted to be a possible trigger for the disintegration of the West Antarctic Ice Sheet, is attributed to enhanced basal melting of the glacier's floating tongue by warm ocean waters," Rignot wrote (Rignot 1998, 549). This glacier is widely believed to be "the ice sheet's weak point" (Kerr, West Antarctica's, 1998, 499). While the accelerated melting of this glacier does not portend an immediate disintegration of the West Antarctic Ice Sheet, Alley wrote that "most models indicate [that the retreat] would speed up if it kept going" (Kerr, West Antarctica's, 1998, 499). One observer was quoted in this context as stating that a quick collapse of the West Antarctic Ice Sheet would "back up every sewer in New York City" (Kerr, West Antarctica's 1998, 500). Rignot speculated that warmer ocean waters were causing Pine Island's rapid bottom melting. "This is one of the most sensitive ice sheets to climatic change. For many, many years we have neglected the importance of bottom melting," Rignot said (Melting 1998).

"The sudden appearance of thousands of small icebergs suggests that the shelves are essentially broken up in place and then flushed out by storms or currents afterward," said Scambos (Britt 1999). The Larsen and Wilkins ice shelves have been melting since the 1950s, and scientists had expected them to fall apart, but the disintegration occurred more quickly than anticipated. "We have evidence that the shelves in this area have been in retreat for 50 years, but those losses amounted to only about 7,000 square kilometers," said David Vaughan, a researcher with the Ice and Climate Division of the British Antarctic Survey. "To have retreats totaling 3,000 square kilometers in a single year is clearly an escalation. Within a few years, much of the Wilkins ice shelf will likely be gone" (Britt 1999).

Speculation regarding the future of the West Antarctic Ice Sheet takes place in a climatic context which includes paleoclimate records and instrument readings indicating that the western Antarctic Peninsula has warmed significantly during the last century, with the trend accelerating at the end of the period. Temperatures in this area rose four to five degrees C. during the century, according to the World Wildlife Fund (Mathews-Amos and Berntson 1999). Temperatures in the area averaged above longer-term averages more than 75 percent of the time during the last third of the twentieth century. The number of annual days above freezing has increased by two to three weeks, mainly during the most recent quarter century. This temperature increase tends to erode coastal ice sheets. "Large areas of ice shatter as the meltwater percolates into fractures, and deep cracks are forced open to the base of the ice sheet by the weight of the water. . . . Once ice sheets weaken to a critical point, they may collapse very suddenly" (Mathews-Amos and Berntson 1999).

Average temperatures have increased by almost five degrees F. during the last 50 years on the Antarctic Peninsula, an arm of the mainland reaching a thousand miles northward toward the southern tip of South America (Woodard 1998). Glaciers are in retreat throughout the peninsula, melting ice that may be tens of thousands of years old. Woodard described the scene atop one Antarctic glacier as chunks of it calved into the sea:

Andy Young is drilling holes into the melting snow on top of the Marr Glacier when it happens. There's a sharp crack followed by a rumbling thunder as garage-size chunks of blue ice calve off a nearby glacial cliff and tumble into the frozen sea 400 feet below, throwing waves across the harbor. "That's a good one," Mr. Young says as he sticks a flag marker into the newly drilled hole. "You can see why we mark the route up the glacier—you just don't want to get too near the edges." (Woodard 1998)

Woodard reported that year by year more rocky surfaces emerge from beneath the melting walls of ice behind Palmer Station, a United States research facility operated by the National Science Foundation which is staffed by about 40 people. Scientists absent for a year find new beaches, outcroppings, even islands that had been hidden for thousands of years under the ice.

For several decades, scientists at the Palmer Station, astride the northwestern edge of Antarctica's Palmer Peninsula, have watched steady changes in the environment, as well as the flora and fauna it supports. Few of them doubt that many of the changes they observe are caused by a warmer climate. Annual average temperatures at the station have risen three to four degrees F. since the 1940s; winter averages have climbed seven to nine degrees during the same period (Petit 2000, 66). Southern elephant seals, which can weigh as much as 8,800 pounds, usually raise their young near the Falkland Islands to the north, but, in recent years, several hundred of them have been living year-round on the Palmer Peninsula. Fur seals, virtually unknown near the Palmer Station before 1950, have colonized the area around the station in the thousands, as areas once covered by ice and snow year-round have been sprouting "low grass, tiny shrubs, and mosses" (Petit 2000, 67).

At the same time, animals that depend on sea ice for food have been migrating southward. Adelie penguins (18-inch birds which look as if they are wearing tuxedoes) have become scarcer near the Palmer Station because they eat krill which arrives with sea ice. Adelie populations within two miles of the Palmer Station have declined by about 50 percent in 25 years, including a 10 percent decline during two years in the late 1990s (Petit 2000, 68). About 1950, four of five winters at Palmer Station produced extensive sea ice; by the end of the century, the average was two winters in five. The Adelie penguins have been moving their range southward toward areas which heretofore had been too cold and barren for them.

Researchers say the climate near the ice shelves has warmed roughly 4.5 degrees F. since the 1940s, causing surface ice to melt. That's just the tip of the iceberg, so to speak. As this surface ice melts, it forms deep pools of water which forces its way into cracks in the ice shelf. "The weight of the water essentially forces the cracks open, so a relatively small amount of climate warming can destroy a large, centuries-old ice shelf" Scambos said (Britt 1999). Scambos' research team is unsure whether the localized warming is part of a global warming trend, or whether it is simply a normal fluctuation in regional climate.

Scientists who study this situation have no firm estimates regarding how much warmer the earth would have to become to provoke a general collapse of the West Antarctic Ice Sheet. Some give it six degrees C. (10 degrees F.) and a century or two. Others believe the West Antarctic Ice Sheet as a whole may not drain into the sea for several centuries, if at all. Uncertainty regarding the future of the West Antarctic Ice Sheet has introduced a great deal of variability into models which estimate much global warming may cause world-wide sea levels to rise. (See also Chapter 5, "Warming Seas.")

A study by James G. Titus and Vijay Narayanan examines some of the reasons why estimates of sea-level rise vary so widely. Titus and Narayanan express this wide range of variables in terms of probabilities:

> The reviewer assumptions imply a 50 percent chance that the average global temperature will rise 2 degrees C., as well as a 5 percent chance that temperatures will rise 4.7 degrees C. by 2100. The resulting impact of climate change on sea level has a 50 percent chance of exceeding 34 centimeters and a 1 percent chance of exceeding one meter by the year 2100, as well as a 3 percent chance of a 2 meter rise and a 1 percent chance of a four-meter rise by the year 2200.
>
> Assuming that nonclimatic factors do not change, there is a 50 percent chance that global sea level will rise 45 cm, and a 1 percent chance of a 112 cm rise by the year 2100; the corresponding estimates for New York City are 55 and 122 cm. (Titus and Narayanan 1996.)

The speed with which Antarctic ice may melt depends not only on how much temperatures rise, but also on the ways in which ice moves within the ice cap. Jonathan L. Bamber, David G. Vaughan, and Ian Joughin have been studying these "rivers" of subsurface Antarctic ice. "It has been suggested," they write, "that as much as 90 percent of the discharge from the Antarctic Ice Sheet is drained through a small number of . . . ice streams and outlet glaciers fed by relatively stable and inactive catchment areas." Their research suggests that "each major drainage basin is fed by complex systems of tributaries that penetrate up to 1,000 kilometers from the grounding line to the interior of the ice sheet." Such "complex flows" are noted throughout the Antarctic ice sheet by these researchers (Bamber et al. 2000).

Bamber, Vaughan, and Jouglin assert that "this finding has important consequences for the modeled or estimated dynamic response time of past and present ice sheets to climate forcing" (Bamber et al. 2000, 1248). The researchers also find evidence of similar (although smaller in scale) ice-sheet dynamics in Greenland. "This evidence," they write, "challenges the view that the Antarctic plateau is a slow-moving and homogenous region" (Bamber et al. 2000, 1250). These researchers also contend that the dynamics of large ice flows are too complex for present-day models to predict, so climate modelers have very little idea how global warming will affect the largest of the Earth's remaining ice masses.

An iceberg twice the size of Delaware (183 by 22 miles) broke off the Ross Ice Shelf during the third week of March 2000, moving the boundary of the Ross Ice Shelf southward by 40 miles. Two and a half weeks later, on April 8, another piece, this one roughly twice the size of Manhattan Island, calved from the Ross Ice Shelf.

While icemelt on the Antarctic Peninsula is definitely being produced by climatic warming, the case for global warming as a factor in the calving of the Ross Ice Shelf is rather tenuous. "I would lean against global warming as an explanation," Michael Oppenheimer of the Environmental Defense Fund was quoted as saying (Williams 2000, 10D). Scambos told *USA Today* that periodic calving of the Ross Ice Shelf has changed the overall size of the ice mass in the area little since it was first mapped early in the twentieth century (Williams 2000, 10D). While the breakup of ice sheets on both sides of the Antarctic Peninsula has been provoked by climatic warming, calving of the Ross Ice Shelf seems to be related to the usual dynamics of glacial formation and movement. The Ross Ice Sheet is much closer to the South Pole than the Antarctic Peninsula, with a climate that remains well below freezing for the entire year, even as climatic warming erodes ice sheets on the Antarctic Peninsula.

COULD GLOBAL WARMING *REDUCE* SEA LEVELS?

A few studies assert that global warming could reduce world sea levels. According to two studies, a doubling of precipitation over Greenland and Antarctica would lower mean sea level between 1.3 millimeters per year (Ohmura and Ohmura 1991) and 4.2 millimeters per year (Bentley and Giovinetto 1990). P. Huybrechts and J. Oerlemans (1990) conclude that additional precipitation at higher temperatures will increase the mass of the Antarctic Ice Sheet and thereby lower sea levels.

Considered alone, Titus and Narayanan write, warmer temperatures in the Antarctic could increase snowfall and slightly lower worldwide sea levels. They expect Antarctic air temperatures to rise by approximately 2.5 degrees C. during the next century, largely as a result of reduced sea ice. For each degree of warming, they expect Antarctic precipitation to increase about eight percent, "equivalent to a 0.4 millimeters per year drop in sea level" (Titus and Narayanan 1996). They conclude,

Antarctica's temperatures are well below freezing; so unlike Greenland, warmer air temperatures will not cause a significant degree of glacial melting. Warmer water temperatures, by contrast, could potentially increase melting of the marine-based West Antarctic Ice Sheet and adjacent ice shelves. (Titus and Narayanan 1996)

Titus and Narayanan caution, however, that assertions linking warming to increasing snowfall fail to examine the possibility that "warmer circumpolar ocean water will intrude beneath the ice shelves, increase their rates of basal melting, decrease the backpressure that they exert on the ice streams, and thereby accelerate the rates at which ice streams convey ice from the Antarctic interior toward the oceans" (Titus and Narayanan 1996). They maintain that some of the recent melting along the edges of the West Antarctic Ice Sheet can be attributed to this mechanism. According to Titus and Narayanan, any warming of the circumpolar ocean is likely to lag behind the general increase in global temperatures by at least 50 years, and perhaps by a century or more.

Even with a large rate of ice-shelf melting in Antarctica, Titus and Narayanan believe,

[T]he Antarctic contribution to sea level may be negligible. Because ice shelves float and hence already displace ocean water, shelf melting would raise sea level only if it accelerates the rate at which ice streams convey ice toward the oceans. Several models suggest, however, that shelf melting will not substantially accelerate ice streams—and even the models that project such an acceleration generally suggest a lag of a century or so. Thus, through the year 2100, we estimate a 60 percent chance that the sea-level drop caused by increased Antarctic precipitation will more than offset the sea-level rise caused by increased ice discharge; this probability declines to 50 percent by 2200. Our analysis suggests that if Antarctica is going to have a major impact on sea level, it will probably be after the year 2100. (Titus and Narayanan 1996)

ARCTIC ICEMELT

Every winter since 1912, when an iceberg claimed the Titantic in the North Atlantic, the U.S. Coast Guard has issued warnings for an average of 500 icebergs which float south of 48 degrees north latitude, calving from Greenland's ice sheet. These icebergs threaten the shipping lanes of "Iceberg Alley," the Grand Banks southeast of Newfoundland. During the El Niño winter of 1998–1999, for the first time in 85 years, the Coast Guard did not issue a single iceberg alert for that area. "The lack of ice is remarkable," said Steve Sielbeck, commander of the International Ice Patrol (Wuethrich 1999, 37).

Rapid icemelt in the Arctic was evident to some observers by the 1980s. By 1989, the Arctic ice was reported to be thinning quickly. The Scott Polar Institute of the United Kingdom reported that ice in an area north of Greenland had thinned from an average of 6.7 meters in 1976 to 4.5 meters in 1987. At about the same time, a British Antarctic survey base reported unusual losses of ice as well (Kelly 1990, 23).

Thomas V. Lowell, writing in the *Proceedings of the National Academy of Sciences*, cites "fieldwork on the western margin of the Greenland Ice Sheet [which] provides evidence . . . [that] within the last 150 years, 94 percent of the lobes along the western edge of the [Greenland] ice sheet withdrew, whereas 70 percent of the independent glaciers retreated" (Lowell 2000, 1351). Lowell, finding that glacial melting began before massive infusions of greenhouse gases into the atmosphere due to human activity, raises the possibility that twentieth-century warming may have many causes, only one of which is human-induced greenhouse forcing (Lowell 2000, 1352). Mark B. Dyurgerow and Mark F. Meier report that glaciers around the world began to melt, on balance, most recently about 1975, with "further acceleration in the last decade [the 1990s]" (Dyurgerow and Meier 2000, 1406).

National Aeronautics and Space Administration scientist Bill Krabill leads a team which began measuring the mass of the Greenland Ice Sheet during the early 1990s. Between 1993 and 1998, Krabill's team, flying as many as 2,000 miles a day over the ice cap, compiled the first detailed "inventory" of the Greenland Ice Sheet, from which they may be able to deduce whether it is changing and, if so, how. The team's surveys indicate, "The ice in southeastern Greenland, a region long thought to be frozen tight, has thinned rapidly. The ice in some regions has shrunk[en] as much as thirty-three feet. The ice balance for the entire southern half of Greenland is in the red" (Sudetic 1999, 104). Snowfall has increased ice mass in some inland locations of Greenland, especially its northern half, as warmer temperatures allow the import and convection of more moisture there.

The studies by Krabill and colleagues indicate that three areas of southern interior Greenland have been accumulating icepack at rates of up to 10 inches per year. In the outer regions of the ice sheets, however, the researchers reported that large areas of ice have been thinning, with the rate of thinning increasing rapidly towards the ocean. The most rapid thinning rates (one to three feet per year) were observed in the lower depths of Greenland's southeast coast outlet glaciers, the researchers reported. Ice is thinning to some degree at most elevations of Greenland below 2000 meters above sea level. At elevations higher than 2,000 meters, an average thickening of the ice sheet of 0.5 to 0.7 centimeters a year was measured from 1993 to 1998. The researchers noted that areas of thinning in the eastern part of Greenland also experienced warmer-than-normal temperatures between 1993 and 1998. "However, we also observe areas of thinning near the West coast, where many locations were cooler than normal," the researchers reported (Murgia 1999).

Krabill's work suggests that Greenland's southeastern glaciers are thinning rapidly and that their lower elevations may be particularly sensitive to potential climate changes. "The results of this study are important in that they could represent the first indication of an increase in the speed of outlet glaciers," said Krabill (Steitz 1999). An outlet glacier acts as a major ice drainage region for an ice sheet.

"The excess volume of ice transported by these glaciers has had a negligible effect on global sea level thus far, but if it accelerates or becomes more widespread, it would begin to have a detectable impact," Krabill said (Steitz 1999). Researchers reported that the glacial thinning is too large to have resulted solely from increased ice-surface melting or decreased snowfall. The researchers believe the thinning is the result of increasing discharge speeds of glaciers flowing into the Atlantic Ocean. Krabill said surface-melt water might be seeping to the bottom of glaciers. Such seepage may be reducing the friction between the ice and the rock below it, enabling the glaciers to slide with less friction across the bedrock and thus allow more ice to slip into the ocean, according to Krabill (Steitz 1999).

Krabill elaborated,

The results of this study are significant because they provide the first evidence of widespread thinning of low-elevation parts of one of the great polar ice sheets. The results also suggest that the thinning outlet glaciers must be flowing faster than necessary to remove the annual accumulation of snow within their basins. (Steitz 1999)

"Why they are behaving like this is a mystery," said Krabill, "but it might indicate that the coastal margins of ice sheets are capable of responding quite rapidly to external changes" (Steitz 1999). One such external stress could be global warming, Krabill said.

A new aerial survey was summarized by Krabill and colleagues in the July 21, 2000 edition of *Science* (Krabill et al. 2000). "[A]pproximately 51 cubic kilometers of ice per year [is melting] from the entire [Greenland] ice sheet, sufficient to raise global sea level by 0.005 inches per year, or approximately seven percent of the observed rise," said Krabill. "Why the ice margins are thinning so rapidly warrants additional study. It may indicate that the coastal margins of ice sheets are capable of responding more rapidly than we thought to external changes, such as a warming climate" (Lazaroff 2000).

An Oregon State University study suggests that North American polar temperatures will rise more than eight degrees C. (14 degrees F.) with a doubling of atmospheric carbon dioxide levels (Bernard 1993, 101–102). The polar regions warm more than the mid-latitudes or tropics for two reasons: Carbon dioxide produces more heat absorption in areas with relatively low amounts of water vapor, where relatively low temperatures do not allow the air to hold as much water vapor as it does in warmer regions. In addition, the melting of snow and ice alters the reflective landscape dramatically; ice and snow, which reflect most of the sunlight reaching the surface, are replaced by radiation-absorbing liquid ocean and bare land (Bernard 1993, 102–103).

By the end of the twentieth century, pronounced warming was leaving its mark in parts of the arctic. Anecdotal evidence of global warming is striking in the area. For example, scientists keeping records at Toolik Lake, north of Alaska's Brooks Range, noted that the lake's surface temperature rose by almost three degrees C. between 1979 and 1994. A team of NOAA researchers found an increase in surface temperatures of 5.5 degrees C. (about nine degrees F.) during 30 years, between 1965 and 1995, at nine stations north of the Arctic Circle (Gelbspan, *The Heat is On*, 1997, 74). One report indicates that average air temperatures directly above the snowy top of the Greenland Ice Sheet rose as much as 18 degrees F. during the 1990s, with the year 2000 the warmest year. "There has been no trend in the summer maximum temperature," said Konrad Steffen of the University of Colorado, "but enormous fluctuations and warming trends in the fall, winter and spring" (Suplee, For 500 Million, 2000, A-9).

Evidence of rapid warming had become part of everyday life in the Arctic by the late 1990s. Point Barrow, Alaska, broke its annual average temperature record in 1998 by three degrees F., a very large number. The old annual record, set in 1940, had been 14 degrees F. Point Barrow's annual average temperature in 1998 was seven degrees F. higher than the 30-year average (1961–1990), which meteorologists use as a benchmark. One day during the summer of 1998, the high temperature at Point Barrow hit a balmy 67. Many of the 4,000 people who live in Point Barrow, most of them Inuit, express fears that lack of ice along the shore may expose their storm-lashed town to flooding. They're also afraid that the bowhead whales, which provide the core of their diet, may swim further offshore, making them harder to kill. Whale meat stored in holes outdoors also may be ruined by an untimely thaw. During the summer of 1998, some melting occurred as liquid water seeped into some storage areas.

Average temperatures at the mouth of the Mackenzie River were nine degrees F. above long-term averages during 1998. Several buildings have been lost to erosion by seawater in the village of Tuktoyaktuk, near the mouth of the Mackenzie. Several decades ago, miners in the area deposited toxic wastes in ponds which were expected to remain frozen (and the toxic materials sealed by the permafrost). With warmer temperatures, some of these toxic dumps may thaw and leak.

A significant rise in Siberian temperatures during the late twentieth century has been provoked by "a speeding up of atmospheric circulation in the Northern Hemisphere [which] results in more warm air being transported deep into the region" (Burroughs 1999, 70). Tundra and taiga (high-latitude forests in Asia) have been moving northward during the late twentieth century, as satellite surveys indicate that the amount of photosynthesis between 45 and 70 degrees north latitude has increased. This trend is linked by William J. Burroughs to a "rise in spring temperatures, which lengthened the growing season at high altitudes" (Burroughs 1999, 74).

In this context, Burroughs raises one of many perplexing questions regarding how the Earth manages carbon:

The possible increase in the growth of the taiga touches on another interesting aspect of the current global warming, which relates to our limited knowledge. . . . Measurements in the early 1990s suggested that the amount of carbon dioxide being absorbed at high altitudes was much greater than previously estimated. . . . This "Great Northern Sink" has been the subject of a great deal of speculation among scientists. (Burroughs 1999, 74–75)

This point of view is not universally accepted among scientists. Witness a nearly opposite assertion by tundra researcher George W. Kling, an associate professor of biology of the University of Michigan, who said, "Our latest data show that the Arctic is no longer a strong [net] sink [absorber of] carbon. In some years, the tundra is adding as much or more carbon to the atmosphere than it removes" (Kling et al. 1992).

The thinning of ice in the Arctic affects life all along the food chain. When the ice breaks up earlier than usual, polar bears have less time to build fat reserves, and must rely on these reduced reserves for a longer period of time before ice forms again in the fall. Calculations indicate that a mean air temperature increase of one degree C. could advance the date of ice breakup nearly a week in western Hudson Bay (Mathews-Amos and Berntson 1999). Hunger among adult bears will be reflected in their offspring, who will be born smaller and more prone to die at an early age.

In the Bering Sea and in Hudson Bay, evidence of stress in polar bear populations is mounting as sea ice retreats (Stirling and Derocher 1993; Stirling and Lunn 1997). Some native people in the Arctic have reported difficulty in hunting marine mammals such as walrus in recent years as sea ice has diminished. (See also Chapter 8, "Global Warming and Indigenous Peoples.")

On October 2, 1997, the Canadian Coast Guard icebreaker *Des Groseilliers* entered the pack ice of the Arctic Ocean about 250 miles due north of Prudhoe Bay, Alaska. Starting at this position, the *Des Groseilliers* began a year-long voyage, drifting with the pack ice. Ice Station SHEBA traveled about 1,000 miles in 13 months, ending its journey about 450 miles northwest of Barrow, Alaska. The icebreaker was a base of operations and home for a group of more than 50 scientists involved in the scientific research program called SHEBA, Surface Heat Balance of the Arctic Ocean. Scientists on the ship conducted tests to determine the physics of icemelt, and were astounded by what they found. The ice was much thinner, as a rule, than their models had led them to believe. Where they had expected to find ice six to eight feet thick, it was three to four feet thick. The ocean underneath the ice also was warmer and less saline than it had been previously.

The lack of salinity in the upper reaches of the ocean under the ice cap indicated that a great deal of ice had melted in a relatively short time. The speed with which the Arctic ice cap is thinning has led some observers to estimate that the entire body of ice (an area the size of the continental United States) will melt during the summer by roughly the year 2020. "If the current trend continues, the Arctic ice pack will disappear," Michael Ledbetter of Ice Station SHEBA said. "The models say it and we didn't find anything at SHEBA that would cause us to modify those predictions. In twenty-five years, we might lose the entire Arctic Ocean ice pack" (Johnson 1999, 11).

During late 1999, a team of scientists reported in *Science* that the Arctic ice cap had lost a Texas-sized measure of its mass during the previous two decades. The team, led by Konstantin Vinnikov of the University of Maryland, had to inventory many retreats in the ice cap across the Arctic Ocean. The trend the scientists found was unmistakable, with its human origins nearly as clear. The team set the chance that such melting is a measure of natural climate variability at two percent. Their study "strongly suggests that the observed decrease in Northern Hemisphere sea ice is related to [human caused] global warming" (Arctic Ice 1999, 8).

During the second week of August 2000, a patch of open water a mile wide was reported at the North Pole. Initial reports indicated that this might have been the first melting of ice at the pole since the Eocene Period, about 50 million years ago. "It was totally unexpected," said Dr. James J. McCarthy, an oceanographer, director of the Museum of Comparative Zoology at Harvard University and the coleader of a group working for the Intergovernmental Panel on Climate Change (Wilfred 2000, 1).

McCarthy took part in a cruise of the Arctic aboard the Russian icebreaker *Yamal*. On a similar cruise six years ago, he recalled, the icebreaker plowed through an icecap six to nine feet thick at the North Pole (Wilfred 2000, 1). The *Yamal* cruised six miles away to find ice thick enough for the 100 passengers to get out and be able to say they had stood on the North Pole, or close to it. They saw ivory gulls flying overhead, the first time ornithologists said they had been sighted at the pole (Wilfred 2000, 1).

The iceless North Pole was widely publicized as a new indicator of global warming, drawing a counterattack from climate-change skeptics. S. Fred Singer said in the *Wall Street Journal* that he had cruised the Arctic with the U.S. Navy, and "I can testify that icebreakers always search for leads [cracks] to make their way through the ice" (Singer 2000, A-18). Peter Wadhams, director of the Scott Polar Institute in Cambridge, England, said, "Claims that the North Pole is now ice-free for the first time in 50 million years [are] complete rubbish, absolute nonsense. . . . What is happening is of concern but it is gradual, not sudden or stupendous" (Nuttall 2000). Wadhams said that the extent of Arctic open water during the summer months is on the rise, as Arctic ice gradually thins. "In an ice pack, there will always be some open water" this time of year—even at the North Pole, said Donald Perovich, who studies links between polar ice and climate at the U.S. Army Corps of Engineers' Cold Regions Research and Engineering Laboratory in Hanover, New Hampshire (Spotts 2000, 4). The trend in the Arctic is rather dramatic over several decades, however. The polar ice cap lost about 40 percent of its volume between 1958 and 1997, according to figures assembled by researcher Drew Rothrock and colleagues. They compared ice soundings taken by U.S. Navy submarines between 1958 and 1976 with soundings taken during scientific submarine cruises from 1993 to 1997 (Spotts 2000, 4).

The same month that the passengers of the *Yamal* found open water at the North Pole, an aluminium-hulled catamaran owned by the Royal Canadian Mounted Police, the *St. Roch II*, crossed the 900-mile channel through Canada's Arctic islands in 30 days, a record. The crossing of the *Roch II*, made with minimal delay by ice, raised speculation that regular summer shipping through the Arctic would soon cut 5,000 miles off the shortest route between Europe and Asia, and open the once-fabled Northwest Passage sought by mariners for 500 years.

Water temperatures in parts of the Arctic Ocean increased an astonishing 1.6 degrees C. during the decade of the 1990s alone (Hodges 2000, 35). If the Arctic continues to warm at anywhere near this rate, by the year 2020 the Northwest Passage may open through the thawing Arctic Ocean year-round for the first time in human history. The passage was sought in vain by early sixteenth century explorers of North America. Construction of the Panama Canal finally allowed shorter oceanic transport of goods between the coasts of North America. Canadian military, police, and customs officials already have begun to plan ways to use the passage, as they examine how to manage the new waterway.

Woodard described the retreat of ice sheets in Iceland during the last century:

From the gravelly, newborn shores of this frigid lagoon, Iceland's Vatnajokull ice cap is breathtaking. The vast dome of snow and ice descends from angry clouds to smother jagged 3,000-foot-tall mountains. Then it spills out from the peaks in a steep outlet glacier 9.3 miles wide and 12.4 miles long—an insignificant appendage of Europe's largest ice cap despite its impressive size. A hundred years ago, however, there was no lagoon here. The shoreline was under one hundred feet of glacial ice. The outlet glacier, known as Breidamerkurjokull, extended to within 250 yards of the ocean, having crushed medieval farms and fields in its path during the preceding centuries, a time now referred to as the Little Ice Age. Today, Breidamerkurjokull's massive snout ends about two miles from the ocean. In its hasty retreat, the glacier has left the rapidly expanding lagoon, which is filled with icebergs calved from its front. The lagoon is 350 feet deep and has nearly doubled its size during the past decade. Every year, it grows larger, threatening to wash out Iceland's main highway. (Woodard 2000, A-1)

THE SPEED OF POLAR CLIMATE CHANGE

Evidence from ice cores taken in Greenland has altered the scientific view regarding the speed with which climatic change takes place, especially at higher latitudes. The older assumption that climate changes very slowly in the polar regions is built into the English language when we say that something moves with "glacial speed." On the contrary, the ice-core record indicates that several rapid warmings and coolings have convulsed the Arctic during the last 100,000 years.

The last ice age may have ended much more quickly than previously thought. Jeffery P. Severinghaus of the Scripps Institute of Oceanography described a new method of analyzing gases trapped in Greenland's ice sheet which indicates that temperatures in the area rose about 16 degrees F. within a decade or two at the end of the last great ice age. "The old idea was that the temperature would change over a thousand years, but we found it was much faster," said Severinghaus (Report 1999, 15). "We know that over the next one hundred years the Earth will probably warm because of the greenhouse effect," Severinghaus continued. "There is a remote possibility that we might trigger one of these abrupt climate changes. This certainly gives us pause" (Report 1999, 15).

Rapid climate changes within a few thousand years are sketched by Jonathan Adams, Mark Masli, and Ellen Thomas:

Ice-core evidence (Dansgaard et al. 1993; Taylor et al. 1993) indicated that the Eemian was punctuated by many short-lived cold events, as shown by variations in electrical conductivity (a proxy for windblown dust, with more dust indicating colder, more arid conditions) and stable oxygen isotopes (a proxy for air temperature) of the ice [that] were used by these workers [to] infer climatic conditions. . . . The cold events seemed to last a few thousand years, and the magnitude of cooling was similar to the difference between glacial and interglacial conditions; a very dramatic contrast in climate. Furthermore, the shifts between these warm and cold periods seemed to be extremely rapid, possibly occurring over a few decades or less. . . . Sudden and short-lived warm events occurred many times during the generally colder conditions that prevailed between 110,000 and 10,000 years ago. (Adams et al. in press)

The last cold "spike" (or "Heinrich Event"), the Younger Dryas, peaked between 13,000 and 11,500 years ago. The sudden beginning and end of the Younger Dryas has been studied in particular detail utilizing ice-core and sediment records. A detailed study of two Greenland ice cores suggests that the main Younger Dryas-to-Holocene warming took several decades in the Arctic, but was marked by a series of warming surges, each taking less than five years. About half of the warming was concentrated in a single period of less than 15 years. A rapid global rise in atmospheric methane concentration that occurred at the same time suggests that the warmer and more humid climate (causing more methane output from swamps and other biotic sources) was a factor in these rapid changes. Scientists (see Adams et al. in press) are at a loss to explain why the climate can change suddenly, as they describe records that indicate such changes:

It is still unclear how the climate on a regional or even global scale can change as rapidly as present evidence suggests. It appears that the climate system is more delicately balanced than had previously been thought, linked by a cascade of powerful mechanisms that can amplify a small initial change into a much larger shift in temperature and aridity. (Rind and Overpeck 1993)

Past rapid changes in regional climates took place without human provocation. Factors contributing to these rapid changes may have included variations in the sun's radiance, changes in angle of the Earth's axis, natural variations in atmospheric levels of carbon dioxide and methane (among other "trace" gases), changes in atmospheric water-vapor levels, and dust ejected from volcanoes and other sources. Jonathan Adams and associates comment,

[C]limate has a tendency to remain quite stable for most of the time and then suddenly "flip," at least sometimes over just a few decades, due to the influence of . . . various triggering and feedback mechanisms. . . . Such observations suggest that even without anthropogenic climate modification there is always an axe hanging over our head, in the form of random very large-scale changes in the natural climate system; a possibility that policy makers should perhaps bear in mind with contingency plans and international treaties designed to cope with sudden famines on a greater scale than any experienced in written history. By starting to disturb the system, humans may simply be increasing the likelihood of sudden events which could always occur. . . . To paraphrase W.S. Broecker; "Climate is an ill-tempered beast, and we are poking it with sticks." (Adams et al. in press)

A COOLER EUROPE IN A WARMER WORLD?

The Gulf Stream emerges in the ocean-atmosphere dynamics of the northern Atlantic Ocean as a crucial regulator of climate, especially for western Europe. Sediment records have produced evidence that a massive outflow of debris north to south across the North Atlantic produced by rapid melting of the Arctic ice cap (including glaciers in Greenland and northern Canada) could radically alter the climate of western Europe, producing colder winters in an otherwise warmer world.

Through most of its recorded history, Europe has experienced an unusual degree of warmth for such a northerly latitude because the Gulf Stream flows near the continent's western shores, carrying relatively warm water from tropical regions near North America. Land-locked areas at 50 degrees north latitude, similar to that of London, include the Canadian prairies and much of Siberia.

Climate models suggest that a rapid outflow of cold, relatively fresh water might divert the Gulf Stream southward, depriving Europe of oceanic warmth that bathes the continent during most winters. During such periods in the past, Europe may have endured a harsher climate, with less precipitation and much colder winters. Summers may have been as warm as or warmer than today, because the Gulf Stream tends to dampen some of continental warming's effects during the summer months.

The northern Atlantic Ocean is a region of particular importance to the global climate system because it contains one of a small number of sites in the world where downwelling creates deep-ocean water. At this site, northeast of Iceland, surface seawater is cooled until, because of its increased density, it sinks and forms a narrow, deep current—the "North Atlantic Deep Water"—which hugs the eastern shore of Greenland.

While a geological-sciences doctoral student at the University of Colorado's Institute of Arctic and Alpine Research in Boulder, Don C. Barber made a case that a sudden surge of cold weather was triggered in Greenland and Europe 8,200 years ago by a massive flow of cold, fresh water from the Hudson Bay region into the North Atlantic Ocean. According to Barber, Hudson Bay (now salt water) housed the world's largest freshwater lake system, which formed from melting ice following the last ice age. An ice dam which had been holding the fresh water broke, spilling much of it into the North Atlantic Ocean, suppressing the Gulf Stream southward. Barber said that Greenland ice cores show that temperatures dropped by as much as 15 degrees F. in central Greenland and by nearly six degrees F. in western Europe following the lake drainage. Barber and colleagues estimated that when the ice dam from a remnant of the Laurentide Ice Sheet collapsed, the lake water flowing through the Hudson Strait into the Labrador Sea flowed out at a rate about 15 times greater than the present flow of the Amazon River (Barber et al. 1999; Schrader 1999, A-28).

According to the *Times of London*, evidence that even a small change in ocean currents may sharply decrease temperatures in Europe has been found in seabed sediment by Nick McCave, a professor of geology at Cambridge University. These deposits, collected south of Iceland, show that changes in currents associated with the Gulf Stream have coincided with large shifts in climate, including the "Little Ice Age," which plagued Europe roughly between 1400 and 1800, a time when persistent cold weather caused many harvests to fail and Londoners held fairs on the frozen Thames.

William J. Burroughs describes the dynamics of open-ocean circulation which could produce a significant cooling in Europe as the rest of the world warms:

Modeling work suggests that circulation patterns are extremely sensitive to the runoff of rainwater from the continents, the number of icebergs calved off Greenland, and the amount of precipitation from low-pressure systems tracking northeastwards past Iceland, and into the Norwegian Sea. Small variations in the total output may be able to trigger sudden switches in global ocean currents. (Burroughs 1999, 118)

THAWING PERMAFROST: THE "DRUNKEN FOREST"

Permafrost is profoundly destablilized by warming above the freezing point of water. Permafrost regions which are threatened by global warming also are being provoked to premature thawing by other human activities, including mining, forestry, and the construction of pipelines to carry oil and other fuels. "When . . . vegetation is removed, the permafrost is far more likely to melt and cause flooding," Burroughs writes (Burroughs 1999, 78).

Warming and eventual thawing of permafrost will have major environmental and societal impacts. As tundra and permafrost thaw, they release stored carbon dioxide and methane which raises the atmosphere's level of greenhouse gases. E.A. Koster and M.E. Nieuwenhuyzen comment in *Greenhouse-Impact on Cold-Climate Ecosystems and Landscapes*,

Where permafrost is ice-rich, or contains massive ground ice, subsidence and settling due to thawing will eventually occur. This could amount to several meters, with severe consequences for roads, buildings, bridges, pipelines, and other structures. . . . This can alter drainage patterns, and change the course of streams. Some areas may become swamp-like. (Koster and Nieuwenhuyzen 1992, 51)

The conditions Koster describes already have been observed in some Alaskan forests. Large patches of forest were described in newspaper reports of the late 1990s as drowning and turning gray as the thawing ground sank under them. Trees and roadside utility poles, destabilized by thawing, lean at crazy angles. The warming has contributed a new phrase to the English language in Alaska— "the drunken forest." Scientists at the University of Alaska say that temperatures in the same region have risen about five degrees F. during the past 30 years.

Human settlements may be damaged as permafrost melts around the Bering Sea. Such thawing could cause the land to subside as much as five meters, or more than 15 feet. The thawing of permafrost is also expected to destroy caribou habitat, cause landslides and erosion, clog salmon spawning rivers with silt, and trigger the loss of some boreal forests.

Between 20 and 25 percent of the land in the Northern Hemisphere experiences at least seasonal permafrost. Of this area, Koster and Nieuwenhuyzen write, "At the moment, Arctic tundras provide a major sink of CO_2. Changes in vegetation, as well as in the organic matter within the active layer, induced by climatic change, may cause this sink to become a major source, thus accelerating climatic changes" (Koster and Nieuwenhuyzen 1992, 38). (See also Chapter 2, "Biotic Feedbacks.") A large amount of carbon is stored in peat deposits, most of which have accumulated since the last glaciation. "A climatic shift to higher temperatures will increase the release of CH_4 [methane] from deep peat deposits, especially from tundra soils. Likewise, the release of CO_2 will increase, but hardly by more than 25 percent of its present level" (Malmer 1992, 97).

Nils Malmer, a plant ecologist in the Department of Ecology, Lun University, Sweden, describes how warmer temperatures gradually convert peatlands from a carbon sink to a carbon source:

Virgin peatlands have generally been suggested to have a net accumulation of organic matter and are therefore treated as carbon sinks in most studies of the global carbon cycle. When peatlands are drained for agriculture or forestry, a net release of carbon takes place through an increasing decomposition of . . . organic matter. For a short period of time (less than 100 years) this release can be at least partially compensated by . . . increased tree biomass on drained sites. The instantaneous release of carbon from a peatland is, of course, highly increased when a peat deposit is exploited and used as fuel. (Malmer 1992, 99)

MOUNTAIN GLACIERS: SLIP-SLIDING AWAY

Ice-core samples taken from high elevations in the Himalayas indicate that the 1990s were the warmest decade of the last 1,000 years at "the roof of the world." A team of scientists drilled holes at roughly 500-foot elevation intervals in the ice on a flank of Xixabangma, which rises to 26,293 feet. "This is the highest climate record ever retrieved, and it clearly shows a serious warming during the late twentieth century, one that was caused, at least in part, by human activity," said Lonnie Thompson, a professor of geological sciences at Ohio State University, who led the study (Connor 2000, 9). The ice cores also contain evidence of major droughts, including one that began in 1790, which caused the deaths of more than 600,000 people in India, where monsoon rains did not arrive for six years. The research team also found evidence that levels of atmospheric dust have quadrupled over the Himalayas during the twentieth century. Concentrations of chlorine have doubled during the same period (Thompson et al. 2000).

Global warming will raise snow levels and generally will cause seasonal snowpacks to melt more quickly. M. Zimmermann and W. Haeberli, who have studied mountain debris flow in their native Switzerland, assert,

Glaciers usually protect subglacial sediments from instability due to increased normal pressure and sheet strength. . . . [G]lacier retreat [has] caused remarkable changes in the cryosphere of the Alps, which resulted in an extension of zones prone to debris-flow initiation. The melting of small glaciers in particular exposes steep . . . moraines. In addition, many small lakes [have] formed behind . . . retreating glacier margins. Breaking of such natural dams can cause disastrous debris flows. . . . Continued glacier shrinkage would expose even more morainic material and, hence, further expand the area of potential debris-flow initiation. The reduction of summer runoff due to decreasing glacier surface is not likely to compensate for this negative effect. (Zimmermann and Haeberli 1992, 69)

Switzerland's Grindewald Glacier has retreated about 1.5 kilometers since 1850, with about half the retreat since 1940 (Gribben, *Hothouse Earth*, 1990, 3). The mean surface temperature of alpine permafrost in Switzerland rose between 1.0 and 1.5 degrees C. from 1880 to 1950, then continued more-or-less stable from 1950 to 1980, rising again after that (Haeberli 1992, 28). Warming mountain temperatures have caused "upward displacement" of plant species on 26 alpine peaks, according to one study (Grabherr et al. 1994), and 30, according to another (Pauli et al. 1996).

As glacial ice melted in the higher elevations of the Alps during the summer of 2000, long-lost debris and human remains surfaced in unusually large amounts. On 15,800-foot Mont Blanc, the highest elevation in the Alps, fragments from an Indian Airlines Boeing 707, including human bones, surfaced at the foot of the Bossons glacier, near the ski resort of Chamonix. The airliner crashed in 1966. The remains of a Dijon woman who disappeared in October 1977 were found at 9,000 feet on the Peclet-Polset glacier during August. According to an account in the *Times of London*, "Bones, climbing equipment and parts from the Boeing, and another Indian airliner that crashed in 1950, had been appearing for years on the glaciers, but the heatwave had accelerated the process" (Bremner et al. 2000).

The Intergovernmental Panel on Climate Change declared in 1995 that between a third and one-half of existing mountain glacier mass could disappear in 100 years (Gelbspan, *The Heat is On*, 1997, 139). Ice cores drilled in the Tibetan plateau indicate that the last half-century probably was the warmest period in the last 12,000 years in that area (Gelbspan, *The Heat is On*, 1997, 139).

Ice-core records indicate that higher elevations in the tropics are warmer than they have been in at least 2,000 to 3,000 years, causing glaciers on the highest peaks to retreat. The edge of the Qori Kalis glacier in the Peruvian Andes retreated at a rate of 13 feet a year from 1963 to 1978, but by 1995, the rate of retreat was 99 feet a year (Environmental Media Services 1997, 3). In the meantime, sea levels have risen four to ten inches during the last century, more than they had changed during the previous thousand years. The sea level is now at its highest level in 5,000 years, and rising (Environmental Media Services 1997, 5). In Peru, the retreat of glaciers in the Andes was raising concern during the late 1990s for irrigation water which usually flows from the mountains to the coastal desert. The same rivers also are dammed for hydroelectric power.

Indian researchers have warned that glaciers in the Himalayas are melting at an alarming rate and could cause a catastrophe if meltwater lakes overflow into surrounding valleys. "All the glaciers in the middle Himalayas are retreating," Syed Hasnain, of Jawaharlal Nehru University in Delhi, told *New Scientist* magazine (Aircraft Pollution 1999). The Gangorti glacier at the head of the Ganges River is receding 100 feet a year, according to Hasnain's four-year study. If current trends continue, all the glaciers in the central and eastern Himalayas could disappear by 2035, Hasnain said.

Glaciologists Lonnie G. Thompson and Ellen Mosley-Thompson of Ohio State University have measured the retreat of the largest glaciers in the Peruvian Andes. They calculated a rate of retreat for 1963 to 1983, and found that the rate had tripled between 1983 and 1991. The loss in volume of ice increased seven-fold in less than half a human lifetime (Gelbspan, *The Heat is On*, 1997, 139). Between 1963 and 1987, the glacial mass on Mount Kenya in tropical east Africa decreased 40 percent. The Thompsons comment, "The loss of these valuable hydrological stores may result in major economic and social disruptions in those areas dependent upon the glaciers for hydrologic power and fresh water" (Gelbspan, *The Heat is On*, 1997, 140).

WILL GREENHOUSE FORCING BREAK THE GLACIAL CYCLE?

Anthony D. Socci has surveyed epochs of natural cooling and warming in the earth's atmosphere, with reference to levels of carbon dioxide in the atmosphere roughly 700 million years into the past, during which he finds that the earth has been both markedly warmer and colder than today. The range of temperature has remained mainly 15 degrees C. warmer or colder than today's levels. According to Socci, "The duration of anthropogenic [human-caused], greenhouse warming may likely exceed 300,000 years, conceivably pre-empting at least three . . . global cooling events normally characterized by thickening and spread of continental glaciers" (Socci 1993, 170).

Socci believes that human-induced warming of the Earth's atmosphere is short-circuiting the glacial cycles which have dominated glacial climate for the last several-score million years (Socci 1993, 170–171). "In other words," Socci comments, "the anthropogenic contribution of carbon dioxide to the atmosphere is in excess of the volume of carbon dioxide which nature calls upon to go from one extreme of climate to the other" (Socci 1993, 171). In raising the atmospheric level of carbon dioxide to about 370 parts per million from the pre-industrial level of 280 p.p.m. in one century, humankind has broken an ages-old natural cycle.

During recent decades, study of ice cores has fundamentally changed many of science's assumptions about paleoclimate, the climate of the past. Analyzing air bubbles in ice cores of various ages, scientists can "read" a number of things, including the levels of greenhouse trace gases (notably carbon dioxide and methane) during past climatic cycles. Early ice-core work established a relationship between fluctuations in greenhouse-gas levels and temperatures. Because of ice-core records, paleoclimatologists now know that climate can change abruptly as the cycle shifts.

By 1999, a team of scientists led by J.R. Petit, drilling ice cores at Vostok, Antarctica, reached a depth of 3,623 meters, and in so doing was able to describe the composition of the Earth's atmosphere during the last four glacial cycles, or about 420,000 years. They found that the carbon dioxide level varied in a range of 180 parts per million (p.p.m.) to 300 parts per million, in a cycle which roughly parallels rises and falls in temperature. The atmospheric level of methane during the same period varied between 320 and 770 parts per billion (p.p.b.), in a similar pattern—low during glacial maxima, and high during interglacial warm periods. By the year 2000, greenhouse-gas levels (about 370 p.p.m. for carbon dioxide, and 1,700 p.p.b. for methane) were much higher than any level reached for as long as humankind has proxy records from ice cores, now 420,000 years before the present time.

The Vostok ice-core investigators have not declared that human-induced warming is short-circuiting the glacial cycle. Their words are more circumspect:

A striking feature of the Vostok . . . record is that the Holocene [the most recent inter-glacial period], which has already lasted 11,000 years is, by far, the longest stable warm period recorded in Antarctica during the last 420,000 years. . . . CO2 and CH4 concentrations are strongly correlated with Antarctic temperatures; this is because, overall, our results support the idea that greenhouse gases have contributed significantly to the glacial-interglacial change. This correlation, together with the uniquely elevated concentrations of these gases today, is of relevance with respect to the continuing debate on the future of the Earth's climate. (Petit et al. 1999, 434, 435)

REFERENCES

Adams, Jonathan, Mark Masli, and Ellen Thomas. In press. "Sudden Climate Transitions During the Quaternary." Article in press in *Progress in Physical Geography.* [http://www.esd.ornl.gov/projects/qen/transit.html]

"Aircraft Pollution Linked to Global Warming; Himalayan Glaciers are Melting, with Possibly Disastrous Consequences." 1999. Reuters. In *Baltimore Sun,* 13 June, 13A.

"Arctic Ice Melting, With Help." 1999. Associated Press. In *Omaha World-Herald,* 3 December, 8.

Bamber, Jonathan L., David G. Vaughan, and Ian Joughin. 2000. "Widspread Complex Flow in the Interior of the Antarctic Ice Sheet." *Science* 287 (18 February): 1248–1250.

Barber, D.C., A. Dyke, C. Hillaire-Marcel, A.E. Jennings, J.T. Andrews, M.W. Kerwin, G. Bilodeau, R. McNeely, J. Southon, M.D. Morehead, and J.-M. Gagnon. 1999. "Forcing of the Cold Event of 8,200 Years Ago by Catastrophic Drainage of Laurentide Lakes." *Nature* 400 (22 July): 344–351.

Barry, R.G. 1978. "Cryospheric Responses to a Global Temperature Increase." In Jill Williams, ed., *Carbon Dioxide, Climate, and Society: Proceedings of a IIASA Workshop Co-sponsored by WMO, UNEP, and SCOPE,* 21–24 February. Oxford, U.K.: Pergamon Press, 169–180.

Bentley, Charles. 1980. "Response of the West Antarctic Ice Sheet to CO2-Induced Global Warming." In *Environmental and Societal Consequences of a Possible CO2-Induced Climate Change.* Vol. 2. Washington, D.C.: Department of Energy, page numbers not available.

Bentley, C.R., and M.B. Giovinetto. 1990. "Mass Balance of Antarctica and Sea Level Change." In *International Conference on the Role of Polar Regions in Global Change.* Fairbanks: University of Alaska, 481–488.

Bernard, Harold W., Jr. 1993. *Global Warming: Signs to Watch For.* Bloomington: Indiana University Press.

Billings, W.D. 1992. "Phytogeographic and Evolutionary Potential of the Arctic Flora and Vegetation in a Changing Climate." In F. Stuart Chapin III, Robert L. Jefferies, James F. Reynolds, Gaius R. Shaver, and Josef Svoboda, *Arctic Ecosystems in a Changing Climate: An Ecophysiological Perspective.* San Diego: Academic Press, 91–109.

Bremner, Charles, Richard Owen, and Mark Henderson. 2000. "Heat Wave Uncovers the Grim Secrets of the Snows." *Times of London,* 26 August, in LEXIS.

Britt, Robert Roy. 1999. Antarctic Ice Shelves Falling Apart. Explorezone.com, 9 April. [http://www.explorezone.com/archives/99_04/09_antarctic_ice.htm]

Burroughs, William J. 1999. *The Climate Revealed.* Cambridge, U.K.: Cambridge University Press.

Canadian Temperatures Obliterate Records. 1998. *Toronto Star*, 15 December. [http://benetton.dkrz.de:3688/homepages/georg/kimo/0254.html]

Connor, Steve. 2000. "Ice Cores From a Himalayan Glacier Confirm Global Warming." *The Independent* (London), 15 September, 9.

Dansgaard, W., S.J. Johnson, H.B. Clausen, D. Dahl-Jensen, S. Gundenstrup, C.U. Hammer, C.S. Hvidberg, C.S. Steffensen, and J.P. Sveinbjrnsdottir. 1993. "Evidence for General Instability of Past Climate From a 250-kyr [thousand-year] Ice-core Record." *Nature* 364:218–220.

Davis, M.B. 1989. "Lags in Vegetation Response to Greenhouse Warming." *Climatic Change* 15:75–81.

Doake, C.S.M., H.F.J. Corr, H. Rott, P. Skvarca, and N.W. Young. 1998. "Breakup and Conditions for Stability of the Northern Larsen Ice Shelf, Antarctica." *Nature* 391:778–780.

Dyurgerow, Mark B., and Mark F. Meier. 2000. "Twentieth-century Climate Change: Evidence from Small Glaciers." *Proceedings of the National Academy of Sciences of the United States of America* 97:4 (15 February): 1351–1354.

Environmental Media Services. 1997. *Understanding the Science of Global Climate Change.* Washington, D.C.: Environmental Media Services.

Epstein, Paul, Georg Grabbher, Tom Karl, Ellen Mosley-Thompson, Kevin Trenberth, and George M. Woodwell. 1996. Current Effects: Global Climate Change. An Ozone Action Roundtable, 24 June, Washington, D.C. [http://www.ozone.org/curreff.html]

"Europe Faces an Ice Age as the World Warms." 1999. *Times of London*, 17 January. [http://geography.rutgers.edu/courses/99spring/370sp99/news01_20_99.html#anchor 33828]

Gelbspan, Ross. 1997. *The Heat is On: The High Stakes Battle Over Earth's Threatened Climate.* Reading, Mass.: Addison-Wesley Publishing Co.

———. 1995. "The Heat is On: The Warming of the World's Climate Sparks a Blaze of Denial." *Harper's*, December. [http://www.dieoff.com/page82.htm]

"Global Warming Seen as Cause of Antarctic Melting." 1999. Associated Press, 8 April. [http://www.gsreport.com/articles/art000101.html]

Grabherr, G., N. Gottfried, and H. Pauli. 1994. "Climate Effects on Mountain Plants." *Nature* 369:447.

Gribben, John. 1990. *Hothouse Earth: The Greenhouse Effect and Gaia.* London: Bantam Press.

Haeberli, W. 1992. "Possible Effects of Climatic Change on the Evolution of Alpine Permafrost." In M. Boer and E. Koster, eds., *Greenhouse-Impact on Cold-Climate Ecosystems and Landscapes.* Selected papers of a European Conference on Landscape Ecological Impact: Impact of Climatic Change, Lunteren, the Netherlands, December 3–7, 1989. CARTENA Supplement 22. Cremlingen, Germany: Cartena Verlag, 23–35.

Helfferich, Carla. 1990. "Alaska Science Forum: Cloudy Picture of a Changing Globe." Geophysical Institute, University of Alaska Fairbanks. 6 June. [http://www.gi.alaska.edu/ScienceForum/ASF9/981.html]

Hodges, Glenn. 2000. "The New Cold War: Stalking Arctic Climate Change by Submarine." *National Geographic*, March, 30–41.

Hollin, J.T. 1970. "Interglacial Climates and Antarctic Ice Surges." *Quatenary Research* 2:401–408.

"A Hot '98, Relatively Speaking, in Nation's Northernmost City." 1999. ABC News, 17 January. [http://geography.rutgers.edu/courses/99spring/370sp99/news01_20_99. html#anchor33828]

Huber, Brian T., Kenneth G. MacLeod, and Scott L. Wing. 2000. *Warm Climates in Earth History.* Cambridge, U.K.: Cambridge University Press.

Hughes, T.J., J.L. Fastook, and G.H. Denton. 1979. *Climatic Warming and the Collapse of the West Antarctic Ice Sheet.* Orono: University of Maine Press.

Huybrechts, P., and J. Oerlemans. 1990. "Response of the Antarctic Ice Sheet to Future Greenhouse Warming." *Climate Dynamics* 5:93–102.

Johnson, Tim. 1999. "World out of Balance: In a Prescient Time, Native Prophecy Meets Scientific Prediction." *Native Americas* 16, no. 3/4(Fall/Winter): 8–25. [http:// nativeamericas.aip.cornell.edu]

Kelly, Mick. 1990. "Halting Global Warming." In Jeremy Leggett, ed., *Global Warming: The Greenpeace Report.* New York: Oxford University Press, 83–112.

Kerr, Richard A. 1998. "West Antarctica's Weak Underbelly Giving Way?" *Science* 281 (24 July): 499–500.

Kling, G. W., G. W. Kipphut, and M. C. Miller. 1992. "The Flux of Carbon Dioxide and Methane from Lakes and Rivers in Arctic Alaska." *Hydrobiologia* 240:23–36.

Koster, E.A., and M.E. Nieuwenhuyzen. 1992. "Permafrost Response to Climatic Change." In M. Boer and E. Koster, eds., *Greenhouse-Impact on Cold-Climate Ecosystems and Landscapes.* Selected papers of a European Conference on Landscape Ecological Impact: Impact of Climatic Change, Lunteren, the Netherlands, December 3–7, 1989. CARTENA Supplement 22. Cremlingen, Germany: Cartena Verlag, 23–35.

Krabill, William. 1999. "Rapid Thinning of Parts of the Southern Greenland Ice Sheet." Abstract of Article in *Science* 283 (5 March): 1152–1154. [http://earth.rice.edu/ MTPE/cryo/cryosphere/topics/greenthin.html]

Krabill, W., W. Abdalati, E. Frederick, S. Manizade, C. Martin, J. Sonntag, R. Swift, R. Thomas, W. Wright, and J. Yungel. 2000. "Greenland Ice Sheet: High-elevation Balance and Peripheral Thinning." *Science* 289 (21 July): 428–430.

Lazaroff, Cat. 2000. "Greenland Ice Sheet Melting Away." Washington, D.C.: Environmental News Service, 21 July. [http://www.cpa.org.au/garchve2/1012grn.html]

Lowell, Thomas V. 2000. "As Climate Changes, So Do Glaciers." *Proceedings of the National Academy of Sciences of the United States of America* 97, no. 4 (15 February): 1351–1354.

Magnuson, John J., Dale M. Robertson, Barbara J. Benson, Randolph H. Wynne, David M. Livingstone, Tadashi Arai, Raymond A. Assel, Roger G. Barry, Virginia Card, Esko Kuusisto, Nick G. Granin, Terry D. Prowse, Kenton M. Stewart, and Valery S. Vuglinski. 2000. "Historical Trends in Lake and River Ice Cover in the Northern Hemisphere." *Science* 289 (8 September): 1743–1746.

Malmer, Nils. 1992. "Peat Accumulation and the Global Carbon Cycle." In M. Boer and E. Koster, eds., *Greenhouse-Impact on Cold-Climate Ecosystems and Landscapes.* Selected papers of a European Conference on Landscape Ecological Impact: Impact of Climatic Change, Lunteren, the Netherlands, December 3–7, 1989. CARTENA Supplement 22. Cremlingen, Germany: Cartena Verlag, 97–110.

Mathews-Amos, Amy, and Ewann A. Berntson. 1999. "Turning up the Heat: How Global Warming Threatens Life in the Sea." World Wildlife Fund and Marine Conservation Biology Institute. [http://www.worldwildlife.org/news/pubs/wwf_ocean.html]

"Melting Antarctic Glacier Could Raise Sea Level." 1998. Reuters, 24 July. [http://bonanza.lter.uaf.edu/~davev/nrm304/glbxnews.htm]

Mercer, J.H. 1970. "Antarctic Ice and Interglacial High Sea Levels." *Science* 168:1605–1606.

———. 1968. "Antarctic Ice and Sangamon Sea Level." *International Association for the Science of Hydrology* 79:217–225.

Murgia, Joe. 1999. "NASA: Greenland's Glaciers Are Shrinking: A New Study Suggests That Rapid Thinning and Excess Run-off from Greenland's Southeastern Glaciers May be Partly Caused by Climate Changes." GS Report, 10 March. [http://www.gsreport.com/articles/art000078.html]

National Academy of Sciences. 1991. *Policy Implications of Greenhouse Warming.* Washington, D.C.: National Academy Press.

Nuttall, Nick. 2000. "Experts are Poles Apart Over Ice Cap." *Times of London*, 21 August, in LEXIS.

Oerlemans, J., and J.P.F. Fortuin. 1992. "Sensitivity of Glaciers and Small Ice Caps to Greenhouse Warming." *Science* 258:115–117.

Ohmura, A., and Reeh Ohmura. 1991. "New Precipitation and Accumulation Maps for Greenland." *Journal of Glaciology* 37:140–148.

Pauli, H., M. Gottfried, and M. Grabherr. 1996. "Effects of Climate Change on Mountain Ecosystems—Upward Shifting of Alpine Plants." *World Resources Review* 8:382–390.

Petit, Charles W. 2000. "Polar Meltdown: Is the Heat Wave on the Antarctic Peninsula a Harbinger of Global Climate Change?" *U.S. News and World Report*, 28 February, 64–74.

Petit, J.R., J. Jouzel, D. Raynaud, N.I. Barkov, J.-M. Barnola, I. Basile, M. Benders, I. Chappellaz, M. Davis, G. Delaygue, M. Delmotte, V.M. Kotlyakov, M. Legrand, V.Y. Lipenkov, C. Lorius, L. Pepin, C. Ritz, E. Saltzman, and M. Stievenard. 1999. "Climate and Atmospheric History of the Past 420,000 Years From the Vostok Ice Core, Antarctica." *Nature* 399 (3 June): 429–436.

"Report: Last Ice Age Had Quick End." 1999. Associated Press. In *Omaha World-Herald*, 29 October, 15.

Rignot, E. J. 1998. "Fast Recession of a West Antarctic Glacier." *Science* 281 (24 July): 549–551.

Rind, D., and J. Overpeck. 1993. "Hypothesized Causes of Decade-to-century-scale Climate Variability: Climate Model Results." *Quaternary Science Reviews* 12:357–374.

Satellite Images Show Chunk of Broken Antarctic Ice Shelf. 1998. Eureka Alert, 16 April. [http://www.eurekalert.org/releases/brkantartice.html]

Schneider, Stephen H., and R.S. Chen. 1980. "Carbon Dioxide Warming and Coastline Flooding: Physical Factors and Climatic Impact." *American Review of Energy* 5:107–140.

Schrader, Ann. 1999. "Ice Dam's Collapse in Canada Cooled Europe for Centuries; Torrent was 15 Times Greater than Amazon, CU Scientist Says." *Denver Post*, 24 July, A-28.

Singer, S. Fred. 2000. "Sure, the North Pole is Melting. So What?" *Wall Street Journal*, 28 August, A-18.

Socci, Anthony D. 1993. "The Climate Continuum: An Overview of Global Warming and Cooling Throughout the History of Planet Earth." In Richard A. Geyer, ed., *A Global Warming Forum: Scientific, Economic, and Legal Overview*. Boca Raton, Fla.: CRC Press, 161–207.

Spotts, Peter N. 2000. "As Arctic Warms, Scientists Rethink Culprits." *Christian Science Monitor*, 22 August, 4.

Steitz, David E. 1999. "NASA Researchers Document Shrinking of Greenland's Glaciers." NASA Press Release 99–33, 4 March. [http://www.earthobservatory. nasa.gov/Newsroom/NasaNews/19990304207.html]

Stirling, I., and A.E. Derocher. 1993. "Possible Impacts of Climate Warming on Polar Bears." *Arctic* 46, no. 3: 240–245.

Stirling, I., and N.J. Lunn. 1997. "Environmental Fluctuation in Arctic Marine Ecosystems as Reflected by Variability in Reproduction of Polar Bears and Ringed Seals." In S.J. Woodlin and M. Marquiss, eds., *Ecology of Arctic Environments*. Oxford: Blackwell Science Ltd., 167–181.

Sudetic, Chuck. 1999. "As the World Burns." *Rolling Stone*, 2 September, 97–106, 129.

Suplee, Curt. 2000. "For 500 Million, a Sleeper on Greenland's Ice Sheet." *Washington Post*, 10 July, A-9.

———. 2000. "Historical Records Provide a Growing Sense of Global Warmth." *Washington Post*, 8 September, A-2.

Taylor, K.C., R.B. Alley, G.A. Doyle, P.M. Grootes, P.A. Mayewski, G.W. Lamorey, J.W.C. White, and L.K. Barlow. 1993. "The 'Flickering Switch' of Late Pleistocene Climate Change." *Nature* 361:432–436.

Thompson, L.G., T. Yao, E. Mosley-Thompson, M.E. Davis, K.A. Henderson, and P.-N. Lin. 2000. "A High-Resolution Millennial Record of the South Asian Monsoon from Himalayan Ice Cores." *Science* 289 (15 September): 1916–1919.

Titus, James G., and Vijay Narayanan. 1996. "The Risk of Sea Level Rise: A Delphic Monte Carlo Analysis in which Twenty Researchers Specify Subjective Probability Distributions for Model Coefficients within their Respective Areas of Expertise." *Climate Change* 33, no. 2: 151–212. (Also U.S. Environmental Protection Agency. N.d. [http://users.erols.com/jtitus/Risk/CC.html])

West Antarctic Ice Sheet Not in Jeopardy. 1998. [http://www.enn.com/news/ennstories/1998/12/120198/antarc.asp]

Wilfred, John Noble. 2000. "Ages-Old Polar Icecap Is Melting, Scientists Find." *New York Times*, 19 August, 1.

Williams, Jack. 2000. "Rising Ocean Temperatures Aren't Breaking the Ice." *USA Today*, 25 April, 10D.

Woodard, Colin. 1998. "Glacial Ice is Slip-sliding Away." *Christian Science Monitor*, 10 December. [http://benetton.dkrz.de:3688/homepages/georg/kimo/0254.html]

———. 2000. "Slowly, but Surely, Iceland is Losing its Ice; Global Warming is Prime Suspect in Meltdown." *San Francisco Chronicle*, 21 August, A-1.

Wuethrich, Bernice. 1999. "Lack of Icebergs Another Sign of Global Warming?" *Science* 285 (2 July): 37.

Zimmermann, M., and W. Haeberli. 1992. "Climatic Change and Debris-flow Activity in High Mountain Areas: A Case Study in the Swiss Alps." In M. Boer and E.

Koster, eds., *Greenhouse-Impact on Cold-Climate Ecosystems and Landscapes*. Selected papers of a European Conference on Landscape Ecological Impact: Impact of Climatic Change, Lunteren, the Netherlands, December 3–7, 1989. CARTENA Supplement 22. Cremlingen, Germany: Cartena Verlag, 59–72.21.

CHAPTER 5

Warming Seas

Because humankind is a narcissistic species, our focus, when considering the implications of global warming, usually is fixed on the third of the planet that comprises dry land. While the land warms, the other two-thirds of the Earth will be warming as well, with profound implications for the species which inhabit it, including a holocaust for coral reefs which already has begun.

According to Peter G. Brewer and associates, writing in *Science*, the chemistry of ocean water already is being changed by human-induced greenhouse gases in the atmosphere: "The invading wave of atmospheric CO2 has already altered the chemistry of surface seawater worldwide, and over much of the ocean, this tracer field has now permeated to a depth of more than one kilometer" (Brewer et al. 1999, 943). Given that the ocean is slowly warming, as its carbon dioxide level rises, its capacity as a carbon sink is probably being compromised.

Ken Caldeira and Philip B. Duffy assert in *Science* that "uptake," or removal, of human-induced carbon dioxide by the oceans is less than many earlier investigators have assumed and that, as temperatures warm, carbon uptake will diminish further. Additionally, Caldeira and Duffy contend that absorption of carbon into the oceans of the Southern Hemisphere (the focus of their study) has been diminishing since 1880 when fossil-fuel effluvia became a factor in the composition of the atmosphere. "Ventilation of the deep Southern Ocean was much more vigorous in the period from about 1350 to 1880 than in the recent past" (Caldeira and Duffy 2000, 620).

Among the important results of warming seas will be coastal erosion, shoreline inundation because of higher tide levels, higher storm surges, and saltwater intrusion into coastal estuaries and groundwater supplies. A report by the World Wildlife Fund said: "Scientific evidence strongly suggests that global climate change already is affecting a broad spectrum of marine species and ecosystems, from tropical coral reefs to polar ice-edge communities" (Mathews-Amos and Berntson 1999).

The level of the world's seas and oceans has been rising slowly for much of the twentieth century, a millimeter or two a year, enough to produce noticeable erosion on 70 percent of the world's sandy beaches, including 90 percent of sandy beaches in the United States (Edgerton 1991, 18). In some areas, such as the United States Gulf Coast between New Orleans and Houston, the withdrawal of underground water and oil is causing some areas to sink as the oceans rise.

More than 20 percent of the world's population lives within 30 kilometers of coastal areas. That population is increasing twice as quickly as aggregate population (Mathews-Amos and Berntson 1999). According to one estimate, a one-meter rise in sea level could displace 300 million people around the world (Edgerton 1991, 70). Thirty of the world's largest cities lie near coasts, and are vulnerable to a one-meter rise in ocean level (Augenbraun et al. N.d.). Most of Australia's population lives in coastal cities. By 1990, more than 100 million people lived within 50 miles of coastlines in the United States.

A large number of the people who could be displaced by modest sea-level rise are residents of broad river deltas, such as the Mississippi near New Orleans, the Nile in Egypt, and the Ganges, which joins with several other rivers in Bangladesh. Half of Bangladesh itself is less than 5 meters above sea level. A one-meter sea-level rise in Bangladesh would displace about 10 to 20 percent of the human population, and cost at least six percent of Gross National Product (GNP). An ocean-level rise of three meters would affect 27 percent of the people there, and cost roughly 15 percent of GNP (Edgerton 1991, 74).

One observer of this situation believes, "Global change within the next century will be particularly hard-felt in densely populated delta areas like Bangladesh, where a substantial area is barely above sea level and vulnerable to frequent natural calamities like droughts, tropical cyclones, and tornadoes" (Mahtab 1992, 28). Fesih Mahtab projects that Bangladesh's population will rise from 115 million in 1990 to 145 million in the year 2000 and 232 million in 2030. A sea-level rise of one meter would put 16 percent of the country under water (Mahtab 1992, 35).

During the twenty-first century, the phrase "environmental refugee" may become more familiar around the world. Lowland residents who could be forced out of their homes by a three-foot rise in sea levels during the next century include 26 million people in Bangladesh, 70 to 100 million in China, 20 million in India, and 12 million in the Nile Delta of Egypt (Gelbspan, *The Heat is On*, 1997, 162). In Egypt, a one-meter sea-level rise could cost 15 percent of the country's GNP, including much of its agricultural base (Edgerton 1991, 72–73). A 14-inch rise in sea levels could flood 40 percent of the mudflats which ring Puget Sound, obliterating a significant habitat for shellfish and waterfowl (Gough 1999, 48).

The locations mentioned above are only a few of a great many examples, because the Earth's junctures of seas and rivers have been important crossroads for human trade (as well as fertile farming areas) throughout human history. Many river deltas are densely populated and very vulnerable to even a small amount of sea level rise.

Forecasts of sea level rise must contend with major uncertainties. Most sea level rise projections do not indicate how much the increase hinges on what happens in Antarctica (see Chapter 4, "Icemelt") or how sudden some rises in sea level could be. William R. Cline calls some of these estimates "innocuous," as he comments,

Consider sea level rise, recently downgraded from one meter to about 40 centimeters for CO2 doubling. The IPCC [Intergovernmental Panel on Climate Change] calculations are premised on no contribution to sea level rise from the Antarctic, on grounds that its temperature remains in a range (below minus 12 degrees C.) where warming increases rather than decreases glacier mass. (One reason is that warmer air is more moist, so it provides more snow for glacier buildup.) But warming of 10 degrees C. or more would cause the Antarctic to move into the domain where it contributes to sea level rise. If so, the increase in sea level would be much higher—particularly if the West Antarctic ice sheet [were to begin] to disintegrate. (Cline, Comments, 1991, 223)

RISING OCEAN TEMPERATURES

In November 1994, researchers announced that they had found a 0.9 degree F. (0.5 degree C.) rise in the temperature of the deep waters of the Indian Ocean, compared to measurements taken 20 years earlier. This was the third ocean observed to be warming. In 1992, Nathan Bindoff reported that the subsurface temperature of the southwestern Pacific Ocean had increased at about the same rate, 0.9 degree F. (0.5 degree C.), over 20 years. In early 1994, a group led by Gregorio Parrilla of the Spanish Oceanographic Institute reported that the North Atlantic also was warming. In December 1994, Ed Carmack of the Institute of Ocean Sciences in Sidney, British Columbia, Canada, reported that the waters 200 to 1000 meters (660 to 3280 feet) beneath the Arctic Ocean had warmed 1.8 degrees F. (1.0 degree C.) in a decade.

Scientists at the National Oceanic and Atmospheric Administration (NOAA) reported during late March 2000 that the oceans had warmed significantly world-wide during the preceding four decades. A broad study of temperature data from the oceans showed that average water temperatures had increased from one-tenth to one-half degree, depending on depth, since the 1950s. The NOAA study uncovered significant warming between 1955 and 1995 as deep as 10,000 feet in the oceans. The greatest warming occurred from the surface to a depth of about 900 feet, where the average heat content increased by 0.56 degree F. Water as far down as 10,000 feet (in the North Atlantic) was found to have gained an average of 0.11 degrees F.

"We've known the oceans could absorb heat, transport it to subsurface depths and isolate it from the atmosphere. Now we see evidence that this is happening," said Sydney Levitus, chief of NOAA's Ocean Climate Laboratory and principal author of the study. "Our results support climate modeling predictions that show increasing atmospheric greenhouse gases will have a relatively large warming influence on the Earth's atmosphere," said Levitus (Herbert 2000). According to Levitus, "The whole-Earth system has gone into a relatively warm state" (Kerr, Globe's 'Missing Warming,' 2000, 2126; Levitus et al. 2000).

The ocean-temperature record compiled by Levitus and associates confirms similar trends in land-surface temperature records, with one notable period of warming between 1920 and 1940, followed by a slight cooling. By the late 1970s, both ocean- and land-surface temperatures began to spike upward to new peaks in the observational record. The magnitude of the oceanic warming surprised some experts. Dr. Peter Rhines, an oceanographer and atmospheric scientist at the University of Washington in Seattle, said it appeared roughly equivalent to the amount of heat stored by the oceans as a result of seasonal heating in a typical year. "That makes it a big number," he said (Stevens, Oceans Absorb, 2000, A-16).

The Environmental Protection Agency (EPA) has stated that warming seas also may be influenced by changes in the Earth's gravitational field provoked by melting ice at the poles. "Removal of water from the world's ice sheets would move the earth's center of gravity away from Greenland and Antarctica; the oceans' water would thus be redistributed toward the new center of gravity. Along the coast of the United States, this effect would generally increase sea level rise by less than 10 percent; sea level could actually drop, however, at Cape Horn and along the coast of Iceland" (Clark and Lingle 1977). Climate change could also influence local sea level by changing winds, atmospheric pressure, and ocean currents, but no one has estimated these impacts (U.S. EPA 1997).

SEA-LEVEL RISE AND SMALL ISLAND NATIONS

On many small islands around the world, rising seas are everyday news, and global warming is on everyone's lips. Nicholas Kristof of the *New York Times* described how Teunaia Abeta, a resident of Kiribati island, in the Pacific, watched in horror, as

a high tide came rolling in from the turquoise lagoon and did not stop. There was no typhoon, no rain, no wind, just an eerie rising tide that lapped higher and higher, swallowing up Abeta's thatched-roof home and scores of others in this Pacific Island nation. "This had never happened before," said Abeta, 73, who wore only his colorful *lava-lava*, a skirt-like garment, as he sat on the raised platform of his home fingering a home-rolled cigarette. "It was never like this when I was a boy." (Kristof 1997)

The Alliance of Small Island States (AOSIS), including the Philippines, Jamaica, the Marshall Islands, the Bahamas, Samoa, and others, has been one of few voices at climate talks favoring swift, worldwide reductions in greenhouse-gas emissions. Their self-interest is evident: with warming already "forced," but not yet fully worked into the world's temperature equilibrium, many small island states will lose substantial territory and economic base within the next few decades.

For every other country on Earth, competing economic interests drive climate negotiations. For the small island nations, the driving force is survival. "For us, it's a matter of death and life, whereas in terms of the citizens of the industrialized countries, [global warming] will affect their lifestyles basically, but not to the extent they will be disappearing," said Bikenibeu Paeniu, Tuvalu's prime minister (Webb 1998).

In no other place is global warming more urgent an issue, in a practical sense, than in equatorial Kiribati (pronounced "Kirabas"), a chain of coral atolls in the Pacific, near the Marshall Islands. Kiribati is a republic, a member of the United Nations, and home to about 79,000 people. Its land area, at any point, is no more than two meters above sea level. Additionally, the living corals that have raised the islands above sea level are threatened with demise by warmer ocean waters. Ierimea Tabai, who became president of Kiribati after its independence from Britain in 1979, has said, "If the greenhouse effect raises sea levels by one meter, it will eventually do away with Kiribati. In fifty or sixty years, my country will not be here" (Webb 1998).

Kiribati's 33 islands, comprising 277 square miles, are scattered across 5.2 million square kilometers (2 million square miles) of ocean, including three groups of islands: 17 Gilbert Islands, eight Line Islands, and eight Phoenix Islands. Kiribati includes Kiritimati (formerly Christmas Island), the world's largest coral atoll (150 square miles). The island of Kiribati, located roughly halfway between California and Australia, more than doubles its surface area at low tide.

During 1997, the area was devastated by El Niño, which brought heavy rainfall, a half-meter rise in sea level, and extensive flooding. About 40 percent of the atolls' coral was killed by overheated water, and nearly all of Kiritimati Island's roughly 14 million birds died or deserted the island.

Sea level rise provoked by global warming could imperil more than 3,000 small isolated islands, grouped into 24 political entities in the Pacific Ocean, with a population of about 5 million people in 800 distinct cultures. Many coral atolls also contain permanent areas of fresh water (lagoons), which are vulnerable to salinization as sea levels rise.

Central lagoons of the small islands are social centers and a source of food. According to Kristof,

Children play in the water from infancy. Many adults fish or sail for a living. The Kiribati men in their loincloths go out in outrigger canoes each day to catch tuna, and the women wade out on the coral reef to dive for shellfish and net smaller fish for dinner. . . . "People here think of the ocean as their source of livelihood, as their friend," said Ross Terubea, a Kiribati radio reporter. "It's hard to think that it would destroy us." (Kristof 1997)

By 1998, rising sea levels already were swallowing some small Pacific islands and contaminating drinking water on others. The rising sea has endangered sacred sites and drowned some small islands near Kiribati and Tuvalu, including the islet of Tebua Tarawa, once a landmark for Tuvalu fishermen. Kiribati already has moved some roads inland on its main island as the rising Pacific Ocean eats into its shores. Rising sea levels are seeping into soils on some islands, as water tables rise. Soils in some areas are becoming too salty for growing most vegetables. In Tuvalu, according to a dispatch from Reuters, farmers "are beginning to grow their taro crops in tin containers filled with compost instead of traditional pits" (Webb 1998). Soil contamination is also an issue in the Bahamas, where fear has been expressed that the limestone that underlies the soil on many islands will absorb saline ocean water like a sponge.

Given expectations that global temperatures will rise further in coming decades, the sea level rise of the twentieth century is expected to be dwarfed by the rise during the twenty-first century. Scientists who attend to this issue typically estimate sea level rise possibilities in wide ranges, because no one knows how much ice will melt given a certain rise in temperatures over the poles. A team led by J.J. Wells Hoffman, for example, in 1983 estimated that sea levels in the year 2100 would be between 58 and 368 centimeters higher than they were in the 1980s (Hoffman et al. 1983). R. Thomas, in 1986, put the estimated range at 56 to 345 centimeters (Thomas 1986). The estimates vary by a factor of roughly seven, an indication of the uncertainty which plagues forecasts of how much the sea will rise under various assumptions about global warming. (See also Chapter 4, "Icemelt.")

A HOLOCAUST FOR CORAL REEFS

Coral reefs, nurtured by an aquatic environment, are among the most productive, diverse ecosystems on earth. Sometimes called the rainforests of the sea, coral reefs cover much less than one percent of the world's ocean floors, while at the same time hosting more than a third of the marine species presently described by science, with many species remaining undocumented. Some of these organisms may provide new sources of anticancer compounds and other medicines. Coral reefs also protect shorelines from erosion by acting as breakwaters which, if healthy, can repair themselves.

Coral reefs are extremely intolerant of small rises in water temperature, which destroy the ecosystem of the reef, leaving only a lifeless, bleached exoskeleton. By the early 1990s, an increasing number of the earth's coral reefs were turning white, a sign that they have been killed because of excessive heat. Most corals die at levels above 30 to 32 degrees C. Bleaching results when symbiotic *zooxanthellae* algae are expelled from coral reefs. The algae provide reefs with most of their color, carbon, and ability to deposit limestone. Once bleached, the coral does not receive adequate nutrients or oxygen. If temperatures fall, the reefs may recover, but if thermal stress continues a large part of any given reef will die.

Pollution, disease, ultraviolet radiation, too much shade, and changes in the ocean's salinity also may contribute to coral bleaching. Warming seas are one among many human-induced changes that threaten coral reefs. Other problems include over-fishing, coastal development, nutrient runoff from agriculture and sewage, and sedimentation from logging on streams which feed into coastal waters. By one estimate, 58 percent of the world's coral reefs were being threatened by human activity by the late 1990s (Bryant et al. 1998).

A scientific team led by Thomas J. Goreau plotted water temperatures at a number of locations in the tropical western Atlantic and Caribbean, finding slow rises beginning during the 1980s in such places as the Bahamas and Jamaica. In some cases, these temperature increases have been significant enough to dramatically accelerate the bleaching of coral reefs. The rate of bleaching correlates to general air temperature rises detected by global surveys of temperatures on land. "The fact that the high water temperatures of the late 1980s and mass coral bleaching have not been seen before in the Greater Caribbean during nearly 40 years of continuous study of coral reef ecosystems implies that temperatures have only recently exceeded tolerance limits of the corals" (Goreau et al. 1993, 251). The authors of this paper cite similar results in Hawaiian waters. This article includes an extensive list of references which detail exactly what is happening to coral reefs around the world (Goreau et al. 1993, 253–255).

Some areas, including the Cook Islands and several locations in the Philippines, have been afflicted with massive coral bleaching since 1983. In 1995, warming in the Caribbean produced coral bleaching for the first time in Belize, as sea surface temperatures surpassed 29 degrees C (84 degrees F.). In 1997, Caribbean sea-surface temperatures reached 34 degrees C. (93 degrees F.) off southern Belize, and coral bleaching was accompanied by the deaths of many starfish and other sea life.

The central barrier reef near Belize suffered a mass coral die-off following record-high water temperatures in 1998. According to Richard Aronson, a marine biologist at the Dauphin Island Sea Laboratory in Alabama, this die-off represents the first complete collapse of a coral reef in the Caribbean Sea from bleaching (Connor, Global Warming, 2000, 12). Richard Aronson, working with colleagues at the Smithsonian Institution in Washington D.C., said in *Nature* that lettuce coral was the most abundant species living in the area until 1998 (Aronson et al. 2000). Later surveys showed that "virtually all living colonies had been bleached white at almost every depth investigated" (Connor, Global Warming, 2000, 12). Core samples from the corals indicate that such a mass bleaching is unprecedented for at least the last 3,000 years. The research team found, "Complete bleaching was also evident in almost all . . . species down to the lagoon floor at 21 meters" (Aronson et al. 2000, 36).

By the late 1990s, coral bleaching had become epidemic in many tropical seas and oceans around the world. A World Wildlife Fund report found that massive coral bleaching occurred during the 1990s, in response to unusually high water temperatures, particularly during the El Niño years of 1997 and 1998. Sites of large-scale coral mortality included the Pacific Ocean, Indian Ocean, Red Sea, Persian Gulf, as well as the Mediterranean and Caribbean seas (Mathews-Amos and Berntson 1999).

With more warming anticipated, the extent and severity of damage to coral reefs worldwide is expected to intensify. An Australian marine scientist, Dr. Terry Done, told Murray Hogarth of the *Sydney Morning Herald* that he regards coral reefs as the "canaries in the coalmine" (Hogarth 1998). During 1998, large coral reefs up to 1,000 years old died; and in some parts of the world up to 90 percent of reefs have been devastated, Done said.

The World Wildlife Fund sketches the worldwide nature of coral bleaching caused by rising ocean temperatures:

According to NOAA, coral bleaching was reported throughout the Indian Ocean and Caribbean in 1998, and throughout the Pacific, including Mexico, Panama, Galapagos, Papua New Guinea, American Samoa, and Australia's Great Barrier Reef starting in 1997. . . . The severity and extent of coral bleaching in 1997–98 was widely acknowledged among coral reef scientists as unprecedented in recorded history. (Mathews-Amos and Berntson 1999)

The International Society for Reef Studies concluded that 1997 and 1998 had witnessed the most geographically widespread coral bleaching in the recorded histories of at least 32 countries and island nations. Reports of bleaching came from sites in all the major tropical oceans of the world, some for the first time in recorded history. The 1997–1998 bleaching episode was exceptionally severe, as a large number of corals died. According to one study, parts of Australia's Great Barrier Reef have been so severely affected that many of the usually robust corals, including one dated over 700 years of age, were badly damaged or had died during 1997 and 1998 (Mathews-Amos and Berntson 1999).

A World Wildlife Fund report indicates that corals are directly harmed not only by rising water temperatures, but also by increasing levels of atmospheric carbon dioxide.

Living coral reefs are composed of great numbers of coral animals covering a rigid skeleton formed by coral secretions of calcium carbonate. High levels of atmospheric CO_2 alter water chemistry and reduce the calcification rate, and hence density, of coral skeletons. Some scientists believe that calcification probably already has decreased on some reefs, and predict that calcification could decrease 17 to 35 percent from pre-industrial levels by 2100. (Mathews-Amos and Berntson 1999)

Coral reefs that suffer from reduced density due to decreased calcification face erosion by storm-tossed seas.

An international team of more than 250 marine scientists warned that conditions during 1998, the warmest year in recorded history on land and sea, had produced a mass breaching of many coral reefs around the world.

Scientists meeting in Townsville, near Australia's Great Barrier Reef, during November of 1998 declared that sea temperature rises predicted during the next fifty years pose a serious threat to nearly all of the world's coral reefs. Global coral bleaching and die-off was unprecedented in 1998 in geographic extent, depth and severity, said the International Coral Reef Initiative, an association of scientists and marine-park managers. The only major reef region spared from coral bleaching appears to be the central Pacific, the scientists said in a report on the status of the world's reefs.

Some of the most severe coral bleaching took place in the tropical Indian Ocean, where water temperatures during the 1998 El Niño reportedly rose to between three and five degrees C. above long-term averages in some areas. According to Clive Wilkinson and colleagues writing in the Swedish environmental journal *Ambio*,

Massive mortality occurred on the reefs of Sri Lanka, [the] Maldives, India, Kenya, Tanzania, and [the] Seychelles, with mortalities of up to 90 percent in many shallow areas. . . . Coral death during 1998 was unprecedented in severity. . . . Coral reefs of the Indian Ocean may prove to be an important signal of the potential effects of global climate change, and we should heed that warning. (Wilkinson et al. 1999, 188)

According to another observer, "In some parts of the Indian Ocean . . . reefs in the Maldives, Sri Lanka, Kenya and Tanzania were devastated with shallow reefs looking like graveyards" (Perry 1998).

In the Asia-Pacific region, the worst-hit coral reefs were near Japan, Taiwan, the Philippines, Vietnam, Thailand, Singapore, Indonesia and the islands of Palau. A scientists' statement, released at the same time in the United States and Australia, relied heavily on new satellite data from America's National Oceanic and Atmospheric Administration which indicated a degree of warming that they had not anticipated, especially in the tropics.

During May 1998, marine scientists said Australia's Great Barrier Reef (the world's largest living reef, stretching 1,300 miles), was experiencing its worst case of coral bleaching in recorded history. Australian marine scientists said bleaching had hit more than 60 percent of its 3,000 coral reefs and aerial surveys showed that 88 percent of inshore reefs were bleached, with 25 percent severely bleached (Perry 1998).

By the year 2000 some coral reefs in the Indian and Pacific oceans had recovered somewhat from the rapid bleachings of 1998. Scientific studies indicated that some young corals in these areas had survived bleaching, and were helping to begin rebuilding some of the reefs. Terry Done, a senior research scientist at the Australian Institute of Marine Science, said that reefs may be "more resilient than we had thought." According to a report in *Science*, Done said that many corals may "not be able to mature and recover from the repeated bleaching forecast to accompany global warming" (Normile 2000, 941). Done is best acquainted with some of the Indian Ocean corals which were character- ized as looking like graveyards after the 1998 bleaching. He noted that most coral reefs require at least a decade to recover from bleaching. "This recovery won't do the reefs much good," Done said. "They'll no sooner get one or two years old before they'll be wiped out again" (Normile 2000, 942).

Corals are not the only living things imperiled by rising ocean temperatures. C.D. Harvell and associates reported in *Science*, "In the past few decades, there has been a world-wide increase in the reports of diseases affecting marine or- ganisms" (Harvell et al. 1999, 1505). This rapid increase in disease-related mor- tality among plants and animals of the seas and oceans is related by Harvell and associates to global warming, which acts to expand the range of many diseases, as well as human pollution and other forms of habitat degradation. Many of the diseases surveyed by these researchers are new to science.

"The current trend toward a warming climate could result in modifications of many marine organisms' basic biological properties, thereby making them more susceptible to disease," Harvell and associates assert (Harvell et al. 1999, 1507). One of the team's primary examples of warming's synergy with disease is the mass bleaching of tropical corals worldwide during the 1997–1998 ENSO (El Niño Southern Oscillation). The authors of this article found that, along with the heat, warmer water often promoted diseases, which weakened the coral: "Demise of some corals is likely to have been accelerated by opportunistic infections" (Har- vell et al. 1999, 1507). Harvell and colleagues call corals an "indicator species of a heightened disease load" throughout the oceans (Harvell et al. 1999, 1509). The same authors also cite increased mortality of oysters in Chesapeake Bay, which they ascribe to warmer winters which have "decreased parasite mortality, resulting in oysters retaining heavy infections" (Harvell et al. 1999, 1507).

WARMING AND DECLINE OF LIFE IN THE OCEAN FOOD WEB

(For a discussion of global warming's effects on coastal fisheries, see Global Warming and Fisheries in Chapter 6.) Warming temperatures will force adap- tation or extinction of many plant and animal species in the world's oceans. Great sea turtles, for example, will come ashore at their usual sites to lay their eggs on sand and rocks which have grown warmer. The temperature determines the sex ratio of the turtles, with female births declining as the environment warms. Scientists do not yet know how much warmth will be required before the last sea turtle mother will die (Bates and Project Plenty 1990, 100). Ac- cording to the World Commission on Environment and Development, one plant or animal species is going extinct somewhere in the world every minute, due mainly to human encroachment on (and pollution of) natural habitats (Bates and Project Plenty 1990, 101).

Warming seas distort marine ecosystems in other ways. According to a report by Climate Solutions of Olympia Washington, a dramatic ocean-temperature increase near North America's West Coast began about 1977. Warming sea surface temperatures may interfere with phytoplankton production, with impacts rippling through the food web. Cooler, upwelling ocean water will break through warm surface waters less frequently, reducing nutrients available for plants and animals living in the oceans. By the 1990s, such decreases in productivity were detected near the California coast, where scientists have documented a measurable decrease in the abundance of zooplankton, the second level in the food web. By the 1990s, the abundance of zooplankton was 70 percent lower than it had been during the 1950s.

The decline of zooplankton raises fears for the survival of several fish species which feed on it. Water temperatures in some areas near the United States West Coast have increased one to two degrees C. during the last four decades (Gelbspan 1995). Ocean seabirds in the California Current have declined 90 percent since 1987, one indication of food web collapse. Reduced primary phytoplankton production usually means less overall fertility in marine ecosystems, including reduced fisheries, according to a report compiled by the World Wildlife Fund (Mathews-Amos and Berntson 1999).

Harvard University epidemiologist Paul Epstein warns that rising temperatures in many ocean areas may be related to increasing outbreaks of algal toxins, bacteria, and viruses which affect large numbers of marine species, as well as shorebirds and mammals. "Of great concern," writes Epstein, "are diseases that attack coral and sea grasses, essential habitats that sustain mobile aquatic species" (Epstein, Profound Consequences, 1999, 66). During the 1997–1998 El Niño, weak winds and strong sunlight created highly stratified, low-nutrient surface waters in the eastern Bering Sea, resulting in a massive bloom of *coccolithophores*, a type of phytoplankton usually associated with low-nutrient areas. According to the World Wildlife Fund, such a large-scale bloom had never before been documented in that area (Mathews-Amos and Berntson 1999).

Scientists also have documented decreased reproduction and increased mortality in seabirds and marine-mammal populations in warming water. The World Wildlife Fund reports that Sooty Shearwaters off the California coast declined 90 percent during the late 1980s and early 1990s, as Cassin's Auklets declined 50 percent. Zooplankton populations declined markedly at the same time. In Alaska, a severe decline in Shearwaters from 1997 to 1998 "was clearly due to starvation," according to the World Wildlife Fund (Mathews-Amos and Berntson 1999).

By the late 1990s, the Mediterranean Sea had become home to at least 110 species of tropical fish for the first time in recorded history. The tropical fish, some of which are crowding out native species, have followed warming water into the Mediterranean through the Suez Canal and the Strait of Gibraltar. Barracuda and other tropical fish are migrating to the Mediterranean as water temperatures warm. "Eighty-five species previously unknown on a significant scale in the waters of Italy or the Levant have set up home there over the past ten years, according to research carried out by Professor Franco Andolaro and his team at the Palermo office of the Italian Marine Research Institute. They are competing for food with 550 indigenous species" (Phillips 2000).

"What is in danger," Professor Andolaro said, "is the ecological balance of the Mediterranean basin. Tropical species now make up almost 20 percent of the fauna and the colonisation is continuing constantly" (Phillips 2000). The Italian Environment Ministry says that water temperatures in the Mediterranean have increased about two degrees C. since 1970. Fifty-five of the new species are from the Red Sea and the remainder from the Atlantic. Ten of the Red Sea migrants are so numerous that they are now being fished commercially in the Mediterranean, including gold-band goatfish, striped-fin goatfish, Haifa grouper, streamlined spinefoot, the wahoo (a small tuna), banded barracuda, and Brazilian lizard fish.

News of the influx has caused considerable excitement in Italy, where tourism to exotic destinations is a burgeoning business. "We won't have to go as far as the Seychelles, Sharm el-Sheikh or the Maldives to enjoy the most fascinating marine beauty," *Il Giornale* of Milan commented. "Today we can admire these curious and highly coloured tropical fish by simply taking a swim in Liguria or Sardinia" (Phillips 2000).

Another potential problem with warming seas is an expected reduction in the pH factor, indicating an increase in relative acidity of the water. Wilfrid Bach writes,

A CO_2 doubling in the atmosphere might reduce the pH of surface sea water from the usual 8.1 to 7.6. It has been shown that a reduction of the pH to 7.6 would increase the copper-ion activity by ten-fold, a change to which the marine phytoplankton would react dramatically . . . a pH of 7.7 would cause the death of fish larvae. (Bach 1984, 183)

Rising sea levels have been killing trees in some of Florida's coastal parks and nature reserves, one indication of the fact that the world's oceans have been rising for about 10,000 years, since the last glacial maximum, producing gradual saltwater encroachment into relatively flat coastal land areas. In some areas, certain species of trees do not die, but do cease to reproduce, causing them to expire as a species within one generation. Red cedars are among the first species of tree to die, while cabbage palms survive saltwater intrusion longer than most.

REDRAWING THE MAPS: DIMENSIONS OF SEA LEVEL RISE

In *It's a Matter of Survival*, Anita Gordon and David Suzuki quote Stephen Leatherman regarding an anticipated worldwide sea level rise of 1.5 meters (five feet) during the twenty-first century: "We're looking at land loss greater than we've seen at any time in human history. . . . There will be so much total immersion that maps will have to be redrawn along many of the low-lying coastal areas, such as along the U.S. Atlantic and Gulf coasts" (Gordon and Suzuki 1991, 14).

Anticipated global warming will raise worldwide sea level not only because of melting ice, but also because the molecular structure of liquid water expands as it is heated. The same amount of liquid water will occupy more volume as it warms. James G. Titus and Vijay Narayanan estimate that about 20 centimeters of the next century's sea level rise will be caused by thermal expansion of existing waters. The amount of expansion depends ultimately on how deeply the warmer surface waters mix with lower levels. "Although global temperatures are projected to rise 20 percent less during the twenty-second century than in the twenty-first, thermal expansion is likely to be 20 to 40 percent more, due to the delayed response of expansion to higher temperatures" (Titus and Narayanan 1996).

Projections of sea level rise are plagued with a large degree of uncertainty, in large part because no one knows how much climatic "forcing" will be necessary before major ice masses, such as the West Antarctic Ice Sheet, begin to contribute significantly to world sea level. (See also Chapter 4, "Icemelt.") A report by the Department of Energy estimated that over a period of two to five hundred years, the West Antarctic Ice Sheet could disintegrate, raising sea level six meters (20 feet) (Hughes 1983; Bentley 1983). Studies after 1983, however, have focused on the next century. Most recent estimates of possible sea level rise due to global warming generally fall in a range of 50 to 200 cm (two to seven feet) by 2100. Studies since 1990, however, generally suggest that a 50 to 200 centimeter rise in mean sea level is more likely to take 150 to 200 years (Mercer 1970).

Global sea level was about 328 feet (100 meters) lower at the peak of the last ice age. Since then, a little more than half the Earth's glacial-maximum ice mass has melted; if the rest were to melt, the sea level would rise an estimated 280 feet from present levels. The earth last witnessed such a state for a sustained period during the age of the dinosaurs, about 65 million years ago, when the planet had no permanent ice, and the ambient global temperature was as much as 20 degrees F. higher than today. Very probably following a cataclysmic collision with an asteroid or comet, the Earth entered a period during which dust obscured the sun, causing notable cooling, and giving rise to the new cycle of ice ages alternating with warmer interglacial periods which has characterized the last several million years. During this period, warm-blooded animals (including Homo sapiens) have become the Earth's dominant species. (The Earth experienced ice ages before the dinosaurs' era; some scientists contend that at least once, many millions of years earlier, nearly the entire planet was covered with ice, a "Snowball Earth.")

Even a minor rise in sea level would inundate wetlands and lowlands, accelerate coastal erosion, exacerbate coastal flooding, threaten coastal structures, raise water tables, and increase the salinity of rivers, bays, and aquifers (Barth and Titus 1984). Most of the wetlands and lowlands of the United States are located along the Gulf of Mexico and Atlantic coasts south of central New Jersey. Shoreline erosion is not limited to the East and Gulf coasts. About 950 miles (86 percent) of California's 1,100-mile coastline eroded during the late twentieth century (Edgerton 1991, 27). Large mudflats also ring parts of Puget Sound, and parts of Washington State's ocean shoreline also are relatively flat.

An Environmental Protection Agency report issued during 1997 said that a one-meter rise in sea level will inundate about 7000 square miles of dry land in the United States, including between 50 and 80 percent of U.S. wetlands. The wetlands, the coastal interface between fresh and salt water, are important to the breeding of many fish species which are particularly sensitive to water temperature, acidity, and salinity. The wetlands of Louisiana, which once hosted stands of oak and cypress, are now losing them at a rate of 55 square miles a year as rising ocean waters increase the salinity of swamplands. In some cases, the increase in salinity is being worsened by construction of canals by oil companies.

The State of South Carolina has enacted a Beachfront Management Act, which curtails shorefront development. The State of Maine also limits development in any area that would be eroded by a 90-centimeter rise in average sea level. Land reclamation projects in San Francisco and Hong Kong now include a safety margin for future sea level rise, as do new seawalls in eastern Britain and the Netherlands (Titus and Narayanan N.d.).

The EPA has compiled reports projecting the costs of protecting ocean-resort communities by pumping sand onto beaches and gradually raising barrier islands as seas rise. The EPA also estimated the costs of protecting developed areas along sheltered waters through the use of levees (dikes) and bulkheads, as well as the anticipated loss of coastal wetlands and undeveloped lowlands. The total cost to mitigate a one-meter sea level rise would be $270 to $475 billion, according to EPA estimates. These reports generally assume the efficacy of human artifice, often ignoring the fact that seawalls built to protect coastal developments from erosion may actually hasten its speed because they interfere with the natural seasonal cycle of sand replenishment.

The net effect of sea level rise in any particular place must be adjusted for the gradual rise or fall of the land itself. Some areas that were covered with ice during the last glacial maximum are rising. An example is Stockholm, Sweden (Silver and DeFries 1990, 94). Parts of Canada and Scotland also are slowly rising following the melting of glaciers several thousand years ago, while much of the United States' Atlantic Coast has subsided about a foot during the twentieth century. Along a relatively level, sandy shoreline, such a change may cause a beach to lose as much as 100 feet (Edgerton 1991, 25). Many Atlantic beaches have been receding as much as three feet per year; in some areas of the Gulf of Mexico coast, the rate averages five feet per year (Edgerton 1991, 78; Silver and DeFries 1990, 92).

Coastal Louisiana is subsiding 0.4 inch per year, an unusually rapid rate, due to removal of oil, gas, and groundwater (Silver and DeFries 1990, 96). Because it is subsiding so quickly, the Mississippi River Delta in Louisiana is probably the area of the United States that is most vulnerable to sea level rises caused by global warming (Silver and DeFries 1990, 97). Land loss in Louisiana due to subsidence and rising waters has been estimated at about one million acres during the twentieth century; by 1990, roughly 50 square miles a year were being lost. The highest point in the city of New Orleans is only 13 feet above mean sea level, as the ground under the city sinks three feet per century (Barth and Titus 1984, 299).

As sea levels rise, cities along the U.S. East and Gulf coasts (and other coastal locales around the world) may find saline water seeping into their drinking-water supplies. Cities most at risk in the United States may be New York City and Philadelphia. In the Philadelphia area, a sea level rise of a third of a meter would require a 12 percent rise in reservoir capacity to prevent saltwater intrusion into system intakes on the Delaware River (Cline 1992, 127). Other cities around the world that lie near seacoasts also may face similar problems: Amsterdam, Rotterdam, Liverpool, Istanbul, Venice, Barcelona, Gothenburg, St. Petersburg, Calcutta, among others.

New York City, Philadelphia, and much of California's Central Valley get their water from areas that are just upstream from areas in which the water is salty during droughts. Farmers in central New Jersey as well as the city of Camden rely on the Potomac-Raritan-Magothy aquifer, which could become salty if sea level rises. The South Florida Water Management District already spends millions of dollars per year to prevent Miami's Biscayne aquifer from becoming salty (Miller et al. 1989). A rise in sea level also would enable salt water to penetrate further inland, as well as upstream into rivers, bays, wetlands, and aquifers, which would be harmful to some aquatic plants and animals, and would threaten human uses of water. Increased salinity probably will threaten oyster harvests in the Delaware and Chesapeake bays (Gunter 1974).

Along the United States Eastern Seaboard, oceanfront property, some of the most valuable real estate in the country, may be vulnerable to sea level rises caused by global warming. "In Massachusetts," writes Bill McKibben, "between three thousand and ten thousand acres of ocean-front land worth between $3 billion and $10 billion might disappear by 2025, and that figure does not include land lost to growing ponds and bogs [created] as the rising sea lifts the water table" (McKibben 1989, 112–113).

Chesapeake Bay "is subject to all the SLR [sea-level rise] impacts: erosion, inundation of low-lying lands and wetlands loss, salt-water intrusion into aquifers and surface waters, higher water tables, and increased flooding and storm damage" (Leatherman 1992, 17). Crab populations that rely on aquatic vegetation for protection during their early life stages are declining in Chesapeake Bay. In the same area, many small islands, which have been important rookeries for several species of birds, are eroding. The Cape Hatteras lighthouse, first built in 1803 (then, as now, the tallest lighthouse in the United States), is being moved inland. It was 1,600 feet from shore when built, but 120 feet from open water when it was moved this year (Cape Hatteras 1999, A-16).

Texas may face special hazards from sea level rise along the Gulf Coast. In the modulated tones of science, a report outlining possible outcomes of the greenhouse effect on Texas sketches severe problems with land subsidence and sea level rise along parts of the state's coast:

The Houston-Galveston urban region, with its high groundwater table, subsidence, and long history of severe flooding along its coastline and bayous, will be particularly vulnerable, and could experience permanent loss of urban land in sensitive areas, necessitating extensive relocation programs. (North et al. 1995, 170)

In addition, a rise in sea level may imperil drinking water supplies in the coastal regions of Texas, the state's fastest-growing region in terms of human population and development. The Houston-Galveston area is additionally vulnerable because the land is relatively flat for several miles inland, and has been experiencing subsidence (sinking) as groundwater has been removed for human consumption. The cities of Beaumont, Port Arthur, Orange, Freeport, and Corpus Christi subsided as much as a foot between 1906 and 1974 (North et al. 1995, 171). Some small areas have sunk as much as ten feet during that period. Subsidence rates could increase as more groundwater is withdrawn.

Coastal erosion is an additional problem in the Houston-Galveston area. At Sargent Beach, the shoreline eroded about 1,000 feet between 1956 and the early 1990s. Galveston built a seawall to protect against oceanic flooding during the early 1900s, following a disastrous hurricane and storm surge there in 1900. When the seawall was constructed, it faced a beach that averaged about 300 feet wide. During ensuing years, the beach has eroded; by about 1940, more rocks were required to protect the original seawall. Most of these rocks sank during the following few years, requiring even more rocks (North et al. 1995, 172).

The EPA estimates that if no measures are taken to hold back the sea, a one-meter rise in sea level will inundate 14,000 square miles of coastal land area in the United States with salt water during the twenty-first century. Roughly 1500 square kilometers (600–700 square miles) of densely developed coastal lowlands could be protected at a cost of approximately $1,000 to $2,000 per year for a typical coastal housing lot. Given high coastal property values, holding back the sea could be cost effective, the report contends. The report adds, however, that the environmental consequences of holding back the sea may be unacceptable. Although the most common engineering solution for protecting the ocean coast—pumping sand—would allow beaches, levees, and bulkheads to be maintained along sheltered waters, the same actions would gradually eliminate most of the nation's wetlands. Based on this probability, the report finds, "To ensure the long-term survival of coastal wetlands, federal and state environmental agencies should begin to lay the groundwork for a gradual abandonment of coastal lowlands [by human inhabitants] as sea level rises" (U.S. EPA 1997).

Barrier islands and sand-spits of land along the Atlantic and Gulf coasts are among the most vulnerable areas to rises in sea level. Most coastal barrier islands are long, narrow spits of sand with ocean on one side and a bay on the other. Typically, the oceanfront shore of any given island usually ranges from two to four meters above high tide, while the bay side is less than a meter above high water. Thus, even a one-meter rise in sea level would threaten much of these lands (and their ranks of hotels, businesses, and homes) with saltwater inundation. Erosion, moreover, threatens the high parts of these islands, and is generally viewed as a more immediate problem than the inundation of barrier islands' bay sides. A rise in sea level can cause an ocean beach to retreat by considerably more than the retreat due to inundation alone, according to an EPA report published in 1997:

The shape of a beach profile is determined by the pattern of waves striking the shore; generally, the visible part of the beach is much steeper than the underwater portion that comprises most of the active "surf zone." While inundation alone is determined by the slope of the land just above the water, Bruun (1962) showed that the total shoreline retreat from a rise in sea level depends on the average slope of the entire beach profile. (U.S. EPA 1997)

Studies suggest that a one-meter rise in sea level would generally cause beaches to erode 50 to 100 meters from the New England to Maryland coasts, about 200 meters along the Carolina coasts, 100 to 1000 meters along the Florida coast, and 200 to 400 meters along the California coast (Everts 1985; Kyper and Sorensen 1985; Kana et al. 1984; Bruun 1962; Wilcoxen 1986). Because most U.S. recreational beaches are less than 30 meters (100 feet) wide at high tide, even a 30 centimeter (one foot) rise in sea level could cause serious loses to large areas of high-priced coastal real estate.

Many coastal areas are becoming more vulnerable to flooding for four reasons, according to the 1997 EPA report:

(1) A higher sea level provides a higher base for storm surges to build upon; a one-meter rise in sea level would thus enable a 15-year storm to flood many areas that today are only flooded by a 100-year storm (Kana et al. 1984).

(2) Beach erosion would leave particular properties more vulnerable to storm waves.

(3) Higher water levels would increase flooding due to rainstorms by reducing coastal drainage (Titus and Narayanan 1996).

(4) A rise in sea level would raise water tables (U.S. EPA 1997).

Possible responses to anticipated sea level rise generally fall into three categories, according to the EPA: erection of walls to hold back the sea, allowing the sea to advance and adapting to it, and raising the land itself. For more than five centuries, the Dutch have used dikes and windmills to prevent inundation from the North Sea. By contrast, some English towns have been rebuilt landward as structures and land were lost to erosion; the town of Dunwich, England, has had to rebuild its church seven times in the last seven centuries. More recently, rapidly subsiding communities such as Galveston, Texas, have used fill to raise land elevations; The U.S. Army Corps of Engineers and coastal states regularly pump sand from offshore to counteract beach erosion. Venice, Italy, has used all three responses: allowing the sea to advance into the canals, while raising some lowlands, and erecting storm protection barriers.

GLOBAL WARMING AND TROPICAL CYCLONES

In 1986, Kerry Emanuel, a hurricane specialist at the Massachusetts Institute of Technology, published an article in the *Journal of the Atmospheric Sciences* (Emanuel 1986) in which he argued that hurricane intensity is governed, in part, by the degree of thermodynamic disequilibrium between the atmosphere and the underlying ocean. Therefore, Emanuel reasoned, warmer ocean waters would breed more intense tropical cyclones. By the year 2000, Emanuel's forecast of "hypercanes" had become gist for at least one best-selling book, *The Coming Superstorm* by Art Bell and Whitley Strieber, which rose to number 15 on the *New York Times* best-seller list for nonfiction hardbound books during January 2000.

In 1987, Emanuel published an article in *Nature* (Emanuel 1987) asserting that a three degree C. increase in sea surface temperature could increase the potential destructive power (as measured by the square of the wind speed) of storms by 40 to 50 percent. Emanuel asserted in 1988 that a warming of the sea surface by 6 to 10 degrees C. (assuming no temperature change in the lower stratosphere) would make a supersized, ultrapowerful "hypercane" theoretically possible (Emanuel, Maximum Intensity, 1988; Toward a General Theory, 1988).

Emanuel's theory has sparked intense debate among hurricane specialists. Critics of his theory responded that even the most pessimistic climate models do not project such a large amount of warming in the tropics.

Hurricanes increase in strength with the warmth of the water over which they travel, as well as the depth of the warmth. If warm water is shallow, the turbulence of the storm will soon mix enough of the surface water to drive its temperature below the 80 degrees F. required to sustain it. Under present conditions, the top wind speed of a hurricane is probably about 200 miles an hour, and the lowest possible central pressure is probably about 885 millibars (McKibben 1989, 95).

While the temperature of the surface water over which a hurricane passes is important to its growth and development, temperature is only one of several variables that influence the career of any given storm. Emanuel, writing in *Nature*, outlines the major factors influencing hurricane intensity: "the storm's initial intensity, the thermodynamic state of the atmosphere through which it moves, and the heat exchange with the upper layer of the ocean under the core of the hurricane" (Emanuel 1999, 665). While climate modelers have had some success forecasting the tracks of hurricanes, intensity is a much tougher theoretical nut to crack. Within a few hours, a hurricane such as Opal (in 1995) may intensify from a category 2, with strongest winds about 100 miles an hour, to a devastating category 5, with winds above 150 m.p.h. Opal's intensification was due mainly to its passage over a patch of unusually warm water in the Gulf of Mexico, together with upper-atmospheric dynamics which were favorable for storm development. The speed and complexity of hurricane intensity in the very short term has thus far eluded climate modelers.

The IPCC's *First Assessment* was rather ambivalent regarding global warming's possible effects on hurricanes (Intergovernment Panel 1990). The IPCC report said, "There is some evidence from model simulations and empirical considerations that the frequency per year, intensity, and area of disturbance of tropical cyclones may increase, though it is not yet compelling" (Cline 1992, 121).

A study by T.R. Knutson, R.E. Tuleya, and Y. Kurihara used a climate-simulation model to study 51 Pacific Ocean cyclones under present day conditions and under simulated conditions which include higher levels of carbon dioxide (Knutson et al. 1999). They found that a warming of sea surface temperatures in the Western Pacific of about 2.2 degrees C. would increase the storms' wind speeds an average of 5 to 12 percent, and decrease their central pressures 7 to 20 millibars. The study suggests that global warming will produce more destructive tropical cyclones and that people living in the possible paths of these storms are courting potential disaster.

A report by Christopher W. Landsea of the National Oceanic and Atmospheric Administration's Miami, Florida, office argues that an increase in carbon dioxide levels may raise the threshold temperature in which hurricanes thrive (presently above 80 degrees F.) in proportion to the rise in temperature (Landsea 1999). This change might nullify the increase in strength that is implied by warmer water surface temperatures. Landsea also speculates that increased frequency of El Niño-type weather patterns in the Pacific tend to dampen hurricane development in the Atlantic, although storm frequency and intensity often rises under El Niño conditions in the eastern Pacific.

Landsea (1993) contends that intensity of Atlantic hurricanes has decreased since the middle of the twentieth century. Landsea and associates (1996) also assert that hurricane frequency and intensity had not increased during the past half-century. Landsea and colleagues wrote that "a long-term (five decade) downward trend continues to be evident primarily in the frequency of intense hurricanes. In addition, the mean maximum intensity (i.e., averaged over all cyclones in a season) has decreased" (Landsea et al. 1996, 1700).

Landsea believes that any warming-induced change in hurricane frequency and intensity will probably be lost in the "noise" of year-to-year hurricane variability. He concludes, "Overall, these suggested changes are quite small compared to the observed large natural variability of hurricanes, typhoons and tropical cyclones. However, more study is needed to better understand the complex interaction between these storms and the tropical atmosphere and ocean" (Landsea 1999).

Three greenhouse skeptics (Sherwood Idso, Robert Balling, and R.S. Cerveny) have challenged Emanuel's theories (Idso et al. 1990). They assembled hurricane data for the central Atlantic, the East Coast of the United States, the Gulf of Mexico, and the Caribbean Sea between 1947 to 1987, comparing their information to estimates of the sea surface temperatures in the Northern Hemisphere. Idso, Balling, and Cerveny concluded, "There is basically no trend of any sort in the number of hurricanes experienced in any of the four regions with respect to variations in temperature" (Idso et al. 1990, 261).

Doubts also have been raised about the relationship of warming seas and intensity of tropical cyclones by scientists whose views are less politicized than those of Idso, Balling, and Cerveny. A.J. Broccoli and S. Manabe (1990) included cloud-related feedbacks in their models and found a 15 percent reduction in the number of days with hurricanes, even assuming a temperature rise caused by a doubling of atmospheric carbon dioxide. Broccoli and Manabe noted that their results were easily skewed by (and highly dependent upon) how they decided to represent cloud processes within their models.

On the other hand, S.T. O'Brien and colleagues assert that doubling the level of atmospheric carbon dioxide (increasing tropical sea-surface temperatures between one and four degrees C.) will double the number of hurricanes, and increase their strength by 40 to 60 percent (O'Brien et al. 1992). O'Brien and associates also believe that warmer average temperatures will extend the hurricane season, because the season effectively ends for any given area when water temperature falls below 26 degrees C. (80 degrees F.) R.J. Haarsma and colleagues also estimate that a doubling of greenhouse-gases levels will increase the frequency of hurricanes by 50 percent and increase the average intensity of the storms by 20 percent (Haarsma et al. 1993).

J. Lighthill and associates argue that while global warming may exert some influence on cyclone formation in the tropics, natural variability is more important (Lighthill et al. 1994). This position drew a reply from Emanuel (1995). L. Bengtsson, a member of the German Max Planck Institut für Meteorologie, contends that global warming will strengthen the upper-level westerlies in areas where hurricanes usually develop, inhibiting storm development and intensity, negating any boost the storms may get from warmer sea-surface temperatures (Bengtsson et al. 1996). (This is also typical of the El Niño pattern that seems to be occurring more frequently as temperatures rise.)

In 1995 the IPCC, in its *Second Assessment*, restated its earlier position that the state of science on the subject does not permit a conclusion regarding whether global warming will affect the number and intensity of tropical cyclones (Bolin et al. 1995). Tom Karl and colleagues wrote two years later in *Scientific American*, "Overall, it seems unlikely that tropical cyclones will increase significantly on a global scale. In some regions, activity may escalate; in others, it will lessen" (Karl et al. 1997, 83).

GLOBAL WARMING AND OCEAN CIRCULATION

Circulation patterns within the oceans carry water around the world at several levels, mixing it laterally as well as vertically. Without such circulation, deep waters would become depleted of oxygen and most marine life there would suffocate. A report by the World Wildlife Fund suggests that changes in the circulation of the ocean "are likely to be affected by global climate change, and in turn are likely to affect future climate change" (Mathews-Amos and Berntson 1999). Scientific attention has focussed on the "thermohaline circulation" of the North Atlantic as one major example of ocean-atmosphere interaction which could shape climate change on nearby land masses, especially Western Europe.

Because thermohaline circulation is driven by density differences of water masses, it is extremely sensitive to influxes of fresh water of the type which are expected to occur with global warming due to melting sea ice and increased precipitation. Fresh water makes ocean water less saline and therefore less dense, so it doesn't sink as quickly. If enough fresh water is added, deep-water formation in the North Atlantic may cease altogether. Relatively small amounts may be sufficient to alter thermohaline circulation; some climate models suggest that, under "business-as-usual" conditions, a complete shutdown of thermohaline circulation in the Atlantic could eventually occur. (Manabe and Stouffer 1993; Stocker and Schmittner 1997)

Regarding the relationship between thermohaline circulation and the climate of Europe, Giancarlo G. Bianchi and I. Nicholas McCave, writing in *Nature*, assert,

The main concern for future climate must be that a possible increase in melting of the Greenland ice sheet resulting from anthropogenically induced atmospheric warming may reach a critical level where the "conveyor belt" will flip to its early Holocene operational mode. The resulting perturbations could conceivably result in climate extremes exceeding those of the Little Ice Age for northern Europe. Without such perturbations, the climate looks likely to warm for several hundred years. (Bianchi and McCave 1999, 517)

Carsten Ruhlemann and colleagues wrote in *Nature* that the Earth's emergence from its last glaciation was punctuated by several short, sharp periods during which glacial conditions returned. These variations seem to have been triggered by the type of changes in North Atlantic thermohaline circulation which are considered likely in a world warmed by increasing emissions of greenhouse gases. "The thermohaline circulation was the important trigger for these rapid climate changes," they write (Ruhlemann et al. 1999, 512). Climate changes which diminish thermohaline circulation tend to divert the Gulf Stream southward, cooling the North Atlantic Ocean (as well as Greenland and Western

Europe), but "warming . . . the western tropical North Atlantic . . . and most of the Southern Hemisphere" (Ruhlmann et al. 1999, 512).

Between 1968 and 1972, an area which came to be known as "The Great Salinity Anomaly" was documented. The anomaly is a large pool of relatively fresh water (vis-à-vis other sea water), resulting from two years of unusually rapid Arctic sea ice melt. This flow of relatively fresh water shut down deep-water formation in the Labrador Sea, an important site for deep-water transport in the North Atlantic. The circulation necessary to transport deep water resumed when the fresher water was removed.

Such impairment of ocean circulation may compound atmospheric increases in greenhouse gases because it impedes the mixture of warmer surface water with cooler water in the depths of the ocean. The same mechanism would decrease the ocean's ability to absorb carbon dioxide from the atmosphere, in part because warmer ocean water has less absorption capacity than cooler water. "Additionally," according to a report by the World Wildlife Fund, "[a] slowing of deep-water formation in the Atlantic would likely reduce the transport of oceanic heat to the European continent. European cities along the Atlantic Seaboard may begin to cool, approaching the cooler temperatures of their latitudinal counterparts in the Pacific" (Mathews-Amos and Berntson 1999).

Scientific attention continues to be concentrated on the future of thermohaline circulation in the North Atlantic. Modeling work on North Atlantic ocean circulation reported in 1999 by R.A. Wood, A.B. Keen, J.F.B. Mitchell, and J.M. Gregory indicates "a dramatic change in the Atlantic occurring over the next few decades: a complete shutdown of one of the two main 'pumps' driving the formation of North Atlantic Deep Water, namely the one in the Labrador Sea" (Wood et al. 1999, 572; Rahmstorf 1999, 523). Britain's Hadley Climate Center projects that global warming will reduce overall thermohaline circulation by 25 percent, with the most notable effects in the area's southern reaches (Wood et al. 1999).

Collapse of the Labrador Current, which Richard A. Wood and colleagues believe is possible between the years 2000 and 2030, "could have serious consequences for marine ecosystems, including seabirds, as they depend not only on specific temperature conditions but also on nutrients supplied by oceanic mixing and currents" (Rahmstorf 1999, 523). As of 2000, observations had not confirmed the predicted breakdown in the Labrador Current; it is believed that decay of the current is being forestalled "by the high phase of the North Atlantic Oscillation" (Rahmstorf 1999, 524). Nevertheless, according to Stefan Rahmstorf, "The simulated ending of the Labrador Sea convection found in the Hadley Center model is perhaps the most convincing demonstration so far of [a] qualitative threshold being crossed because of global warming" (Rahmstorf 1999, 524).

Changes in ocean circulation are not limited to the Atlantic. Northward flow of ocean water from the Pacific Ocean through the Bering Strait into the Arctic Ocean also appears to have slowed, according to the World Wildlife Fund report.

This change reduces northward flow of nutrient-rich North Pacific Ocean water over the Bering Sea shelf, "potentially reducing the overall primary productivity in the region" (Mathews-Amos and Berntson 1999; Bering Sea Task Force 1999).

Oceanographers have found similar trends in parts of the oceans bordering Antarctica. Marine geochemist Wallace Broecker of Columbia University's Lamont-Doherty Earth Observatory and colleagues have found, "The renewal of deep waters by sinking surface waters near Antarctica has slowed to only one-third of its flow a century or two ago" (Kerr, Oceanography, 1999, 1062). The work of Wallace Broecker, Stewart Sutherland, and Tsung-Hung Peng suggests that the slowing of deep-water formation may be related to warming attending the end of Europe's "Little Ice Age" (roughly 1400 to 1800 A.D.). Furthermore, they posit, "A see-sawing of deep water production between the northern Atlantic and the Southern oceans may lie at the heart of the 1,500-year ice-rafting cycle" (Broecker et al. 1999, 1132).

"This huge, climate-altering change in the oceans—if it's real—would greatly complicate attempts to understand how the ocean and climate are responding ... to the buildup of greenhouse gases in the atmosphere," writes Richard A. Kerr in *Science* (Kerr, Oceanography, 1999, 1062). Kerr quoted Jorge Sarmiento, an ocean-circulation modeler at Princeton University, as saying that the work of Broecker and colleagues is "really interesting and provocative." Sarmiento also said, however, that he was "very uneasy about the calculations. I find the paper more of a stimulation to further work than what I would accept as proven fact" (Kerr, Oceanography, 1999, 1062).

Variations in the thermohaline circulation of the North Atlantic can have marked climatic consequences in Mesoamericam as well as in Europe. In *The Great Maya Droughts*, for example, Richardson B. Gill (2000) postulates that a southward shift in the Gulf Stream forced by cold deep-water flow from the north cooled climate in northwestern Europe between 800 and 1000 A.D. The same shift is said by Gill to have displaced the North Atlantic High southwestward, causing many years of severe drought in regions populated by the classic Maya civilization in Mesoamerica. Gill makes a case that these droughts played a major role in collapse of the classic Maya civilization (Gill 2000). (Possible effects of changes in North Atlantic thermohaline circulation on the climate of Europe are described in Chapter 4, "Icemelt.")

GLOBAL WARMING AND EL NIÑO/LA NIÑA

Will a warmer world experience more frequent El Niño-type ocean temperature patterns, with associated extreme weather events? As global temperatures have risen since 1980, the El Niño Southern Oscillation (ENSO) has occurred more frequently than at any other time in the century and a half that detailed worldwide records have been kept. Climate modeling by A. Timmermann and associates tends to support the idea that further warming will enhance the prominence of El Niño-type events in world oceanic and terrestrial circulation and climate patterns. The forecasts of Timmermann and associates come with an added twist: the La Niña side of this oscillation may be rarer, but possibly colder as well. In brief, their models indicate that weather may become generally wilder as well as warmer as levels of greenhouse gases increase. The researchers believe that in a warmer world climate will change in several ways.

The climatic effects will be threefold. First, the mean climate in the tropical Pacific region will change toward a state corresponding to present-day El Niño conditions. It is therefore likely that events typical of El Niño will also become more frequent. Second, a stronger inter-annual variability will be superimposed on the changes in the mean state, so year-to-year variations may become more extreme under enhanced greenhouse conditions. Third, the inter-annual variability will be more strongly skewed, with cold events (relative to the warmer mean state) becoming more frequent. (Timmermann et al. 1999, 696)

Global warming may be giving El Niño a "jump start" by providing ocean water at appropriate latitudes with increased warmth. The two most severe El Niño events since 1850 occurred late in the twentieth century—the first in 1982–1983, the second in 1997–1998. The second severe El Niño caused 23,000 deaths and $33 billion worth of property damage to human infrastructure around the world (Kerr, Big El Niños, 1999, 1108).

Researchers have speculated that the strength of these two El Niño events may have been influenced not only by warmer ocean waters in general, but also by peak activity in the "interdecadal ENSO," another climate cycle in the eastern tropical Pacific in which "the temperature swings slowly from warm to cold and back over ten to twenty years" (Kerr, Big El Niños, 1999, 1109) Yet another cycle in Pacific Ocean temperature (as yet unnamed), which involves the central reaches of the tropical Pacific Ocean, began to warm in the late 1970s and, like the "interdecadal ENSO," seems to have reinforced the better-known El Niño cycle from the 1980s into the 1990s. Another possible influence on the strength of El Niño events is the Pacific Decadal Oscillation (PDO), in the colder waters of the central northern Pacific. Researchers have proposed that a cooler-than-usual northern Pacific Ocean may help intensify storms spawned by El Niño conditions, especially in North America (Kerr, North American Climate, 1999, 1109).

El Niño conditions do not always occur in warm weather, however. The lead author of an article in *Science*, Tammy Rittenour, said that traces of El Niño-like conditions have been found in the last ice age, 13,500 to 17,500 years ago (Rittenour et al. 2000). The work of Rittenour and colleagues suggests that temperature may be only one of several factors affecting the ebb and flow of several El Niño cycles. "The results fit well with new climate models that suggest that periods of weakened El Niños rhythmically alternate with the current mode of strong El Niños. Driving these swings, at least in the models, are periodic variations of solar heating as Earth wobbles on its spin axis" (Kerr, Viable, 2000, 945).

Rittenour was surprised by indications that her climate research was indicating El Niño-type events during a glacial period. She had begun by investigating the rate at which melting drained glacial Lake Hitchcock, in present day southern New England. "It kind of threw me back," Rittenour told a reporter. "Coming up with El Niño was kind of a shock" (Anderson 2000, 12B). (For more information on global warming and increased frequency of El Niño events, see Chapter 9, "Greenhouse Gases and the Weather.")

REFERENCES

Anderson, Julie. 2000. "UNL Student Helps Shed New Light on El Niño." *Omaha World-Herald*, May 14, 12B.

Aronson, Richard B., William F. Precht, Ian G. MacIntyre, and Thaddeus J.T. Murdoch. 2000. "Coral Bleach-out in Belize." *Nature* 405 (4 May): 36.

Augenbraun, Harvey, Elaine Matthews, and David Sarma. N.d. The Greenhouse Effect, Greenhouse Gases, and Global Warming. [http://icp.giss.nasa.gov/research/methane/greenhouse.html]

Bach, Wilfrid. 1984. *Our Threatened Climate: Ways of Averting the CO2 Problem Though Rational Energy Use*. Trans. Jill Jager. Dordrecht, Germany: D. Reidel.

Balling, Robert C., Jr. N.d. "The Spin on Greenhouse Hurricanes." Fraser Institute. [http://www.fraserinstitute.ca/publications/books/g_warming/hurricanes.html]

Barth, M.C., and J.G. Titus, eds. 1984. *Greenhouse Effect and Sea Level Rise: A Challenge for this Generation*. New York: Van Nostrand Reinhold Company.

Bates, Albert K., and Project Plenty. 1990. *Climate in Crisis: The Greenhouse Effect and What We Can Do*. Summertown, Tenn.: The Book Publishing Co.

Bell, Art, and Whitley Strieber. 1999. *The Coming Global Superstorm*. New York: Pocket Books.

Bengtsson, L., M. Botzet, and M. Esch. 1996. "Will Greenhouse Gas-induced Warming over the Next 50 Years Lead to a Higher Frequency and Greater Intensity of Hurricanes? *Tellus* 48A:57–73.

Bentley, C.R. 1983. *West Antarctic Ice Sheet: Diagnosis and Prognosis*. Washington, D.C.: U.S. Department of Energy.

Bering Sea Task Force. 1999. *Status of Alaska's Oceans and Marine Resources: Bering Sea Task Force Report to Governor Tony Knowles*. Juneau: State Government of Alaska, March.

Bianchi, Giancarlo, and I. Nicholas McCave. 1999. "Holocene Periodicity in North Atlantic Climate and Deep-ocean Flow South off Iceland." *Nature* 397 (11 February): 515–517.

Bolin, Bert, et al. 1995. *Intergovernmental Panel on Climate Change. Second Assessment Synthesis of Scientific-Technical Information Relevant to Interpreting Article 2 of the United Nations Framework Convention on Climate Change*. Approved by the IPCC at its eleventh session, 11–15 December, Rome. [http://www.unep.ch/ipcc/pub/sarsyn.htm].

Brewer, Peter G., Gernot Friederich, Edward T. Peltzer, and Franklin M. Orr, Jr. 1999. "Direct Experiments on the Ocean Disposal of Fossil Fuel CO2." *Science* 284 (7 May): 943–945.

Broccoli, A.J., and S. Manabe. 1990. "Can Existing Climate Models be Used to Study Anthropogenic Changes in Tropical Cyclone Intensity?" *Geophysical Research Letters* 17:1917–1920.

Broecker, Wallace S., Stewart Sutherland, and Tsung-Hung Peng. 1999. "A Possible 20th-Century Slowdown of Southern Ocean Deep Water Formation." *Science* 286 (5 November): 1132–1135.

Bruun, P. 1962. "Sea-level Rise as a Cause of Shore Erosion." *Journal of Waterways and Harbor Division* 88:117–130. American Society of Civil Engineers.

Bryant, D., L. Burke, J. McManus, and M. Spaulding. 1998. *Reefs at Risk: A Map-based Indicator of Threats to the World's Coral Reefs.* Washington, D.C.: World Resources Institute.

Caldeira, Ken, and Philip B. Duffy. 2000. "The Role of the Southern Ocean in the Uptake and Storage of Anthropogenic Carbon Dioxide." *Science* 287 (28 January): 620–622.

"Cape Hatteras, N.C. Lighthouse Lights Up Sky From New Perch." 1999. Associated Press. In *Omaha World-Herald,* 14 November, A-16.

Clark, J.A., and C.S. Lingle. 1997. "Future Sea-level Changes Due to West Antarctic Ice Sheet Fluctuation." *Nature* 269:206–209.

Cline, William R. 1992. *The Economics of Global Warming.* Washington, D.C.: Institute for International Economics.

———. 1991. "Comments." In Rudiger Dornbusch and James M. Poterba, eds., *Global Warming: Economic Policy Reponses.* Cambridge, Mass.: MIT Press, 222–228.

Connor, Steve. 2000. "Global Warming is Blamed for First Collapse of a Caribbean Coral Reef." *The Independent* (London), 4 May, 12.

Corson, Walter H., ed. 1990. *The Global Ecology Handbook: What You Can Do About the Environmental Crisis.* Washington, D.C.: The Global Tomorrow Coalition.

Cowen, Robert C. 1998. "New Research Shows that Hurricanes Pump More CO2 into the Air by Roiling Oceans." *Christian Science Monitor,* 3 September. [http://www.csmonitor.com/durable/1998/09/03/fp4s1-csm.htm]

Edgerton, Lynne T., and Natural Resources Defense Council. 1991. *The Rising Tide: Global Warming and World Sea Levels.* Washington, D.C.: Island Press.

Emanuel, Kerry A. 1987. "The Dependence of Hurricane Intensity on Climate." *Nature* 326, no. 2 (April): 483–485.

———. 1999. "Thermodynamic Control of Hurricane Intensity." *Nature* 401 (14 October): 665–669.

———. 1986. "An Air-sea Interaction Theory for Tropical Cyclones. Part I: Steady-state Maintenance." *Journal of the Atmospheric Sciences* 43:585–604.

———. 1988. "The Maximum Intensity of Hurricanes." *Journal of the Atmospheric Sciences* 45:1143–1156.

———. 1988. "Toward a General Theory of Hurricanes." *American Scientist* 76:370–379.

———. (1995). "Comments on 'Global Climate Change and Tropical Cyclone': Part I." *Bulletin of the American Meteorological Society* 76:2241–2243.

Epstein, Paul. 1999. "Profound Consequences: Climate Disruption, Contagious Disease and Public Health." *Native Americas* 16, no. 3/4(Fall/Winter): 64–67. [http://nativeamericas.aip.cornell.edu/fall99/fall99epstein.html]

Everts, C.H. 1985. "Effects of Sea Level Rise and Net Sand Volume Change on Shoreline Position at Ocean City, Maryland." In J.G. Titus, ed., *Potential Impacts of Sea*

Level Rise on the Beach at Ocean City, Maryland. Washington, D.C.: U.S. Environmental Protection Agency.

Gelbspan, Ross. 1997. *The Heat is On: The High Stakes Battle Over Earth's Threatened Climate.* Reading, Mass.: Addison-Wesley Publishing Co.

———. 1995. "The Heat is On: The Warming of the World's Climate Sparks a Blaze of Denial." *Harper's*, December. [http://www.dieoff.com/page82.htm]

Gill, Richardson Benedict. 2000. *The Great Maya Droughts: Water, Life, and Death.* Albuquerque: University of New Mexico Press.

Glick, Patricia. 1998. *Global Warming: The High Costs of Inaction.* San Francisco: Sierra Club. [http://www.sierraclub.org/global-warming/inaction.html]

Gordon, Anita, and David Suzuki. 1991. *It's a Matter of Survival.* Cambridge: Harvard University Press.

Goreau, Thomas J., Raymond L. Hayes, Jenifer W. Clark, Daniel J. Basta, and Craig N. Robertson. 1993. "Elevated Sea-surface Temperatures Correlate With Caribbean Coral Reef Bleaching." In Richard A. Geyer, ed., *A Global Warming Forum: Scientific, Economic, and Legal Overview.* Boca Raton, Fla.: CRC Press, 225–262.

Gough, Robert. 1999. "Stress on Stress: Global Warming and Aquatic Resource Depletion." *Native Americas* 16, no. 3/4(Fall/Winter): 46–48. [http://nativeamericas.aip.cornell.edu]

Gunter, G. 1974. "An Example of Oyster Production Decline with a Change in the Salinity Characteristics of an Estuary—Delaware Bay, 1800–1973." *Proceedings of the National Shellfish Association* 65:3–13.

Haarsma, R.J., J.F.B. Mitchell, and C.A. Senior. 1993. "Tropical Disturbances in a G[lobal] C[limate] M[odel]." *Climate Dynamics* 8:247–257.

Harvell, C.D., K. Kim, J.M. Buckholder, R.R. Colwell, P.R. Epstein, D.J. Grimes, E.E. Hofmann, E.K. Lipp, A.D.M.E. Osterhaus, R.M. Overstreet, J.W. Porter, G.W. Smith, and G.R. Vasta. 1999. "Emerging Marine Diseases—Climate Links and Anthropogenic Factors." *Science* 285, no. 3 (3 September): 1505–1510.

Herbert, H. Josef. 2000. "Study: World's Oceans Warming." Associated Press, 24 March, in LEXIS.

Hoffman, J.S., D. Keyes, J.J. Wells, and J. Titus. 1983. *Projecting Future Sea-level Rise.* Washington, D.C.: U.S. Environmental Protection Agency.

Hogarth, Murray. 1998. "Sea-warming Threatens Coral Reefs." *Sydney Morning Herald* (Australia), 26 November. [http://www.smh.com.au/news/9811/26/text/national13.html]

Hughes, T. 1983. *The Stability of the West Antarctic Ice Sheet: What Has Happened and What Will Happen.* Washington, D.C.: U.S. Department of Energy.

Idso, S.B., R.C. Balling, Jr., and R.S. Cerveny. 1990. "Carbon Dioxide and Hurricanes: Implications of Northern Hemispheric Warming for Atlantic/Caribbean Storms." *Meteorology and Atmospheric Physics* 42:259–263.

Intergovernmental Panel on Climate Change (IPCC). 1990. *Climate Change: The IPCC Scientific Assessment.* Report prepared for IPCC by Working Group I. John T. Houghton et al., eds. Cambridge: Cambridge University Press.

Kana, T.W., J. Michel, M.O. Hayes, and J.R. Jenson. 1984. "The Physical Impact of Sea Level Rise in the Area of Charleston, South Carolina." In M.C. Barth and J.G. Titus, eds., *Greenhouse Effect and Sea Level Rise: A Challenge for this Generation.* New York: Van Nostrand Reinhold Company, 105–150.

Karl, T.R., N. Nicholls, and J. Gregory. 1997. "The Coming Climate." *Scientific American* 276:79–83.

Kerr, Richard A. 1999. "Big El Niños Ride the Back of Slower Climate Change." *Science* 283 (19 February): 1108–1109.

———. 1999. "In North American Climate, a More Local Control." *Science* 283 (19 February): 1109.

———. 1999. "Oceanography: Has a Great River in the Sea Slowed Down?" *Science* 286 (5 November): 1061–1062.

———. 2000. "Globe's 'Missing Warming' Found in the Ocean." *Science* 287 (24 March): 2126–2127.

———. 2000. "Viable But Variable Ancient El Niño Spied." *Science* 288 (12 May): 945.

Knutson, T.R., R.E. Tuleya, and Y. Kurihara. 1999. "Simulated Increase in Hurricane Intensities in a CO2-Warmed Climate." *Science* 279 (13 February): 1018–1020.

Kristof, Nicholas. 1997. "For Pacific Islanders, Global Warming Is No Idle Threat." *New York Times*, 2 March. [http://sierraactivist.org/library/990629/islanders.html]

Kyper, T., and R. Sorensen. 1985. "Potential Impacts of Selected Sea-level Rise Scenarios on the Beach and Coastal Works at Sea Bright, New Jersey." In O.T. Magoon, H. Converse, D. Miner, D. Clark, and L.T. Tobin, eds., *Coastal Zone '85*. New York: American Society of Civil Engineers.

Landsea, Christopher W. 1993. "A Climatology of Intense (or Major) Atlantic Hurricanes." *Monthly Weather Review* 121:1703–1713.

———. 1999. NOAA: Report on Intensity of Tropical Cyclones. Miami, Fla., 12 August. [http://www.aoml.noaa.gov/hrd/tcfaq/tcfaqG.html#G3]

Landsea, C.W., N. Nicholls, W.M. Gray, and L.A. Avila. 1996. "Downward Trends of Atlantic Hurricanes During the Past Five Decades." *Geophysical Research Letters* 23:1697–1700.

Leatherman, Stephen P. 1992. "Coastal Land Loss in the Chesapeake Bay Region: An Historical Analog Approach to Global Change Analysis." In Jurgan Schmandt and Judith Clarkson, eds., *The Regions and Global Warming: Impacts and Response Strategies*. New York: Oxford University Press, 17–27.

Levitus, Sydney, John I. Antonov, Timothy P. Boyer, and Cathy Stephens. 2000. "Warming of the World Ocean." *Science* 287:2225–2229.

Lighthill, J., G. Holland, W. Gray, C. Landsea, G. Craig, J. Evans, Y. Kurihara, and C. Guard. 1994. "Global Climate Change and Tropical Cyclones." *Bulletin of the American Meteorological Society* 75:2147–2157.

Mahtab, Fasih Uddin. 1992. "The Delta Regions and Global Warming: Impact and Response Strategies for Bangladesh." In Jurgan Schmandt and Judith Clarkson, eds., *The Regions and Global Warming: Impacts and Response Strategies*. New York: Oxford University Press, 28–43.

Manabe, S., and R. J. Stouffer. 1993. "Century-scale Effects of Increased Atmospheric CO2 on the Ocean-atmosphere System." *Nature* 364:215–218.

Mathews-Amos, Amy, and Ewann A. Berntson. 1999. "Turning up the Heat: How Global Warming Threatens Life in the Sea." World Wildlife Fund and Marine Conservation Biology Institute. [http://www.worldwildlife.org/news/pubs/wwf_ocean.htm]

McKibben, Bill. 1989. *The End of Nature*. New York: Random House.

Mercer, J.H. 1970. "Antarctic Ice and Interglacial High Sea Levels." *Science* 168:1605–1606.

Miller, T., J.C. Walker, G.T. Kingsley, and W.A. Hyman. 1989. "Impact of Global Climate Change on Urban Infrastructure." In J.B. Smith and D.A. Tirpak, eds., *Potential Effects of Global Climate Change on the United States: Appendix H, Infrastructure*. Washington, D.C.: U.S. Environmental Protection Agency.

National Academy of Sciences. 1991. *Policy Implications of Greenhouse Warming*. Washington, D.C.: National Academy Press.

Normile, Dennis. 2000. "Some Coral Bouncing Back From El Niño." *Science* 288 (12 May): 941–942.

North, Gerald R., Jurgen Schmandt, and Judith Clarkson. 1995. *The Impact of Global Warming on Texas*. Austin: University of Texas Press.

O'Brien, S.T., B.P. Hayden, and H.H. Shugart. 1992. "Global Climatic Change, Hurricanes, and a Tropical Forest." *Climatic Change* 22:175–190.

Perry, Michael. 1998. "Global Warming Devastates World's Coral Reefs." Reuters, 26 November. [http://www.gsreport.com/articles/art000023.html]

Phillips, John. 2000. "Tropical Fish Bask in Med's Hot Spots." *London Times*, 15 July.

Rahmstorf, Stefan. 1999. "Shifting Seas in the Greenhouse?" *Nature* 399 (10 June): 523–524.

Revelle, Roger, and Hans S. Suess. 1957. "Carbon Dioxide Exchanges Between Atmosphere and Ocean and the Question of an Increase of Atmopsheric CO2 During the Past Decades." *Tellus* 9:18–27.

Rittenour, Tammy M., Julie Brigham-Grette, and Michael E. Mann. 2000. "El Niño-Like Climate Teleconnections in New England During the Late Pleistocene." *Science* 288 (12 May): 1039–1042.

Ruhlemann, Carsten, Stefan Mulitza, Peter J. Muller, Gerold Wefer, and Rainer Zahn. 1999. "Warming of the Tropical Atlantic Ocean and Slowdown of Thermohaline Circulation During the Last Deglaciation." *Nature* 402 (2 December): 511–514.

Silver, Cheryl Simon, and Ruth S. DeFries. 1990. *One Earth, One Future: Our Changing Global Environment*. Washington, D.C.: National Academy Press.

Stevens, William K. 2000. "The Oceans Absorb Much of Global Warming, Study Confirms." *New York Times*, 24 March, A-16.

Stocker, T.F., and A. Schmittner. 1997. "Influence of CO2 Emission Rates on the Stability of the Thermohaline Circulation." *Nature* 388:862–864.

Thomas, R. 1986. "Future Sea-level Rise and its Early Detection by Satellite Remote Sensing." In *Effects of Changes in Atmospheric Ozone and Global Climate*. Vol. 4. New York: United Nations Environment Programme/United States Environmental Protection Agency.

Timmermann, A., J. Oberhuber, A. Bacher, M. Esch, M. Latif, and E. Roeckner. 1999. "Increased El Niño Frequency in a Climate Model Forced by Future Greenhouse Warming." *Nature* 398 (22 April): 694–696.

Titus, J.G., C.Y. Kuo, M.J. Gibbs, T.B. LaRoche, M.K. Webb, and J.O. Waddell. 1997. "Greenhouse Effect, Sea-level Rise, and Coastal Drainage Systems." *Journal of Water Resources Planning and Management* 113:2–21.

Titus, J.G., R.A. Park, S.P. Leatherman, J.R. Weggel, M.S. Greene, P.W. Mausel, S. Brown, C. Gaunt. M. Trehan, and G. Yohe. 1991. "Greenhouse Effect and Sea Level Rise: The Cost of Holding Back the Sea." *Coastal Management* 19:171–210.

Titus, James G., and Vijay Narayanan. 1996. "The Risk of Sea Level Rise: A Delphic Monte Carlo Analysis in which Twenty Researchers Specify Subjective Probability Distributions for Model Coefficients within their Respective Areas of Expertise." *Climate Change* 33, no. 2: 151–212. (Also U.S. Environmental Protection Agency. N.d. [http://users.erols.com/jtitus/Risk/CC.html])

U.S. Department of Energy. 1983. *Proceedings to Carbon Dioxide Research Conference: Carbon Dioxide, Science, and Consensus.* Washington, D.C.: U.S. Department of Energy.

U.S. Environmental Protection Agency (EPA). 1997. "The Cost of Holding Back the Sea." [http://users.erols.com/jtitus/Holding/NRJ.html#causes]

U.S. Environmental Protection Agency (EPA) and Louisiana Geological Survey. 1989. *Saving Louisiana's Coastal Wetlands: The Need for a Long-term Plan of Action.* Washington, D.C.: U.S. Environmental Protection Agency.

Webb, Jason. 1998. "Small Islands Say Global Warming Hurting Them Now." Reuters. [http://bonanza.lter.uaf.edu/~davev/nrm304/glbxnews.htm]

Weiner, Jonathan. 1990. *The Next One Hundred Years: Shaping the Fate of Our Living Earth.* New York: Bantam Books.

Wilcoxen, P.J. 1986. "Coastal Erosion and Sea Level Rise: Implications for Ocean Beach and San Francisco's Westside Transport Project." *Coastal Zone Management* 14, no. 3: 173–191.

Wilkinson, Clive, Olof Linden, Herman Cesar, Gregor Hodgson, Jason Rubens, and Alan E. Strong. 1999. "Ecological and Socioeconomic Impacts of 1998 Coral Mortality in the Indian Ocean: An ENSO Impact and a Warning of Future Change?" *Ambio* 28, no. 2 (March): 188–196.

Wood, Richard A., Anne B. Keen, John F.B. Mitchell, and Jonathan M. Gregory. 1999. "Changing Spatial Structure of the Thermohaline Circulation in Response to Atmospheric CO2 Forcing in a Climate Model." *Nature* 399 (10 June): 572–575.

CHAPTER 6

Flora and Fauna

David Quammen, an environmental writer, believes that the Earth during the next century will become a "planet of weeds," where human domination forces the extinction of most undomesticated living things. Quammen expects that the flora and fauna of the Earth eventually will include food crops, animals raised to be eaten or petted, and a few stubborn weed species which will benefit from a harsher, hotter world. By 2050, Quammen believes that deforestation will cause half the world's wild birds and two-thirds of other wild animal species to become extinct (Quammen 1998, 61, 69). "Wildlife," he writes, "will consist of pigeons, coyotes, rats, roaches, house sparrows, crows, and feral dogs" (Quammen 1998, 67). Human beings—"remarkably widespread, prolific, and adaptable," are "the consummate weed" (Quammen 1998, 68).

Albert K. Bates supports Quammen's opinion: "Sixty-five million years ago, 60 to 80 percent of the world's species disappeared in a cataclysmic mass extinction, possibly caused by an asteroid's impact with Earth. Human population, not an asteroid, will cut the remaining number of species in half again, in just the next few years" (Bates and Project Plenty 1990, 137). Another observer, Robert L. Peters, factors climate change into a similar picture of the Earth's biological future:

Habitat destruction in conjunction with climate change sets the stage for an even larger wave of extinction than previously imagined, based on consideration of human encroachment alone. Small, remnant populations of most species, surrounded by cities, roads, reservoirs, and farmland, would have little chance of reaching new habitat if climate change makes the old unsuitable. Few animals or plants would be able to cross Los Angeles on the way to the promised land. (Peters 1989, 91)

Global warming could destroy or fundamentally alter a third of the world's plant and animal habitats within a century, bringing extinction to thousands of species, according to a study by Great Britain's World Wide Fund for Nature, an affiliate of the World Wildlife Fund (Clover 2000). The study said that the most vulnerable plant and animal species will be in Arctic and mountain areas, where as many as 20 percent could be driven to extinction. In the north of Canada, Russia and Scandinavia, where warming was predicted to be most rapid, up to 70 percent of habitat could be lost, according to this study. The report, *Global Warming and Terrestrial Bio-diversity Decline*, was written by Jay Malcolm, professor of forestry at Toronto University, and Adam Markham, the executive director of Clean Air/Cool Planet. Pests and weedy species would fare best, they said, concluding, "If past fastest rates of migration are a good proxy for what can be attained in a warming world, then radical reductions in greenhouse gas emissions are urgently required to reduce the threat of bio-diversity loss" (Clover 2000, 9). To adapt and survive the expected rate of warming during the next century, the report said that plants may need to move 10 times more quickly than they did when recolonizing previously glaciated land at the end of the last ice age. Few plant species can move at a rate of one kilometer per year, the speed that will be required in many parts of the world.

THE DANGERS OF TROPICAL DEFORESTATION

The Earth had about 5.2 million square miles of old-growth forest remaining by about 1995. An estimated deforestation rate of 62,000 square miles a year, the amount being logged in the late 1990s, could reduce that figure to zero within two human lifetimes.

More than 50 percent of the world's forests have been destroyed within the past 100 years. An area equivalent to 57 soccer fields falls each minute. (Webb, World Forests, 1998). Trees are roughly 50 percent carbon. Alive, they remove carbon dioxide from the air and replace it with oxygen. Dying, dead, or being burned as fuel, trees become sources, not absorbers ("sinks") of carbon dioxide and other greenhouse gases. The destruction of forests around the world is no trivial matter for the atmosphere's carbon budget. While the atmosphere contains about 750 billion tons of carbon dioxide, forests contain about 2,000 billion tons. Roughly 500 billion tons is stored in trees and shrubs and 1,500 billion tons in peat bogs, soil, and forest litter (Jardine 1994).

Stephen Schwartzman, a senior scientist with the International Program of the Environmental Defense Fund, sketches the scope of tropical deforestation:

An area of forest bigger than Belgium, Holland and Austria put together, or about 40 percent of California, was cut down and burned every year between 1980 and 1995, some 62,000 square miles per year. NASA's Landsat satellite photographs show that more than 200,000 square miles, an area about the size of France, has been cleared and burned in Brazil alone. All of this has happened since the 1970s. (Schwartzman 1999, 60)

Deforestation increased rapidly during the 1980s. Three million acres of forest were being lost per year in Indonesia, providing 70 percent of the world's plywood and 40 percent of its tropical hardwoods. By the early 1980s in the tropical forests of Indonesia, roughly 1.2 percent of remaining coverage was being felled each year. In 1988, 48,000 square miles of Amazon rain forest were burned to clear land for farming. Nearly overnight, most of the burned land was turned from a net producer of oxygen to a source of carbon dioxide and methane (Gordon and Suzuki 1991, 118–119).

As Indonesia's forests were being felled, government policy encouraged a homesteading program on Sumatra and Borneo for farmers moving from overcrowded Java. Indonesia was using some of the same government incentives Brazil was using to "develop" the Amazon Valley. During the 1980s, hundreds of thousands of Brazilians moved to the Amazon Valley in a fashion resembling the homesteading of the North American West a century earlier. Some of this migration was financed by international development agencies, such as the World Bank.

In Indonesia and Brazil, the new landowners began altering the landscape by felling portions of the rain forest, then burning what they had cut. The rainforest canopy, which had heretofore produced a surplus of oxygen, was replaced by cattle, human beings, and motor vehicles. A survey conducted by the United Nations in 1982 estimated that tropical forests were being felled at a rate of 70,000 square miles (180,000 square kilometers) per year. The report estimated that 300 million subsistence farmers around the world, most of them in the tropics, were turning an area of forest the size of Denmark into a net producer of carbon every three months (Bates and Project Plenty 1990, 92). In 1987, 25,000 square miles of Brazilian rainforest were felled; a year later, the area that was deforested doubled (Bates and Project Plenty 1990, 93). In the 1980s, lowland tropical forests in Indonesia, the Philippines, Malaysia, and parts of West Africa shrank quickly as well.

According to the Brazilian government's own figures, the amount of Amazonian forest lost to fire doubled between 1997 and 1998. The Woods Hole Research Center, in Massachusetts, said that 7,800 square miles of Amazonian rain forest caught fire in 1997. About 80 percent of the fires were set on purpose, according to the United Nations; 42 million acres of rain forest (an area the size of Florida) were being lost per year by the 1990s.

Given deforestation and other land-use changes in the Amazon Valley, the area no longer functions generally as "the lungs of the world." The area acts as a carbon sink (absorber) only when temperature and precipitation patterns are healthy for the forest. In years of unusual drought (such as during the El Niño episodes of the middle and late 1990s), unusually hot, dry weather turns the Amazon into a net producer of carbon dioxide. In a usual year, the Amazon absorbs about 700 million tons of carbon dioxide, but in the recent El Niño year of 1995, the Amazon forest *added* 200 million tons of carbon dioxide to the atmosphere. According to A. Lindroth, A. Grelle, and A.S. Moren, writing in

Nature, a decrease in soil moisture can turn a carbon dioxide consuming forest into one that adds the gas to the atmosphere; this variation occurs in temperate zone forests as well as those of the tropics, the authors assert (Lindroth et al. 1998).

By the 1990s, according to R.A. Houghton and colleagues, "This large area of tropical forest is nearly balanced with respect to carbon" (Houghton et al. 2000, 301).

The combined effects of deforestation, abandonment, logging, and fire may thus yield sources of carbon. . . . These fluxes are similar in magnitude (but opposite in sign) to the sink calculated recently for natural ecosystems in the region. Taken together, the sources (from land-use change and fire) and the sinks (in natural forests) suggest that the net flux of carbon between Brazilian Amazonia and the atmosphere may be nearly zero, on average. (Houghton et al. 2000, 303–304)

Threats to the last remnants of a nation's forests sometimes provoke action. According to Stephen Schwartzman, "China, not a world leader in green consciousness, last year [1998] banned all logging in its few remaining natural forests after disastrous flooding wreaked havoc along heavily populated rivers. In so doing, China hoped to save remnants of forest cover on the upper headwaters. But so much forest is already gone that it may not make much difference" (Schwartzman 1999, 61).

Deforestation has taken the trees from the world's tropical forests to some notable venues. Vintage ships of the British Royal Navy, for example, were refurbished during the 1990s with large amounts of mahogany harvested illegally from an indigenous reserve in the Amazon River basin. The harvest raised a ruckus among English environmentalists, including Friends of the Earth, as the defense ministry's mahogany purchase was questioned on national television. More than 7,000 cubic feet of mahogany was cut, moving by ten to twelve trucks a day from the Kayapo indigenous reserve in Para state to the town of Redencao, home of the Juary logging company, harvester of the lumber. Each log was harvested illegally under Brazilian law. Despite the illegality of the harvest (aided by slack enforcement) the mahogany was purchased for the British ministry of defense by an English supplier, Parker Kislingbury.

In February 1999, deforestation was implicated as a major culprit when dozens of people died and hundreds of millions of dollars in property was destroyed in massive floods which shut down the industrial capital of South America, São Paulo. The same year flooding, also aggravated by deforestation, provoked unprecedented destruction in Caracas, Venezuela. Deforestation also intensified the death and destruction wrought in Central America by Hurricane Mitch.

About half of the rain which falls on a rainforest is produced by the forest's own humidity. Deforestation over large areas influences regional climate because there are fewer plants to produce the moisture that returns to the Earth as rain. More water runs off, carrying more topsoil, leaving less forest cover. Deforestation thus breaks a forest's hydrology cycle. As a tropical forest declines, regional weather becomes hotter, drier, and more prone to fire within surviving stands of trees. Increasing wildfires (such as those during the late 1990s in Mexico, Brazil, and Indonesia) add even more carbon to the atmosphere. By the late 1990s, the burning of tropical forests was contributing about 20 percent of the human-induced carbon dioxide buildup in the atmosphere. The burning of the Amazon rainforest alone contributes about a quarter of the worldwide total. The "lungs" of the world are being turned, within living memory, into yet another anthropogenic source of greenhouse gases.

According to Schwartzman,

The Woods Hole Research Center has found that for every acre cleared and burned in the Amazon, at least another acre burns in ground fires under the forest canopy or is degraded by selective logging (not picked up by the satellites). The frequency and extent of these ground fires skyrocket in El Niño events, which can then cause drought in some tropical forests.... [S]uch fires are likely to increase in frequency and intensity with global warming. (Schwartzman 1999, 63)

Deforestation is especially dangerous because once an area is denuded, local soil conditions (and sometimes weather) changes in such a way as to make the regrowth of trees more difficult. In many areas where deforestation has been extensive, previously forested areas have been replaced by sparse grasses, stunted shrubs, and bare, eroded earth.

"The scale of the problem is mind-boggling," said Daniel C. Nepstad of the Woods Hole Research Center, who has been surveying deforestation in the Amazon Valley. Nepstad estimated that 400,000 square kilometers of rainforest, an area twenty times that of Massachusetts, became vulnerable to fire during 1998, compared to an average of 15,000 square kilometers a year which had been cleared and burned during the previous few years. "Once burned," wrote Nepstad, "Amazonian forests become more susceptible to future burning. Wildlife is killed or dispersed (except for some species that thrive on disturbance). Timber, forest medicines and forest fruits are destroyed, and a large part of the carbon contained in the trees is gradually released into the atmosphere" (Nepstad 1998).

A major problem, according to Nepstad, "is that the rainforest is available in abundance and is therefore very cheap. When forest is cheap and labor and capital are scarce, it is the forest itself that becomes the fertilizer, the pesticide, the herbicide, the plow" (Nepstad 1998). Nepstad suggested "[a] new model of rural development . . . in Amazonia, which restricts access to most (about 70 percent) of the region's forests, and which creates a situation of land scarcity, higher land prices, and higher investments in agricultural production systems" (Nepstad 1998). Nepstad believes that people will value the forest only when it becomes an economic good in human financial terms. People will quit burning the forests when they can earn more money by sustaining it. About half of the area burned each year in Amazonia is accidental, according to Nepstad, who recommends prohibition of burning at times when fires may easily spread out of control.

Destruction of the Amazon rainforest accelerated toward the end of the twentieth century. According to studies by Nepstad, the forest was being destroyed two to three times more quickly than previous estimates. Nepstad conducted his research with scientists at the Institute of Environmental Research in Belem, Brazil. They measured forest losses at 1,104 sample points from a light plane and on the ground. They also interviewed more than 1,500 mill operators and landholders in the Amazon region. Nepstad and his associates assert that satellite imagery, on which most previous estimates have been based, "often fails to distinguish between pristine forest and burned or cut forest land newly covered with fast-growing brush" (Amazon Destruction 1999).

Nepstad said that roughly 17,000 square miles of Amazon forest were lost to cutting and burning in 1998, about three times the official Brazilian government estimate of 5,700 square miles. Nepstad also said that the total amount of rainforest already lost to human encroachment is about 217,000 square miles, or about 16 percent of the entire Amazon Valley. Nepstad cautioned that these figures are conservative and that "the real numbers could be significantly larger" (Amazon Destruction 1999).

Research published during October 1998 by the Hadley Center, part of the United Kingdom Meteorological Office, anticipates that large areas of "closed" tropical forests will die as the world warms during the next century. Instead of soaking up pollution, these forests will dump more carbon dioxide into the atmosphere than all the world's power stations and cars have produced during the past 30 years. "Closed" tropical forests, according to the Hadley Center report, contain trees covering a large proportion of the ground where grass does not form a continuous layer on the forest floor. Forests of this type declined from 289,700 thousand hectares (1980) in Africa to 241,500 in 1990. In Latin America, tropical forests declined from 825,000 to 753,000 hectares during the same period, and in Asia, tropical forest coverage fell from 334,500 hectares in 1980 to 287,500 in 1990 (National Academy 1991, 9).

Deforestation sometimes places additional debits on the world account of greenhouse gases. When tropical forests are cut and slashed, for example, the resulting piles of dead wood attract termites, whose populations explode as they feast on a food supply that has suddenly increased. Termites very efficiently digest carbon, turning it into methane (McKibben 1989, 17).

At the November 1998 conference on climate change in Buenos Aires, Britain's Hadley Center for Climate Prediction and Research presented a report indicating that large parts of the Amazon Valley could become desiccated by 2050. It was said that this desiccation may threaten the world with an unstoppable greenhouse effect. The Hadley report asserts that land temperatures will rise by an average 10 degrees F. over the next century. In addition to intensifying aridity in parts of tropical South America, large parts of tropical Africa are expected to become desiccated by 2050, according to the Hadley Center researchers. These climate changes will force roughly 30 million people to face intense hunger, the report asserts (Vrolijk 1999).

By about the year 2050, according to one study, "a decrease in annual rainfall of up to 500 millimeters in some key areas, combined with temperature rises of up to seven degrees C., will begin to kill tropical forests, creating deserts" (Nuttall 1998). At this point, the study says that these forests will become carbon sources rather than absorbers (sinks), as their remains burn or rot. "Up to 2050, the land surface takes up carbon through an increased growing season, which is being measured at the moment. But when we move to 2050, the tropical dieback releases so much carbon to the atmosphere it causes the concentrations to increase. This may enhance the build-up of carbon dioxide in the atmosphere" (Nuttall 1998).

HURRICANE MITCH AND DEFORESTATION

As early as 1984, economist Ian Cherrrett, working with a Dutch nongovernmental organization in Honduras, witnessed "total destruction of the environment going on at an accelerated rate," and warned that if forest cover was not protected, "as far as I'm concerned, within 20 years it's going to be too late." Orin Langelle of the Vermont-based Action for Community and Ecology in the Rainforests of Central America (ACERCA) observed, "Fourteen years later, Hurricane Mitch confirmed his prediction" (Weinburg 1999, 54).

Five years before flooding from Hurricane Mitch devastated Honduras, J. Almendares, a Honduran medical doctor, warned readers of the British medical journal *Lancet* that deforestation was making the country more vulnerable than ever to deadly flooding. Almendares presented evidence that "desiccation and soil erosion caused by cattle grazing and sugarcane and cotton cultivation have altered the regional hydrological cycle" (Almendares and Sierra 1993, 1401). These changes have led to fewer rainy days, but more intense downpours.

Almandares presented temperature statistics from one deforested area of Honduras indicating that the average ambient air temperature had risen 7.5 degrees C. between 1972 and 1990 (Almendares and Sierra 1993, 1403). In neighboring Nicaragua, during the final months of the Sandinista revolution (1979), similar temperature rises were reported in Managua after dictator Anastasio Somoza ordered the wholesale destruction of trees in the city to deny Sandinistas places to hide during gun battles.

Central America had 200,000 square miles of forest in 1900. By the 1980s, only 36,000 square miles survived, and the rate of deforestation was increasing, especially in the Miskito region of Nicaragua and Honduras. A 1998 report by United Nations agencies and nongovernmental organizations documented a regional deforestation rate of 958,360 acres (1,500 square miles) a year (Weinburg 1999, 51).

Honduras lost a third of its forests between 1964 and 1990. Honduran forests continued to be felled at a rate of 80,000 hectares a year during the 1990s, a rate which, if sustained, amounts to a quarter of remaining forested land per decade. In the meantime, people who can no longer wrest a living from denuded (or corporate-controlled) land have been moving to Honduran cities, where malaria has become endemic. "Blood transfusions have now become a significant means of . . . malaria transmission in Honduras," Almendares and Sierra wrote (Almendares and Sierra 1993, 1402).

During October, 1998, Hurricane Mitch made landfall on Central America's Miskito Coast with winds as strong as 178 miles an hour, dropping as much as three feet of rain. Crossing the mountains, Mitch turned northwest, through El Salvador, Guatemala, and southern Mexico. By the time the storm reached the sea again, it had killed 10,000 people and left nearly three million others homeless. In Honduras, thousands of people who survived the storm lost their jobs in devastated banana plantations. (For more information on Hurricane Mitch, see Greenhouse Weather in Chapter 9.)

The devastation wrought by Mitch raised questions in Central America not only about potential increases in hurricane severity due to global warming, but also about changes in land use across the region which makes many areas more prone to severe flooding during heavy rains. The same questions were raised in Caracas, Venezuela, after devastating floods occurred there late in 1999.

Bill Weinburg described the conditions which have made hurricane flooding so devastating in Central America:

Fire and flood fuel each other in a vicious cycle. Landless peasants colonize the agricultural frontier, or clear forested slopes for their *milpas* (fields). The more forest is destroyed, the more the hydrologic cycle is disrupted; with no canopy for transpiration, local rainfall and cloud cover decline; aridity makes the surviving forest vulnerable to wildfires. Then, when the rains do come, sweeping in from the Caribbean on the trade winds, there are no roots to hold the soil and absorb the water. Millennias' accumulated wealth of organic matter is swept from the mountainsides in deluges of mud. Tlaloc, the revered Nahua rain god who the Maya called Chac Mool, brings destruction instead of abundance. (Weinburg 1999, 51–52)

In Nicaragua, 2,000 people died in the Chinandega municipality of Posoltega, as ten communities were buried in mudslides when the Casita volcano crater collapsed. Three-quarters of a million people were left homeless in the area. Posoltega lies in an area that has been almost completely deforested. Hurricane Mitch's trail of death and damage in eastern Nicaragua also was intensified by deforestation of the upper Rio Coco watershed. More than 20 inches of rain caused the river to rise more than 60 feet within a few days.

Preservation of forests became an environmental issue in Eastern Nicaragua following Hurricane Mitch's devastation. Local native peoples and ecologists united that same year to evict Korean-owned timber giant Solcarsa, which had won a government contract in a large area of the Miskito rainforest. According to Weinburg, at least 370,500 acres were being deforested annually in Nicaragua at the time of Hurricane Mitch. Nicaragua has lost 60 percent of its forest cover within the last two generations.

Hurricane Mitch left Nicaragua with an enduring legacy, as described by Weinburg:

[I]n September 1999, almost a year after the disaster ... President Aleman declared a national emergency over a plague of rats in ... Nicaragua. Their numbers exploded from an overabundance of dead meat—both human and animal—following the hurricane. The rats overran fields and homes, decimating crops. Poison had to be distributed in mass quantities to beat the infestation. (Weinburg 1999, 54)

Former Honduran President Rafael Callejas attributed the extreme death and damage wrought by Hurricane Mitch to "mudslides that were the result of un-controlled deforestation and therefore could have been prevented" (Weinburg 1999, 54). Ecological activism can be as dangerous in Honduras as the winds, rains, and mudslides of a major hurricane. A few weeks before Mitch made landfall, Carlos Luna, a local opponent of logging in the central mountains near Tegucigalpa, the country's capital and largest city, was gunned down by un-known assailants.

Deforestation has become a major problem (and political issue) in Mexico as well as in Central America. During the spring of 1999, fires raged out of control throughout the mountains of southern Mexico, sweeping through parts of the Chipas highlands, as well as the Sierra Tarahumara of Chihuahua. The fires cloaked Mexican skies in an acrid haze from its borders with Guatemala and Texas. Among the causes of these fires are private timber operations on *ejido* (Native-held) lands that compound the deforestation caused by campesinos clearing lands for their *milpas*. In October 1997, deforestation took a deadly toll when Hurricane Paulina hit the Sierra Madre del Sur in the Mexican states of Guerrero and Oaxaca. Mudslides cascading down denuded mountainsides left scores dead and thousands homeless, especially in Zapotec country.

The main cause of tropical deforestation—perhaps two-thirds of it—is slash-and-burn agriculture. Much of the deforestation is caused, in turn, by small farmers and ranchers who are threatened by the spread of commercial farming and ranching enterprises. Driven out of more populous areas, subsistence farmers are forced to destroy large amounts of tropical forest to create new farmland. According to a study by the *Global Futures Bulletin* (Much Deforestation 1999), the pattern is the same in much of Africa, Asia and Latin America. Increasing human population and pressure on the land is the ultimate cause of the defor-estation of slash-and-burn agriculture, according to the report.

FIRE, CARBON, AND BOREAL (NORTHERN) FORESTS

A report by Greenpeace International suggests that between 50 and 90 percent of the Earth's existing boreal forests are likely to disappear if atmospheric levels of carbon dioxide and other greenhouse gases double (Jardine 1994). These forests comprise a third of the Earth's remaining tree cover, about 15 million square kilometers, across Russia (where they are called "taiga"), Canada, the United States, Scandinavia, and parts of the Korean Peninsula, China, Mongolia, and Japan. Large forests also clothe many mountain ranges outside of these zones. In total, boreal forests cover about 10 percent of the world's land area.

The Greenpeace report indicates that global warming's toll on the boreal forests had begun by the early 1990s. The report warns that decaying forests may provide an extra boost to rising carbon dioxide levels, causing warming to feed upon itself. (See also Biotic Feedbacks in Chapter 2.) The decline of boreal forests also endangers more than one million indigenous people who live in them, including the Dene and Cree of Canada, the Sami (Lapplanders) of Norway, Sweden, and Finland, the Ainu of northern Japan, and the Nenets, Yakut, Udege, and Altaisk of Siberia (Jardine 1994). Some of the forests' larger animals, such as the Siberian tiger, already are near extinction. The Greenpeace report concludes that rapid logging of the boreal forests is intensifying pressure on animal life, and accelerating the release of even more carbon dioxide and other greenhouse gases into the atmosphere.

Climate change could imperil New England's maple syrup industry, as a gradual rise in global temperatures strips New England of its native sugar maples. A report by the U.S. Office of Science and Technology Policy said that expected climate changes within the present century are likely to shift the ideal range for some North American forest species northward by as much as 300 miles, exceeding the species' ability to migrate naturally. If emissions of greenhouse gases continue to increase, the report projects that maples will recede from all areas of the continental United States except the northern tip of Maine.

The Sierra Club also has described the possible effects of global warming on sugar maples:

Regions dependent on forests for various commodities, such as the maple-syrup-producing states of New England and the Midwest, could also face devastating losses. EPA projects that by 2050 the range for the sugar maple could shift north of all but the northernmost tip of New England. This possibility has serious implications for the maple syrup industry, which currently provides up to $40 million annually to the regional economy. (Glick 1998)

The same Sierra Club report traces increasing temperatures and insect infestations in boreal forests to a rise in temperatures which began in 1976.

With rising temperatures have come larger and more frequent forest fires, according to Greenpeace. "Unless atmospheric concentrations of greenhouse gases are quickly stabilized," the report states, "climate-vegetation models predict that large areas of boreal forest will be reduced to patchy open woodland and grassland, resulting in lowered biological diversity and a reduced ability to store carbon" (Jardine 1994).

If boreal forests continue to decline, their burning and rotting could contribute to the release of as much as 225 billion tons of extra carbon dioxide into the atmosphere, raising current levels by a third, accelerating the pace of warming. Trees could have difficulty colonizing the thawing tundra to the north of their present ranges because they simply cannot migrate quickly enough, and because the treeless tundra cannot evolve quickly enough to sustain them.

Kevin Jardine sketches the role of insect outbreaks in the anticipated destruction of boreal forests:

Pests which may invade boreal forests under warming conditions include the western spruce budworm, the Douglas-fir tussock moth, and the mountain pine beetle. (Kurz et al. 1995, 127) Kurz [and colleagues] conclude that "biospheric feedbacks from temperate and boreal forest ecosystems will be positive feedbacks that further enhance the carbon content of the global atmosphere" (Kurz et al. 1995, 129). By 1998, spruce budworms had devoured 50 million acres (20 million hectares) of Alaskan woodland.

In moderation, the forest usually benefits from insect outbreaks because they reduce the likelihood of catastrophic fire by helping to eliminate older stands and diseased trees. However, populations of insects such as bark beetles, the Siberian silkworm and the spruce budworm can explode and devastate millions of acres of forest. The life cycles of the spruce budworm (*Choristoneura fumiferana*) and the spruce bark beetle (*Ips typographus*) are strongly influenced by climate, with both species likely to increase in numbers during the kind of warmer, drier weather predicted in a global warming world. . . . A Canadian government study released in 1987 showed that conifers were growing on average up to 65 percent slower than in the 1940s and 1950s, because of spruce budworm outbreaks, and possibly acid rain. (Jardine 1994)

William K. Stevens of the *New York Times* described unprecedented destruction of boreal forests in Alaska by spruce bark beetles that have been whipped into a reproductive frenzy by a warming environment:

Once these purple-gray stretches of tree skeletons were a green and vital part of the spruce-larch-aspen tapestry that makes up the taiga. . . . Today, in a stretch of 300 or 400 miles reaching westward from the Richardson Highway, north of Valdez, past Anchorage, and down through the Kenai Peninsula, armies of spruce bark beetles are destroying the spruce canopy or have already done so. Often the trees are red instead of gray—freshly killed but not yet desiccated. (Stevens 1999, 178)

On the south-central coast of Alaska, cool temperatures have heretofore kept the spruce bark beetle under control. As temperatures have warmed, however, the beetles have killed much of the tree cover across three million acres, one of the largest insect-caused forest devastations in North America's history.

While warming increases many insects' reproductive energies, warming destroys many trees' reproductive capacities. According to Jardine, rising temperatures can cause boreal trees' pollen and seed cones to develop too rapidly because of higher-than-usual spring temperatures, leading to reproductive failure. Boreal seeds also germinate within a specific range of soil temperatures. For example, writes Jardine, black spruce seeds germinate between 15 and 28 degrees C. "If the soil temperature falls below 15 degrees," writes Jardine, "the processes that cause germination come to a halt. If soil temperatures rise above 28 degrees C., bacteria and fungi can attack and consume seeds. The probability of germination also declines rapidly for higher temperatures" (Jardine 1994).

Valerie A. Barber, Glen Patrick Juday, and Bruce P. Finney studied tree-ring data from Alaskan forests for 90 years during the twentieth century, and found that warmer temperatures inhibit the growth of Alaskan white spruce, thereby decreasing boreal forests' ability to draw carbon dioxide from the atmosphere.

The tree-ring records show a strong and consistent relationship over the last 90 years, and indicate that, in contrast with earlier predictions, radial growth has decreased with increasing temperature. Our data show that temperature-induced drought stress has disproportionally affected the most rapidly growing white spruce. . . . If this limitation in growth due to drought stress is sustained, the future capacity of the northern latitudes to sequester carbon may be less than currently expected. (Barber et al. 2000, 668)

R. Suffling correlates increases in temperature to a rise in the amount of land lost to forest fires in the boreal forests of Scandinavia and Canada. Given the unplanned nature of forest fires, policy responses will be limited, Suffling warns, and he speculates whether increasing pressure to harvest remaining forests for human use will accelerate this process (Suffling 1992, 128). He describes a wave of forest fires in the boreal forests of Canada: "Since the mid-1970s, there has been a massive fire outbreak in response to a series of warm, dry summers" (Suffling 1992, 111). Work by W.R. Emanuel and colleagues indicates that boreal forests on Earth may become nearly extinct within roughly a century if the world warms to the extent forecast by many climate models (Emanuel et al. 1985).

During the summer of 1995, large fires scorched forests in parts of northern and central Canada, consuming as many as 240,000 acres a day. A study by the Canadian Forest Service concluded, "The northern forest has lost almost a fifth of its biomass over the last twenty years because of enormous increases in fires and insect outbreaks" (Gelbspan, *The Heat is On*, 1997, 21). In 1970, these forests were absorbing 118 million tons of carbon dioxide each year, but during the 1990s, the same area had become a net producer of 57 million tons of carbon dioxide per year (Gelbspan, *The Heat is On*, 1997, 21).

Canada's boreal forests may be reduced to a fraction of their current range by warming temperatures, according to a governmental report. The area covered by boreal forests in Canada declined 20 percent between 1975 and 1995 (Rolfe 1996). The report, compiled by Environment Canada in 1994, predicted an average winter warming on the coast of British Columbia of four degrees C. in the winter and 2.5 degrees C in summer by 2050. Such a change could raise snow levels as much as 3,000 feet and reduce mountain snowpack to between one-half and one-sixth its size in 1995, according to the report (Environment Canada 1994). If, over the next century, there is a doubling of atmospheric carbon dioxide and a warming of roughly 9 to 10 degrees F., yellow birch, sugar maple, hemlock, spruce, beech, and other trees which now flourish in the temperate zones would be forced to move northward (in the northern hemisphere) 300 to 600 miles to maintain a constant environment. Short of growing legs, many trees will not be able to travel so quickly (Bates and Project Plenty 1990, 97; Corson 1990, 233).

W.D. Billings describes some of the problems which confront plants because of the speed of anthropogenic warming:

If the predicted rates of climatic warming during the twenty-first century hold in the Arctic, the change may be far more rapid than any climatic changes during deglaciation, or even during the Little Ice Age. Will this warming be much faster than the ability of plant migration and soil formation to keep up? Perhaps. . . . For tree taxa in the Great Lakes states, future range extensions will have to occur at least ten times as fast as the average Holocene rate, or 200 kilometers percentury, to track the expected temperature rise. (Billings 1992, 105)

The flora of the future will favor plants best adapted to increased carbon dioxide levels as well as a warmer environment. Swift changes in temperatures also will favor plants which can adapt to weather extremes and those which can move quickly. These attributes describe most "weedy shrubs and grasses which directly compete with young trees in unforested areas. New forests will not be established easily, and the world will become much weedier" (Bates and Project Plenty 1990, 98). In addition, mid-latitude forests which perish because they cannot beat the heat to higher latitudes will release their own carbon when they die (McKibben 1989, 33).

According to George M. Woodwell, director of the Woods Hole Research Institute, a one degree C. change in temperature is equivalent in the mid-latitudes to between 60 and 100 miles of latitudinal change (Epstein et al. 1996). Woodwell sketches the speed of the climate changes suggested by many climate models: "With warming in the range of tenths of a degree per decade and the highest warming rates in the higher latitudes, the changes in climatic zones will be of the order of miles per year" (Epstein et al. 1996).

Insect infestations also have been increasing in forests that are accustomed to sustained heat, as well as in cooler boreal forests. By 1998, the southeastern United States was suffering through its worst infestation of armyworms in recent memory, as the pests chewed through thousands of acres of crops, pasture, and turf. The armyworm population explosion was caused by a mild winter following a summer drought, according to Richard Sprenkel, pest management specialist with the University of Florida's Institute of Food and Agricultural Sciences (Greening the Planet 1998).

Armyworms balloon from the width of two hairs to the thickness of a pencil during a life span of three to four weeks. A field of crops can be wiped clean by the insects within 48 hours. Armyworms have been eating corn, cotton, peanuts, and grasses used for pasturing animals. Armyworms also threaten lawns, golf courses, and athletic fields. The insects do most of their damage during a frenzy which consumes the last two days of their lives. Insecticides do little good at this stage, since the armyworms have already laid the eggs which will hatch into a new generation. The rise of the armyworms has been linked to the increasing rarity of hard winter freezes in the Southeast (Greening the Planet 1998).

ANIMAL ADAPTATIONS TO WARMING

By the late 1990s, the migration and breeding patterns of many animals were responding to warmer temperatures. For example, Jerram Brown has charted the breeding seasons of Mexican jays in the Chiricahua Mountains of southern Arizona for 31 years. By 1998, Brown found that the jays were laying their eggs an average of 10 days earlier than they did in 1971. Camille Parmesan has analyzed records tracking the distribution patterns of 57 nonmigratory butterfly species across Europe. She found that, during the last century, two-thirds of the species have shifted their ranges northward, some of them by as much as 240 kilometers (Wuethrich 2000, 795). "We ruled out all other obvious factors, such as habitat change," said Parmesan. "The only factor that correlated was climate" (Wuethrich 2000, 795).

Parmesan and colleagues' tracking of butterfly ranges provided evidence of poleward shifts in entire species' ranges. In a sample of 35 nonmigratory European butterflies, 63 percent have ranges that shifted to the north by 35 to 240 kilometers during the twentieth century, while only three percent of ranges have shifted to the south (Parmesan et al. 1999, 579). The study team's evaluation of its data ends with a warning about other species:

Given the relatively slight warming in this century compared [with anticipated temperature] increases of 2.1 to 4.6 degrees C. for the next century, our data indicate that future climate warming could become a major force in shifting species distributions. But it remains to be seen how many species will be able to extend their northern range margins substantially across the highly fragmented landscapes of northern Europe. This could prove difficult for all but the most efficient colonizers. (Parmesan et al. 1999, 583)

By the year 2000, warmer temperatures were bringing butterflies to England two weeks to a month earlier than during the 1970s. Scientists studying 35 of the estimated 60 species of British butterflies say that some, such as the red admiral, can now be seen a month earlier. Others, "[s]uch as the peacock and the orange tip, are appearing between 15 and 25 days earlier than two decades ago," according to a report in the *Times of London* (Nuttall 2000). Roy and Tim Sparks, both of the Centre for Ecology and Hydrology at Monks Wood, Cambridgeshire, analyzed data from 1976 to 1998 provided by the Butterfly Monitoring Scheme, whose members check more than 100 sites each week from April to September. According to a news report in *The Guardian* of London, "One [red admiral] was monitored after crossing the [English] Channel on New Year's Day" (Vincent and Brown 2000, 9).

Some birds, like butterflies, have been extending their ranges northward in Europe since 1970. Chris D. Thomas and Jack J. Lennon of the University of Leeds' School of Biology analyzed the breeding distribution for birds in Britain, finding that their northern ranges had expanded northward an average of 18.9 kilometers between roughly 1970 and 1990. The authors note that temperatures warmed in Britain during this period, "which we propose might be casually related" (Thomas and Lennon 1999, 213).

GLOBAL WARMING AND FISHERIES

Drastic declines in some western Alaskan salmon populations during 1997 and 1998 led some observers to ask whether rising water temperatures had played a role, because salmon are very sensitive to temperature. According to a report by the World Wildlife Fund,

While salmon can withstand higher temperatures in summer when food is abundant, in the winter their tolerance drops considerably. As cold-blooded creatures, their metabolism increases in warmer water and keeping up with this high metabolism requires large amounts of food. If sufficient food is not available, salmon can starve. (Mathews-Amos and Berntson 1999)

The decline of Pacific salmon runs during the late 1990s suggests that global climate change could devastate fish populations on which millions of people rely for food. In Alaska during 1997 and 1998, few salmon returned from the ocean to spawn at their birthplaces. Those which did return were smaller than average and arrived later than usual. The key factor in the decline of salmon runs appears to have been the fact that water temperatures in 1997 and 1998 were much higher than usual (Kruse 1998).

For Canada's western regions, climate models forecast an increase in precipitation, water runoff, and flooding in winter and a decrease in precipitation and runoff during summer. Higher winter river flows are expected to damage salmon spawning grounds, reduce survival and growth of fish because of increased stream temperatures, and damage Fraser River salmon due to increased predation by warm-water species (Rolfe 1996). In 1995, the Canadian Department of Fisheries and Oceans blamed a collapse of Fraser River salmon runs on predation by mackerel, which invaded the salmon spawning grounds along with warmer-than-average ocean waters provoked by El Niño conditions. At the same time (and according to the Canadian government, for the same reasons) the Queen Charlotte chinook salmon runs declined about 80 percent.

The salmon catch in Scotland was 35 percent less in 1999 than during 1998, which itself was the second worst year since comprehensive records began being kept in 1952. The reduced catch, due in part to warming water temperatures, is threatening the livelihood of river proprietors, country hotels and bed and breakfasts, as well as ghillies (guides) who help the angler find his fish. Warmer sea water decreases the population of krill which the salmon eat. Anglers also say that sea lice from salmon reared on fish farms are infesting wild salmon, killing them. The fish farms also produce a polluting slurry. In 8 of 32 rivers on the west coast of Scotland, salmon were virtually extinct by the year 2000. Seals are killing salmon in some rivers as well (Buxton 2000, 11).

Cod (a cold-water fish similar in some ways to salmon) also has been declining as habitats warm. In mid-July, 2000, the World Wildlife Fund (WWF) placed North Sea Cod, staple of British fish and chips (as well as a $60 million annual fishery), on its endangered-species list. Populations of North Sea Cod have declined 90 percent in 30 years, according to the WWF (Brown 2000, 3). The cod have been over-fished, and their population is falling because they do not breed well in warmer water. The North Sea in the year 2000 is as much as three degrees C. warmer than it was in 1970 (Brown 2000, 3).

Global warming could result in an estimated eight percent decrease in fish yields worldwide (Fankhauser 1995, 39). Global warming also may significantly damage recreational fishing. An Environmental Protection Agency (EPA) report issued in 1995 shows that several cool-water fish species, especially trout, would diminish as waters warm. This study projects that between eight and ten states (of the United States of America) could lose all of their cool-water sports fisheries within 50 to 60 years. Between 11 and 16 other states could lose half of their cool-water sports fisheries, according to the study (U.S. EPA 1995, ix).

Rising temperatures in British Columbia's largest sockeye salmon spawning river, the Fraser, provoked a government shutdown of commercial salmon fishing at the height of the season late in September 1998. According to an Environmental News Service dispatch from Victoria, British Columbia (Thomas 1998), above-average temperatures in the river impaired the salmon's swimming and jumping abilities, and afflicted them with "proliferating pathogens and an invasion of warm-water predators such as squaw fish" (Thomas 1998). The article also described fears that "seawater temperatures 1.5 degrees C. above average in the adjacent Georgia Strait could trigger toxic algae blooms" (Thomas 1998).

Between a quarter and two-thirds of salmon returning to various locations in the Fraser river system were dying before they could spawn, many of them from conditions related to rapid warming of their aquatic environment. The water warmed because of the below-average snowpack and above-normal temperatures in the interior of British Columbia, the Fraser's drainage basin. A large aluminum smelter also was delivering large amounts of heated water to an upstream tributary of the Fraser, adding yet another human provocation to the salmon's warming environment. Clear-cutting of upstream forests is also blamed for some of the warming of the river's waters.

Above-average water temperatures in this area also had caused salmon to die in unusually large numbers during 1992 and 1994. If the temperature rises two degrees more (from a late-summer peak of 21 degrees C. to 23 degrees C.) none of the salmon will survive the swim upstream to spawn. At 23 degrees C., most salmon will die of heat prostration. The warming of the river and resulting fishing shutdown in September, 1998, brought out angry fishermen, who threatened to block cruise ships in Vancouver's busy harbor.

Robert Gough comments,

Global (primarily marine) fisheries provide the world with about 20 percent of its animal protein. Worldwide fisheries are pressed to their natural limits and international fishing fleets are at . . . capacity. Fishing fleets of various nations exceed their own boundaries and encroach on the territorial waters of other countries. (Gough 1999, 46)

According to the Intergovernmental Panel on Climate Change (IPCC), global warming may cause some fisheries to collapse, while others expand. As Gough comments,

Climate changes can exacerbate the effects of over-fishing at a time of inherent instability in world fisheries. In addition, over-fishing creates an increased imbalance in the age composition of a stock, and may reduce the resiliency of the population. Further, changes in ocean currents may result in changes in fish population location and abundance and the loss of certain fish populations. (Gough 1999, 47)

Fish populations in many areas also probably will be affected by interruptions in their breeding cycles caused by saltwater intrusion into estuaries as seas rise. Roughly 70 percent of the world's fish use shoreline waters in their breeding cycles. Writes Gough, "Fish production will thus suffer when such nursery habitats are lost" (Gough 1999, 47). Gough warns that many species in the sea and on the land which have adapted to specific environments may not be able to adapt to the speed of coming temperature changes: "Surviving species may succumb to predatory pressure and competition from more exotic species better adapted to the new conditions" (Gough 1999, 48).

GLOBAL WARMING AND AGRICULTURE

Many people who do not farm for a living share a stereotype of agriculture as a family affair, a builder of character, and a style of employment which evokes, for tillers of the soil, a basic sense of enjoyment from communion with nature. During the twentieth century, however, agriculture has become progressively more mechanized on a massive scale suited to large, worldwide markets. Agriculture, like other modes of production in our machine culture, has come to demand less human labor and an increasing amount of fossil-fuel energy. Industrial-scale agriculture also requires copious amounts of synthetic fertilizers. For all except a few remaining (and often struggling) family farmers, agriculture has become as industrialized as factory work.

Ecologist Barry Commoner described how the American farm has changed:

Between 1950 and 1970, the total U.S. crop output increased by 38 percent, although the acreage decreased by 4 percent and the labor [number of people employed] fell by 58 percent. This sharp increase in productivity was accomplished by an 18 percent increase in the use of machinery and a 295 percent increase in the application of synthetic pesticides and fertilizer. (Commoner 1990, 49)

How will agriculture fare in a warmer world? The MINK [Missouri-Iowa-Nebraska-Kansas] Study surveyed potential climate change in the central United States, North America's agricultural heartland. Under certain circumstances, the authors found, higher levels of carbon dioxide might enhance growth of some crops, but as a whole, "[u]nder the best of these scenarios . . . the productivity of the region's agriculture would be significantly diminished" (Rosenburg et al. 1992, 151). Agriculture would be severely affected not only by heat stress, but also by reduced surface-water supplies, since most global climate models predict that as the atmosphere warms the interiors of continents would become not only hotter, but also drier, especially during the growing season. An additional problem facing farmers in Nebraska and Kansas is depletion and salinization of aquifers which already support a large part of agricultural production in both states, especially their drier western areas. (For more information on global warming and the American Midwest, see Heat, Drought, and the Great Plains of North America in Chapter 9.)

Craig Benjamin, writing in *Native Americas*, describes how large scale monocultural farms make themselves vulnerable to a rising risk of pest attack in a warmer, more humid world.

This impressive vulnerability of industrial agriculture is key to understanding how climate change will likely have an impact on global agriculture and on the relationship between industrial agriculture and indigenous farming communities. Faced with rapid and dramatic climate change, the impressively vulnerable industrial farm can conceivably continue to use large-scale irrigation and artificial fertilizers to counter the effects of changing temperature and precipitation. (Benjamin 1999, 80)

While some skeptics argue that global warming will benefit agriculture by providing plants with a higher level of carbon dioxide (see Carbon Dioxide "Enrichment" in Chapter 3), Pim Martens contends that increased growth may be counterbalanced by molds and other parasites which thrive best in hot, humid weather. "It is generally believed that a climate change will have negative effects for global food production," Martens writes (Martens 1999, 540). To cite one of many examples, the Mediterranean Fruit Fly could expand into Northern Europe during the next century with the degree of global warming projected by the IPCC.

Research by Fakhri A. Bazzaz and Eric D. Fajer casts doubt on the skeptics' assertions that a carbon dioxide enriched atmosphere will lead to more plant growth and greater agricultural yields.

Studies have shown that an isolated case of a plant's positive response to increased CO_2 levels does not necessarily translate into increased growth for entire plant communities. . . . [P]hotosynthetic rates are not always greatest in CO_2-enriched environments. Often plants growing under such conditions initially show increased photosynthesis, but over time this rate falls and approaches that of plants growing under today's carbon-dioxide levels. . . . When nutrient, water, or light levels are low, many plants show only a slight CO_2 fertilization effect. (Bazzaz and Fajer 1992, 68–71)

Bazzaz and Fajer assert, "[W]e do not expect that agricultural yields will necessarily improve in a CO_2-rich future" (Bazzaz and Fajer 1992, 68). William R. Cline adds that scarcity of water (which is forecast for many continental interiors as the atmosphere warms) also may reduce agricultural yields (Cline 1992, 91).

Studies by Martin Parry (1990; Parry and Jiachen 1991) estimate that the European corn borer could move 165 to 200 kilometers northward (in the northern hemisphere) with each one-degree C. rise in temperature. The potato leaf-hopper, a major pest for soybeans, presently spends its winters along the Gulf Coast. Global warming could move this range northward. The range of the hornfly, which caused about $700 million in damage to beef and diary cattle across the United States during the late 1980s, could be similarly affected.

N.C. Bhattacharya has found that while enriched carbon dioxide causes accelerated growth in most plants, others respond negatively. Increasing the growth rate of plants also tends to accelerate depletion of soils, stunting later growth. Heat also may be detrimental to some plants even as their growth is being stimulated by rising carbon dioxide levels in the atmosphere (Bhattacharya 1993).

Y.A. Izrael summarizes the rough road of agriculture in a warmer world:

Estimates of the impact of doubled CO_2 on crop potential have shown that in the northern mid-latitudes summer droughts will reduce potential production by 10 to 30 percent. The impact of climate change on agriculture in all, or most, food-exporting regions will entail an average cost of world agricultural production [of] no less than 10 percent. (Izrael 1991, 83)

Martin Parry elaborates, "While global levels of food production can probably be maintained in the face of climate change, the cost of this could be substantial" (Parry 1990, 279). Gains in production at higher latitudes are unlikely to balance reductions in the hotter mid-latitudes, which are major grain exporters today (Parry 1990, 279).

Harold W. Bernard, in *Global Warming: Signs to Watch For*, describes global warming's anticipated role in the spread of wheat rust, a fungus which thrives in dry heat. Wheat rust destroyed millions of tons of wheat in North America during the Dust Bowl decade of the 1930s. At the same time, the pale western cutworm, another pest which favors hot and dry conditions, damaged thousands of acres of wheat in Canada and Montana (Bernard 1993, 47).

Cline models global warming three centuries into the future. By that time, he expects that agricultural production in many contemporary breadbaskets will have been devastated. Indicating an average July maximum for Iowa between 100 and 108 degrees F., Cline scoffs at the idea that higher carbon dioxide levels in the atmosphere may enhance agricultural yields. By the year 2275, Cline indicates that temperatures may become so hot that most staple grain crops in present day Iowa (and surrounding states) may die of heat stress. Wheat, barley, oats, and rye simply will not grow in the climate which Cline projects for the United States Midwest roughly three centuries from today. In Eurasia, rice, corn, and sorghum will be pressed to their limits by such temperatures as far north as Moscow. By 2275, according to Cline, the atmosphere's carbon dioxide load may be eight times the levels of the 1990s (Cline 1992).

In the much shorter range (at twice contemporary levels of greenhouse gases, a level expected by many climate models by the end of the twenty-first century), the United States is forecast to experience a decline of perhaps 20 percent in agricultural production, while some areas of Russia and northern Europe may see production increases. A beneficiary of moderate global warming could be Iceland, which, according to Cline, is expected to see its agricultural production rise 50 percent (Cline 1992, 99–100). For comparative purposes, Cline remarks that the average temperature in the United States during the 1930s Dust Bowl years was only 0.9 degrees C. higher than the 1951–1980 average range (Cline 1992, 95). After the twenty-first century, following Cline's analysis (which assumes increasing use of fossil fuels), farmers will be dealing with a degree of warmth that probably will be outside human historical experience.

According to Thomas R. Karl and associates, increases in minimum temperatures are important because of their effect on agriculture. Observations over land areas during the latter half of the twentieth century indicate that average minimum temperatures have increased at a rate more than 50 percent greater than that of maximums.

The rise in minimum temperatures has lengthened the growing (frost-free) season in many parts of the United States; in the Northeast, for example, Karl and associates write that the frost-free season began an average of 11 days earlier during the 1990s than during the 1950s. The compression of daily high and low temperatures may be related to increasing cloud cover and evaporative cooling in many areas, Karl and associates propose. Clouds depress daytime temperatures because they reflect sunlight, as they also warm night-time temperatures by inhibiting loss of heat from the surface. "Greater amounts of moisture in the soil from additional precipitation and cloudiness inhibit daytime temperature increases because part of the solar energy goes into evaporating this moisture," they write (Karl et al. 1997).

In a warmer world, erosion and drought will probably pose greater dangers to agriculture. This is not the paradox that it seems, because, according to Karl and associates "[N]ot only will a warmer world be likely to have more precipitation, but the average precipitation event is likely to be heavier" (Karl et al. 1997). Karl cites data indicating that heavy precipitation events increased roughly 25 percent from the beginning to the end of the twentieth century. Karl describes how a generally wetter world also will be a place in which drought may threaten flora and fauna more often:

As incredible as it may seem with all this precipitation, the soil in North America, southern Europe, and in several other places is actually expected to become drier in the coming decades. Dry soil is of particular concern because of its far-reaching effects, for instance, on crop yields, groundwater resources, lake and river ecosystems. . . . Several models now project significant increases in the severity of drought. (Karl et al. 1997)

Karl and associates temper this statement by citing studies indicating that increased cloud cover may reduce evaporation in some of the areas which are expected to become drier. All in all, however, most of the world's flora and fauna will suffer markedly in a significantly warmer world.

REFERENCES

Almendares, J., and M. Sierra. 1993. "Critical Conditions: A Profile of Honduras." *Lancet* 342 (4 December): 1400–1403.

Amazon Destruction More Rapid Than Expected. 1999. GS Report, 10 April. [http://www.gsreport.com/articles/art000099.html]

"Antarctic Meltdown." 1998. *World Press Review*, November, 36.

Barber, Valerie A., Glen Patrick Juday, and Bruce P. Finney. 2000. "Reduced Growth of Alaskan White Spruce in the Twentieth Century from Temperature-induced Drought Stress." *Nature* 405 (8 June): 668–673.

Bates, Albert K., and Project Plenty. 1990. *Climate in Crisis: the Greenhouse Effect and What We Can Do*. Summertown, Tenn.: The Book Publishing Co.

Bazzaz, Fakhri A., and Eric D. Fajer. 1992. "Plant Life in a CO2-rich World." *Scientific American*, January, 68–74.

Benjamin, Craig. 1999. "The Machu Picchu Model: Climate Change and Agricultural Diversity." *Native Americas* 16, no. 3/4(Summer/Fall): 76–81.

Bernard, Harold W., Jr. 1993. *Global Warming: Signs to Watch For*. Bloomington: Indiana University Press.

Bhattacharya, N.C. 1993. "Prospects of Agriculture in a Carbon-Dioxide-Enriched Environment." In Richard A. Geyer, ed., *A Global Warming Forum: Scientific, Economic, and Legal Overview*. Boca Raton, Fl.: CRC Press, 487–505.

Billings, W.D. 1992. "Phytogeographic and Evolutionary Potential of the Arctic Flora and Vegetation in a Changing Climate." In F. Stuart Chapin III, Robert L. Jefferies, James F. Reynolds, Gaius R. Shaver, and Josef Svoboda, eds., *Arctic Ecosystems in a Changing Climate: An Ecophysiological Perspective.* San Diego: Academic Press, 91–109.

Brown, Paul. 1999. "Global Warming: Worse Than We Thought." *World Press Review,* February, 44.

———. 2000. "Overfishing and Global Warming Land Cod on Endangered List." *The Guardian* (London), 20 July, 3.

Buxton, James. 2000. "Suspects in the Mystery of Scotland's Vanishing Salmon: Fish Farms, Seals and Global Warming are All Blamed for What Some See as a Crisis." *Financial Times* (London), 13 June, 11.

Cline, William R. 1992. *The Economics of Global Warming.* Washington, D.C.: Institute for International Economics.

Clover, Charles. 2000. "Thousands of Species 'Threatened by Warming.' " *London Daily Telegraph,* 31 August, 9.

Commoner, Barry. 1990. *Making Peace with the Planet.* New York: Pantheon.

Corson, Walter H., ed. 1990. *The Global Ecology Handbook: What You Can Do About the Environmental Crisis.* Washington, D.C.: The Global Tomorrow Coalition.

Devitt, Terry. 1998. "Landscape Changes Seen As Bad As Greenhouse Gases." *Uniscience Research News,* 9 December. [http://benetton.dkrz.de:3688/homepages/georg/kimo/0254.html]

Emanuel, W.R., H.H. Shugart, and M.P. Stevenson. 1985. "Climatic Change and the Broad-scale Distribution of Terrestrial Eco-system Complexes." *Climatic Change* 7:29–43.

Environment Canada. 1994. *Potential Impacts of Global Warming on Salmon Production in the Fraser River Watershed.* Ottawa: Environment Canada.

Epstein, Paul, Georg Grabbher, Tom Karl, Ellen Mosley-Thompson, Kevin Trenberth, and George M. Woodwell. 1996. Current Effects: Global Climate Change. An Ozone Action Roundtable, 24 June, Washington, D.C. [http://www.ozone.org/curreff.html]

Fankhauser, Samuel. 1995. *Valuing Climate Change: The Economics of the Greenhouse.* London: Earthscan Publications, Ltd.

Gelbspan, Ross. 1997. *The Heat is On: The High Stakes Battle Over Earth's Threatened Climate.* Reading, Mass.: Addison-Wesley Publishing Co.

Glick, Patricia. 1998. "Global Warming: The High Costs of Inaction." San Francisco: Sierra Club. [http://www.sierraclub.org/global-warming/inaction.html]

Gordon, Anita, and David Suzuki. 1991. *It's a Matter of Survival.* Cambridge: Harvard University Press.

Gough, Robert. 1999. "Stress on Stress: Global Warming and Aquatic Resource Depletion." *Native Americas* 16, no. 3/4(Fall/Winter): 46–48. [http://nativeamericas.aip.cornell.edu]

Greening the Planet. 1998. Climate Change Debate, 8 December. [http://climatechange-debate.com/archive/12-08_12-15_1998.txt]

Houghton, R.A., D.L. Skole, Carlos A. Nobre, J.L. Hackler, K.T. Lawrence, and W.H. Chomentowski. 2000. "Annual Fluxes of Carbon From Deforestation and Regrowth in the Barazilian Amazon." *Nature* 403 (20 January): 301–304.

Izrael, Yu A. 1991. "Climate Change Impact Studies: The IPCC Working Group II Report." In J. Jager and H.L. Ferguson, *Climate Change: Science, Impacts, and*

Policy. Proceedings of the Second World Climate Conference. Cambridge: Cambridge University Press, 83–86.

Jardine, Kevin. 1994. "The Carbon Bomb: Climate Change and the Fate of the Northern Boreal Forests." Ontario, Canada: Greenpeace International. [http://dieoff.org/page129.html]

Karl, Thomas R., Neville Nicholls, and Jonathan Gregory. 1997. "The Coming Climate: Meteorological Records and Computer Models Permit Insights Into Some of the Broad Weather Patterns of a Warmer World." *Scientific American* 276:79–83. [http://www.scientificamerican.com/0597issue/0597karl.html]

Kruse, Gordon H. 1998. "Salmon Run Failures in 1997–1998: A Link to Anomalous Oceanic Conditions?" *Alaska Fishery Research Bulletin* 5, no. 1: 55–63.

Kurz, Werner A., Michael J. Apps, Brian J. Stocks, and Jan A. Volney. 1995. "Global Climate Change: Disturbance Regimes and Biospheric Feedbacks of Temperate and Boreal Forests." In George M. Woodwell and Fred T. MacKenzie, eds., *Biotic Feedbacks in the Global Climate System: Will the Warming Feed the Warming?* New York: Oxford University Press, 119–133.

Lindroth, A., A. Grelle, and A.S. Moren. 1998. "Long-term Measurements of Boreal Forest Carbon Balance Reveal Large Temperature Sensitivity." *Global Change Biology* 4 (April): 443–450.

Malcolm, Jay R., and Adam Markham. 2000. *Global Warming and Terrestrial Biodiversity Decline*. Washington, D.C.: World Wildlife Fund. [http://www.library.adelaide.edu.au/cgi-bin/director?id=1379565]

Manning, Anita. 2000. "Ragweed Warms to Climate Change; Increasing CO2 has Doubled Pollen, Misery." *USA Today*, 15 August, 8 D.

Martens, Pim. 1999. "How Will Climate Change Affect Human Health?" *American Scientist* 87, no. 6 (November/December): 534–541.

Mathews-Amos, Amy, and Ewann A. Berntson. 1999. "Turning up the Heat: How Global Warming Threatens Life in the Sea." World Wildlife Fund and Marine Conservation Biology Institute. [http://www.worldwildlife.org/news/pubs/wwf_ocean.htm]

McKibben, Bill. 1989. *The End of Nature*. New York: Random House.

"Much Deforestation Driven By Population, Poverty." 1999. *Global Futures Bulletin* 84 (15 May). [http://www.gsreport.com/articles/art000149.html]

National Academy of Sciences. 1991. *Policy Implications of Greenhouse Warming*. Washington, D.C.: National Academy Press.

Nepstad, Daniel C. 1998. "Report from the Amazon: May, 1998." Woods Hole Research Center. [http://terra.whrc.org/science/tropfor/fire/report2.htm]

"New England May Lose Sugar Maples to Global Warming." 1998. Reuters. [http://bonanza.lter.uaf.edu/~davev/nrm304/glbxnews.htm]

Nuttall, Nick. 1998. "Global Warming 'Will Turn Rainforests into Deserts.' " *London Times*, 3 November. [http://bonanza.lter.uaf.edu/~davev/nrm304/glbxnews.htm]

———. 2000. "Climate Change Lures Butterflies Here Early." *London Times*, 24 May.

Parmesan, Camille, Nils Ryrholm, Constanti Stefanescu, Jane K. Hill, Chris D. Thomas, Henri Descimon, Brian Huntley, Lauri Kaila, Jaakko Kulberg, Toomas Tammaru, W. John Tennent, Jeremy A. Thomas, and Martin Warren. 1999. "Poleward Shifts in Geographical Ranges of Butterfly Species Associated with Regional Warming." *Nature* 399 (10 June): 579–583.

Parry, Martin. 1990. *Climate Change and World Agriculture*. London: Earthscan.

Parry, Martin, and Zhang Jiachen. 1991. "The Potential Effect of Climate Changes on Agriculture." In J. Jager and H.L. Ferguson, *Climate Change: Science, Impacts, and Policy*. Proceedings of the Second World Climate Conference. Cambridge: Cambridge University Press, 279–289.

Peters, Robert L. 1989. "Effects of Global Warming on Biological Diversity." In Edwin Abrahamson, ed., *The Challenge of Global Warming*. Washington, D.C.: Island Press, 82–95.

Quammen, David. 1998. "Planet of Weeds: Tallying the Losses of Earth's Animals and Plants." *Harpers*, October, 57–69.

Rolfe, Christopher. 1996. "Comments on the British Columbia Greenhouse Gas Action Plan." A Presentation to the Air and Water Management Association, 17 April, West Coast Environmental Law Association. [http://www.wcel.org/wcelpub/11026.html]

Rosenburg, Norman J., Pierre R. Crossman, William E. Easterling III, Mary S. McKenney, Kenneth D. Frederick, and Michael Bowes. 1992. "Methodology for Assessing Regional Economic Impacts of and Responses to Climate Change: The MINK [Missouri-Iowa-Nebraska-Kansas] Study." In Jurgan Schmandt and Judith Clarkson, *The Regions and Global Warming: Impacts and Response Strategies*. New York: Oxford University Press, 132–153.

Schemo, Diana Jean. 1998. "Amazon Fires Threaten Intact Forest, Indigenous People." *New York Times*. In *The Age* (Melbourne, Australia), 14 September.

Schneider, Stephen H. 2001. "No Therapy for the Earth: When Personal Denial Goes Global." In Michael Aleksiuk and Thomas Nelson, eds., *Nature, Environment & Me: Explorations of Self In A Deteriorating World*. Montreal: McGill-Queens University Press.

Schwartzman, Stephen. 1999. "Reigniting the Rainforest: Fires, Development and Deforestation." *Native Americas* 16, no. 3/4(Fall/Winter): 60–63.

Stevens, William K. 1999. *The Change in the Weather: People, Weather, and the Science of Climate*. New York: Delacorte Press.

Suffling, R. 1992. "Climate Change and Boreal Forest Fires in Fennoscandia and Central Canada." In M. Boer and E. Koster, eds., *Greenhouse Impact on Cold-climate Ecosystems and Landscapes*. Selected Papers of a European Conference on Landscape Ecological Impact: Impact of Climatic Change, Lunteren, the Netherlands, December 3–7, 1989. CARTENA Supplement 22. Cremlingen, Germany: Cartena Verlag, 111–132.

Thomas, Chris D., and Jack J. Lennon. 1999. "Birds Extend Their Ranges Northwards." *Nature* 399 (20 May): 213.

Thomas, William. 1998. "Salmon Dying in Hot Waters." ENS [Environmental News Service], 22 September. [http://www.econet.apc.org/igc/en/hl/9809244985/hl11.html]

U.S. Environmental Protection Agency. 1995. *Ecological Impacts From Climate Change: An Economic Analysis of Freshwater Recreational Fishing*. Washington, D.C.: EPA.

Vincent, John, and Paul Brown. 2000. "Swoop to Conquer: Global Warming Brings Butterflies to Britain Earlier." *The Guardian* (London), 24 May, Home Pages, 9.

Vrolijk, Christiaan. 1999. "The Buenos Aries Climate Conference: Outcome and Implications." The Royal Insitute of International Affairs. 2 April. [http://www.riia.org/briefingpapers/bp53.html]

Webb, Jason. 1998. "World Forests Said Vulnerable to Global Warming." Reuters. 4 November. [http://bonanza.lter.uaf.edu/~davev/nrm304/glbxnews.htm]

Weinburg, Bill. 1999. "Hurricane Mitch, Indigenous Peoples and Mesoamerica's Climate Disaster." *Native Americas* 16, no. 3/4(Fall/Winter): 50–59. [http://nativeamericas.aip.cornell.edu/fall99/fall99weinberg.html]

Wuethrich, Bernice. 2000. "How Climate Change Alters Rhythms of the Wild." *Science* 287 (4 February): 793–795.

CHAPTER 7

Human Health

When climate-change scientists and diplomats met in Buenos Aires during 1998, they were greeted by news that mosquitoes carrying dengue fever had invaded more than a third of the homes in Argentina's most populous province, with 14 million people. The *aedes aegypti* mosquito appeared in Argentina in 1986; within 12 years, it was found in 36 percent of homes in Buenos Aires province, according to Dr. Alfredo Seijo of the Hospital Munoz. "*Aedes aegypti* now exists from the south of the United States as far as the south of Buenos Aires province and this is obviously due to climatic changes which have taken place in Latin America over the past few years," Seijo told a news conference organized by the World Wildlife Fund at the United Nations climate talks in Argentina (Webb 1998). Dengue, a common disease in tropical regions, is a prolonged, flu-like viral infection which can cause internal bleeding, fever, and sometimes death. Dengue, which is sometimes called "breakbone fever," may be accompanied by headache, rash, and severe joint pain. The World Health Organization lists dengue fever as the tenth deadliest disease worldwide.

During 1995, an explosion of termites, mosquitoes, and cockroaches hit New Orleans, following an unprecedented five years without a killing frost. "Termites are everywhere. The city is totally, completely, inundated with them," said Ed Bordees, a New Orleans health official, who added, "The number of mosquitoes laying eggs has increased tenfold" (Gelbspan, *The Heat is On*, 1997, 15). The situation in New Orleans was aggravated not only by unusual warmth, but also by above-average rainfall totaling about 80 inches the previous year. Some of the 200-year-old oaks along New Orleans' St. Charles Avenue were found to have been eaten alive from the inside by billions of tiny, blind, Formosan termites. The same year, dengue fever spread from Mexico across the border into Texas for the first time since records have been kept. Dengue fever, like malaria, is carried by a mosquito that is limited by temperature. At the same time, Colombia was experiencing plagues of mosquitoes and outbreaks of the diseases they carry, including dengue fever and encephalitis, triggered by a record heat wave followed by heavy rains.

Mild winters with a lack of freezing conditions allow many disease-carrying insects to expand their ranges. "Indeed," comments Harvard University epidemiologist Paul Epstein, "fossil records indicate that when changes in climate occur, insects shift their range far more rapidly than do grasses, shrubs, and forests, and move to more favorable latitudes and elevations hundreds of years before larger animals do. 'Beetles,' concluded one climatologist, 'are better paleo-thermometers than bears' " (Epstein 1998).

GLOBAL WARMING AND DISEASE

John T. Houghton, author of *Global Warming: The Complete Briefing*, believes that global warming will accelerate the spread of many diseases from the tropics to the middle latitudes. Malaria could increase from its present level, Houghton warns. "Other diseases which are likely to spread for the same reason are yellow fever, dengue fever, and . . . viral encephalitis," he wrote (Houghton 1997, 132). After 1980, small outbreaks of locally-transmitted malaria occurred in Texas, Georgia, Florida, Michigan, New Jersey, New York, and California, usually during hot, wet spells. Worldwide, according to Epstein, as many as 500 million people contract malaria every year, and between 1.5 and 3 million die, mostly children. Mosquitoes and parasites that carry the disease have evolved immunities to many drugs.

Epstein stated, "If tropical weather is expanding it means that tropical diseases will expand. We're seeing malaria in Houston, Texas" (Glick 1998). Epstein suggests that a resurgence of infectious disease may be one result of global warming. Warming may appear beneficial at first, Epstein says. Initially, plants may be fertilized by warmth and moisture, an earlier spring, and more carbon dioxide and nitrogen. "But," he cautions, "warming and increased CO2 can also stimulate microbes and their carriers" (Epstein 1998).

Since 1976, Epstein reports, 30 diseases have emerged which are new to medicine. Old ones, such as drug-resistant tuberculosis, have been given new life by new diseases (such as HIV/AIDS) which compromise the human immune system. By 1998, tuberculosis was claiming three million lives annually around the world. "Malaria, dengue, yellow fever, cholera, and a number of rodent-borne viruses are also appearing with increased frequency," Epstein reports (Epstein 1998). During 1995, mortality from infectious diseases attributed to causes other than HIV/AIDS rose 22 percent above the levels of 15 years before in the United States. Adding deaths complicated by HIV and AIDS, deaths from infectious diseases have risen 58 percent in 15 years (Epstein 1998).

Dengue fever is one of a number of mosquito-vector diseases which have been increasing their coverage in many areas of the Earth—climbing in altitude in the tropics and rising in latitude in temperate zones—as global temperatures have warmed during the last quarter of the twentieth century. Rising temperatures and humidities increase the range of many illnesses spread by insects, including mosquitoes, warm-weather insects which die at temperatures below a range of 50 to 61 degrees F., depending on species.

In the tropics, elevation has long been used to shield human populations from diseases that are endemic in the lowlands. With global warming, mosquito-borne diseases are expected to reach higher altitudes, affecting peoples with little or no immunity. Writes Pim Martens of the Netherlands' Center for Integrative Studies (and a senior scientist at the University of Maastricht): "A minor temperature rise will be sufficient to turn the populated African highlands into an area that is suitable for the malaria mosquito and parasite" (Martens 1999, 537).

During 1997, a large outbreak of malaria ravaged large areas of Papua, New Guinea, at an elevation of 2,100 meters, notably higher than the 1,200 to 2,000 meters which had heretofore provided a barrier to the disease in different parts of Central and Southern Africa. In northwestern Pakistan, according to Martens, a rise of about half a degree C. in the mean temperature was a factor in a rising incidence of malaria there, from a few hundred cases a year in the early 1980s to 25,000 in 1990 (Martens 1999, 537). While most strains of malaria can be controlled, drug-resistant strains were proliferating late in the twentieth century.

Martens writes that while the overall impact of global warming on human health is expected to be markedly negative, human beings may experience a few positive outcomes. Some diseases that thrive in cold weather (such as influenza) may find their ranges reduced in a warmer world. The elderly might die less frequently of cardiovascular and pulmonary ailments which peak during cold weather. "Whether the milder winters could offset the mortality during the summer heat waves is one of the questions that demands further research," Martens writes (Martens 1999, 535).

Dengue fever, for which no vaccine exists, had nearly disappeared from the Americas by the 1970s. During the 1980s, however, the disease increased dramatically in South America, infecting over 300,000 people there by 1995. Also during 1995, Peru and the Amazon Valley were especially hard hit by the area's largest epidemic of yellow fever since 1950, which is carried by the same mosquito that transmits dengue fever. The annual world incidence of dengue fever, which averaged about 100,000 cases between 1981 and 1985, averaged 450,000 cases a year between 1986 and 1990 (Gelbspan, *The Heat is On*, 1997, 149).

Writing in the *Bulletin of the American Meteorological Society*, Paul R. Epstein and seven co-authors described the spread of malaria and dengue fever to higher altitudes in tropical areas of the Earth because of warmer temperatures. Rising winter temperatures have allowed disease-bearing insects to survive in areas previously closed to them. According to Epstein, frequent flooding which is associated with warmer temperatures also promotes the growth of fungus and provides excellent breeding grounds for large numbers of mosquitoes. The flooding caused by Hurricane Floyd and other storms in North Carolina during 1999 are cited by some as a real-world example of global warming promoting conditions ideal for the spread of diseases imported from the tropics (Epstein et al. 1998).

Countering the views of Epstein and others, some health researchers contend that global warming will do little to increase incidence of tropical diseases. "For mosquito-borne diseases such as dengue, yellow fever, and malaria, the assumption that warming will foster the spread of the vector is simplistic," contends Bob Zimmerman, an entomologist with the Pan American Health Organization (PAHO). Zimmerman points out that in the Amazon basin, over 20 species of *Anopheles* mosquitoes can transmit malaria, and all are adapted to different habitats: "All of these are going to be impacted by rainfall, temperature, and humidity in different ways. There could actually be decreases in malaria in certain regions, depending on what happens" (Taubes 1997). Virologist Barry Beaty of Colorado State University in Fort Collins, Colorado, agrees with Zimmerman: "You don't have to be a rocket scientist to say we've got a problem," he says. "But global warming is not the current problem. It is a collapse in public-health measures, an increase in drug resistance in parasites, and an increase in pesticide resistance in vector populations. Mosquitoes and parasites are efficiently exploiting these problems" (Taubes 1997).

Countering the majority view that a warmer world will spread malaria, David J. Rogers and Sarah E. Randolph, using their own models, wrote in *Science* that even extreme rises in temperature will not spread the disease. They argue that the spread of malaria is too poorly understood to base a forecast several decades into the future on one variable, temperature. For example, the "Dengie marshes" of Essex, in England, a breeding ground for malaria-carrying mosquitoes in the seventeenth century, have dried up, making an increase in temperatures not a factor vis-à-vis malaria's transmission (Rogers and Randolph 2000).

According to the Intergovernmental Panel on Climate Change's projections for human health, a rise in global average temperatures of three to five degrees C. by 2100 could lead to 50 to 80 million additional annual cases of malaria worldwide, "primarily in tropical, subtropical and less well-protected temperate-zone populations" (Bolin et al. 1995). Italy experienced a brief outbreak of malaria during 1997. Hadley Climate Center researchers expect the same disease to reach the Baltic states by 2050. In parts of the world where malaria is now unknown most people have no immunity (Brown 1999). The World Health Organization projects that warmer weather will cause tens of millions of additional cases of malaria and other infectious diseases. The Dutch health ministry anticipates that more than a million people may die annually as a result of the impact of global warming on malaria transmission in North America and Northern Europe (Epstein, Profound Consequences, 1999, 7).

The United Nations' Intergovernmental Panel on Climate Change (IPCC) included a chapter on public health in an update of its 1990 assessment. The IPCC's public-health chapter concluded that "climate change is likely to have wide-ranging and mostly adverse impacts on human health, with significant loss of life" (Taubes 1997).

One observer asserts,

Although winter bronchitis and pneumonia may be reduced [by global warming], it is quite likely that hay fever and perhaps asthma could increase. A combination of increase in temperature with increasing levels of tropospheric ozone could have clinically important effects, particularly in patients with asthma and chronic obstructive airways disease. (Haines 1990, 154)

Malaria is not a new disease in the temperate zones. It was common in the Roman Empire. A British invasion of Holland in 1806 failed to drive out French troops because large numbers of the Britons became ill with malaria. Malaria was a public health problem in most of the Eastern United States during warm, humid summers before medications were developed for it about a century ago.

By the late 1990s, malaria had been transmitted by mosquitoes as far north as Toronto, Canada, according to Epstein. "The extreme events we are seeing today in Nicaragua and Honduras [as a result of Hurricane Mitch in 1998] are spawning outbreaks of cholera and dengue fever with new breeding sites for mosquitoes and increased water-born diseases," Epstein said (Webb 1998).

Further north, nighttime and winter temperatures have warmed twice as fast as overall global temperatures since 1950, Epstein said, meaning that fewer pests are being killed by frost in the southern reaches of the temperate zones. Humidity also has increased in many regions, including much of the eastern United States, helping mosquitoes to breed. Disease-carrying mosquitoes usually require a certain level of temperature *and* humidity to survive.

In May, 1995, researchers in the Netherlands and in England estimated the increase in malaria's geographic range which could occur if the IPCC's projections for global warming prove correct. These researchers concluded that, in tropical regions, the epidemic potential of the mosquito population would double. In temperate climates, according to these projections, the epidemic potential could increase a hundred-fold. Furthermore, this study said, "There is a real risk of reintroducing malaria into non-malarial areas, including parts of Australia, the United States, and southern Europe." All told, the study estimated that a temperature rise of 5.4 degrees F. (3 degrees C.) during the twenty-first century could cause an additional 50 million to 80 million new cases of malaria each year worldwide (Rachel's 1995).

By the late 1970s, dwindling investments in public health programs, growing insecticide resistance, and prevalent environmental changes (such as deforestation) contributed to a widespread resurgence of malaria, according to Epstein. By the late 1980s, he reports, large epidemics of malaria were being associated with warm, wet weather. Between 1993 and 1998, adds Epstein, worldwide incidence of malaria quadrupled.

Malaria is now found in higher-elevations in central Africa and could threaten cities such as Nairobi, Kenya (at about 5,000 feet, roughly the elevation of Denver, Colorado), as freezing levels have shifted higher in the mountains. In the summer of 1997, for example, malaria took the lives of hundreds of people in the Kenyan highlands, where populations had previously been unexposed. (Epstein 1998)

Between 1970 and 1995, the lowest level at which freezing occurs has climbed about 160 meters higher in mountain ranges from 30 degrees north to 30 degrees south latitude, based on radiosonde data analyzed at the National Oceanic and Atmospheric Administration's Environmental Research Laboratory. This shift corresponds to a warming at these elevations of about one degree C. (almost two degrees F.), which is nearly twice the average warming that has been documented over the Earth as a whole (Epstein 1998). As higher elevations warm, mosquito-vector diseases are ascending tropical mountainsides around the world. Bill Weinburg, in *Native Americas*, describes changes provoked by warming in the mountains of Mexico:

Dr. Juan Blechen Nieto, a Cuernavaca physician, traveled through the Sierra del Sur on a survey of local health conditions in November 1998, and found an alarming incidence of dengue fever and malaria. "These are diseases that are traditionally associated with lowland coastal regions, and are now appearing in the Sierra del Sur," he told me. . . . "Indians in highland Oaxaca communities tell me they have mosquitoes now, for the first time. This has to do with deforestation impacting local and regional climate. It gets hotter, and the undergrowth that comes up after forests are destroyed provides a habitat for pests." (Weinburg 1999, 58–59)

Epstein identifies three tendencies in global climate change, and relates each of them to an increasingly virile environment for infectious diseases. The three indicators are

(1) Increased air temperatures at altitudes of two to four miles above the surface in the Southern Hemisphere;

(2) A disproportionate rise in minimum temperatures, in either daily or seasonally-averaged readings;

(3) An increase in extreme weather events, such as droughts and sudden heavy rains. (Epstein 1998)

"There is growing evidence for all three of these tell-tale 'fingerprints' of enhanced greenhouse warming," says Epstein (Epstein 1998).

A Sierra Club study indicates that a lengthy El Niño event during the middle 1990s provided an indication of how sensitive some diseases can be to changes in climate. The Sierra Club study cites evidence that warming waters in the Pacific Ocean contributed to a severe outbreak of cholera which led to thousands of deaths in Latin American countries during the 1990s. According to health experts quoted by the Sierra Club study, "The current outbreak [of dengue fever], with its proximity to Texas, is at least a reminder of the risks that a warming climate might pose" (Sierra Club 1999). The Sierra Club study concludes, "While it is difficult to prove that any particular outbreak was caused or exacerbated by global warming, such incidents provide a hint of what might occur as global warming escalates" (Sierra Club 1999).

Willem Martens and colleagues writing in *Climatic Change*, attempt to sketch how a warmer, wetter climate would affect transmission of three vector-borne diseases: malaria, schistosomiasis, and dengue fever. Martens anticipates that "the periphery of the currently endemic areas" will expand with global warming, with diseases notable at higher elevations in the tropics, an expectation that has been borne out by several observers. Martens and colleagues expect, "The increase in epidemic potential of malaria and dengue transmission may be estimated at 12 to 27 percent and 31 to 47 percent respectively." In contrast, they forecast that the transmission potential of schistosomiasis may decrease 11 to 17 percent (Martens et al. 1997, 145).

The incidence of eastern equine encephalitis, which attacks both horses and humans, has been increasing in parts of the United States, although transmission to humans is still considered rare. Prince Georges County, adjacent to Washington, D.C., reported five cases in 1996, with the origins of contraction unknown (Bloomfield and Showell 1997). Early symptoms include fever, headache, drowsiness, and muscle pain, followed by disorientation, weakness, seizures, and coma. Sixty percent of cases are fatal, and most survivors suffer permanent neurological damage. Mild winters and wet springs, which are expected to become more likely with global warming, are associated with increased risk of eastern equine encephalitis.

Another disease which may become more common in the temperate zones of a warmer world is diarrhea, which during the 1990s was killing more than three million children a year worldwide, mainly in the tropics of Asia, Africa, and the Americas. The bacteria which cause diarrhea thrive in warm weather, especially after heavy rainfall. Warm, moist weather also promotes the growth and activity of flies and cockroaches.

The effects of warming-induced disease extend to the oceans. Comments Epstein,

Warming—when sufficient nutrients are present—may also be contributing to the proliferation of coastal algal blooms. Harmful algal blooms of increasing extent, duration, and intensity—and involving novel, toxic species—have been reported around the world since the 1970s. Indeed, some scientists feel that the worldwide increase in coastal algal blooms may be one of the first biological signs of global environmental change. (Epstein 1998)

TEMPERATURE, HUMIDITY, AIR POLLUTION, AND HUMAN MALADIES

Aside from insect-vector diseases, warmer, more humid weather may aggravate urban air pollution (especially tropospheric ozone) which is a factor in many human maladies, such as asthma. Pim Martens points to a number of studies which indicate that air pollution's effect on human health rises as temperatures go up: "Simultaneous exposure to heat and pollution appears too be more harmful than the sum of the individual effects" (Martens 1999, 535). Allergenic pollens and spores are more readily dispersed during hot, dry summers, Martens says.

Asthma is aggravated by heat (which increases the pollen production of many plants) as well as air pollution. According to the American Lung Association, more than 5,600 people died of asthma in the United States during 1995, a 45.3 percent increase in mortality over ten years, and a 75 percent increase since 1980. Roughly a third of those cases occurred in children under the age of 18. Since 1980, children under age five have experienced a 160 percent increase in asthma.

"What happens to asthmatics in the heat and humidity? Well, we can't breathe," says Barbara Mann, an author, teacher, and asthmatic who lives in Toledo, Ohio. "I'm permanently on three different prescription inhalers that I must use at regular intervals, four times a day—and that's just when I'm well. When I'm sick, there are antibiotics to bust up the hardened mucus in my lungs, along with other regimens of pills to aid the process. When my lungs are irritated by pollutants or by natural 'triggers' such as pollens or infected, my air passages swell shut, suffocating me" (Mann 1999).

A study by the Sierra Club indicated that air pollution which will be enhanced by global warming could be responsible for a number of human health problems, including respiratory diseases such as asthma, bronchitis, and pneumonia. According to Dr. Joel Schwartz, an epidemiologist at Harvard University, current air pollution concentrations are responsible for 70,000 early deaths per year and more than 100,000 excess hospitalizations for heart and lung disease in the United States. This could increase 10 to 20 percent in the United States as a result of global warming, with significantly greater increases in countries that are more polluted to begin with, according to Schwartz (Sierra Club 1999).

In addition to aggravating specific diseases, persistent heat and humidity has a general debilitating influence on most warm-blooded animals, including human beings. Farm animals are especially adversely affected when the air temperature remains higher than usual throughout the night. During the latter half of the twentieth century, according to Epstein, nighttime minimum temperatures over land areas have risen 1.86 degrees C., while maximum temperatures have risen 0.88 degrees C. (Epstein 1998).

MORTALITY FROM HEAT WAVES

Laurence S. Kalkstein has estimated that a doubling of the carbon dioxide level in the atmosphere could increase heat-related mortality to seven times present levels if acclimatization is not factored in. With acclimatization (human adaptation to higher temperatures), the estimated increase in heat wave mortality estimated by Kalkstein is four times the present rate (Kalkstein 1993, 1397). Kalkstein observes that each urban area has its own "temperature threshold," at which the death rate from heat prostration rises rapidly. Seattle, for example, has a lower threshold than Dallas. "Mortality rates in warmer cities seemed to be less affected no matter how high the temperature rose," Kalkstein (1993, 1398) wrote. He suggests that residents of urban areas in poor countries will find adaptation more difficult because of limited access to air conditioning. A 1988 Environmental Protection Agency study estimates that heat wave mortality in 15 large U.S. cities would rise from 1,200 a year to 7,500 a year if carbon-dioxide levels double (Schneider 1989, 182). Another EPA report indicated in the late 1980s that a four degree C. rise in San Francisco's temperature would raise ozone levels there by about 20 percent, with attendant health effects (Schneider 1989, 183).

According to Anthony J. McMichael, Professor of Epidemiology at the London School of Hygiene and Tropical Medicine, higher summer temperatures in both temperate and tropical regions could increase the rates of serious illness and death from heat-related causes by as much as six times the current level, with the greatest impact falling on the sick and elderly (McMichael 1993, 143).

The urban heat-island effect was first identified by a meteorologist, Luke Howard, in 1818. Extra heat is produced in urban areas by a city's many sources of waste heat, from building heating and air conditioning, as well as from motor vehicles, among other sources. Heating also increases when open fields and forests become streets, sidewalks, parking lots, and buildings. The dark colors of city structures, especially asphalt streets and parking lots which may make up as much as 30 percent of many urban surfaces, have a very low albedo (reflectivity), so most of the sun's heat energy is absorbed, not reflected.

Cities also warm more rapidly than surrounding countryside because they are usually drier and have less surface water and plant mass (both of which cool the air through evaporation) than most rural areas. Furthermore, as new housing and businesses spread from urban areas, some of the cities' urban heat follows with them, spreading in widening suburban circles. In Japan, suburban areas near Tokyo have seen temperatures rise between two to three degrees C. in 10 years, following urbanization. In a compact urban area such as Manhattan Island, the total heat generation of the city can add quite substantially to solar radiation. By one estimate, the heat energy generated by motor vehicles and space heat in Manhattan during an average winter day sometimes exceeds that of incoming solar radiation (Weiner 1990, 262).

Since World War II, when many Phoenix residents slept on porches outdoors, average summertime lows in Phoenix have risen above the human comfort zone. Average summertime lows have risen from 73 degrees to more than 80 degrees F. during the last half of the twentieth century. During the same fifty years, the Phoenix area's human population has increased nearly twenty times, from roughly 150,000 to 2.8 million. Average daytime summer high temperatures in Phoenix have remained roughly the same during the same half-century, at between 102 and 104 degrees F. Dale Quattrochi, senior research scientist at NASA's Global Hydrology and Climate Center in Huntsville, Alabama, estimated that Phoenix temperatures likely will increase as much as 15 degrees, and possibly up to 20 degrees, over historic averages the next several decades (Yozwiak 1998). (For more information on urban heat islands, see Chapter 2.)

REFERENCES

Bloomfield, Janine, and Sherry Showell. 1997. *Global Warming: Our Nation's Capital at Risk*. Washington, D.C.: Environmental Defense Fund. [http://www.edf.org/pubs/Reports/WashingtonGW/index.html]

Bolin, Bert, et al. 1995. *Intergovernmental Panel on Climate Change. Second Assessment Synthesis of Scientific-Technical Information Relevant to Interpreting Article 2 of the United Nations Framework Convention on Climate Change*. Approved by the IPCC at its eleventh session, 11–15 December, Rome. [http://www.unep.ch/ipcc/pub/sarsyn.htm].

Brown, Paul. 1999. "Global Warming: Worse Than We Thought." *World Press Review*, February, 44.

Epstein, Paul R. 1998. "Climate, Ecology, and Human Health." 18 December. [http://www.iitap.iastate.edu/ gccourse/ issues/ health/ health.html]

———. 1999. "Profound Consequences: Climate Disruption, Contagious Disease, and Public Health." *Native Americas* 16, no. 3/4(Fall, Winter): 64–67.

Epstein, Paul R., Henry F. Diaz, Scott Elias, Georg Grabherr, Nicohlas E. Graham, Willem J.M. Martens, Ellen Mosley-Thompson, and Joel Susskind. 1998. "Biological and Physical Signs of Climate Change: Focus on Mosquito-borne Diseases." *Bulletin of the American Meteorological Society* 79, part 1: 409–417.

Gelbspan, Ross. 1997. *The Heat is On: The High Stakes Battle Over Earth's Threatened Climate*. Reading, Mass.: Addison-Wesley Publishing Co.

Glick, Patricia. 1998. "Global Warming: The High Costs of Inaction." San Francisco: Sierra Club. [http://www.sierraclub.org/global-warming/inaction.html]

Haines, Andrew. 1990. "The Implications for Health." In Jeremy Leggett, ed., *Global Warming: The Greenpeace Report*. New York: Oxford University Press, 149–162.

Houghton, John. 1997. *Global Warming: The Complete Briefing*. Cambridge, England: Cambridge University Press.

Kalkstein, Laurence S. 1993. "Direct Impacts in Cities." *Lancet* 342 (4 December): 1397–1400.

Mann, Barbara. 1999. Personal communication, 3 August.

Martens, Pim. 1999. "How Will Climate Change Affect Human Health?" *American Scientist* 87, no. 6 (November/December): 534–541.

Martens, Willem J.M., Theo H. Jetten, and Dana A. Focks. 1997. "Sensitivity of Malaria, Schistosomiasis, and Dengue to Global Warming." *Climatic Change* 35:145–156.

McMichael, A. J. 1993. *Planetary Overload: Global Environmental Change and the Health of the Human Species.* Cambridge: Cambridge University Press.

"Rachel's #466: Warming & Infectious Diseases." 1995. *Rachel's Environment and Health Weekly*, 2 November. Annapolis, Md.: Environmental Research Foundation. [http://www.igc.apc.org/awea/wew/othersources/rachel466.html]

Reany, Patricia. 1997. " 'Millions Will Die' Unless Climate Policies Change." Reuters, 6 November. [http://benetton.dkrz.de:3688/homepages/georg/kimo/0254.html]

Rogers, David J., and Sarah E. Randolph. 2000. "The Global Spread of Malaria in a Future, Warmer World." *Science* 289 (8 September): 1763–1766.

Schneider, Stephen H. 1989. *Global Warming: Are We Entering the Greenhouse Century?* San Francisco: Sierra Club Books.

Sierra Club. 1999. "Global Warming: The High Costs of Inaction." [http://www.sierraclub.org/global-warming/resources/innactio.htm]

Taubes, Gary. 1997. "Apocalypse Not." [http://www.junkscience.com/news/taubes2.html]

Webb, Jason. 1998. "Mosquito Invasion as Argentina Warms." Reuters. [http://bonanza.lter.uaf.edu/~davev/nrm304/glbxnews.htm]

Weinburg, Bill. 1999. "Hurricane Mitch, Indigenous Peoples and Mesoamerica's Climate Disaster." *Native Americas*, 16, no. 3/4(Fall/Winter): 50–59. [http://nativeamericas.aip.cornell.edu/fall99/fall99weinberg.html]

Weiner, Jonathan. 1990. *The Next One Hundred Years: Shaping the Fate of Our Living Earth.* New York: Bantam Books.

Yozwiak, Steve. 1998. " 'Island' Sizzle, Growth May Make Valley an increasingly Hot Spot." *The Arizona Republic* (Phoenix), 25 September. [http://www.sepp.org/reality/arizrepub.html]

CHAPTER 8

A Fact of Daily Life: Global Warming and Indigenous Peoples

For indigenous peoples who live above the Arctic Circle, global warming by the 1990s had passed from the realm of theory into day-to-day reality. For peoples whose lives are entwined with natural processes, daily life by the turn of the Christian millennium pulsed with evidence that climate and life were changing swiftly, and sometimes dangerously.

GOVERNMENTAL REPORTS CONFIRM NATIVE OBSERVATIONS

Tim Johnson writes in *Native Americas* that indigenous peoples' observations vis-à-vis global warming are dovetailing with a growing body of scientific evidence:

Now scientists are conducting an extensive array of studies in the north that support Native observations. This time a spirit of cooperation and collaboration is evolving between the Western-trained researcher and the Native-trained specialist. And this time, both are offering blunt assessments of the foreboding impacts of global warming. (Johnson 1999, 11)

Statements of Native American residents in the Arctic confirm observations of scientists on Ice Station SHEBA, who were surprised to observe unusually heavy rains during the warmer periods of their journey. Scientists who had worked decades in the Arctic said they had never experienced rain on the Arctic Ocean ice pack.

During December 1998, Canada's Environment Department released a study of global warming's potential effects, including displaced wildlife, increased pollution risks, and a thaw of the permafrost that could destabilize infrastructure across the Canadian sub-Arctic. The study suggests that a warming trend could push the edge of the permafrost zone 300 miles further north, causing "massive slumping of terrain" in the thawed area (Crary 1998). The thawing permafrost could destabilize roads, bridges, buildings, and oil pipelines. The Canada Country Survey, compiled by dozens of Canadian government, academic, and industry experts, also expresses concern for the Inuit, the region's indigenous people, "who could find their hunting prey out of reach, their water supplies contaminated and their coastal communities subjected to erosion from seas no longer covered by ice. . . . Northern indigenous people would be affected by ecosystem shifts that may be outside the limits of historical experience" (Crary 1998). Some Arctic coastal communities already are experiencing damage from an ocean that generates increased wave action because it is no longer covered with ice. In Tuktoyaktuk, a town near the mouth of the Mackenzie River, several buildings have been lost to erosion by the sea.

Scientists at the University of Alaska have documented carbon dioxide and temperature increases in Alaska during the 1990s which confirm what Natives have seen in the natural systems on which they rely. In addition to a reduction in Bering Sea ice cover, more precipitation is falling in many areas. Native statements and scientific observations note increasing rates of thawing in subsiding permafrost, increased coastal erosion, later river freezings, and earlier thaws. During the fall of 1998, sea ice formed in northern Alaska more than a month later than usual, postponing the annual seal and walrus hunt.

Sarah James, an environmental monitor in Arctic Village, a small hamlet of about 100 people in the Yukon-Koyukuk region of Alaska, said that the tree line is now about 35 miles north of where it used to be. She said, "The glaciers are melting and we are losing lakes to the widening flow of rivers. The flow cleans out the lakes and more new vegetation is moving northward. With this, in summer has come more insects that are eating and killing many trees" (Johnson 1999, 24). Warmer winters and more extensive thawing of permafrost are upsetting the animals. "Fur animals are dropping their fur when it is too warm at the wrong time of the year," said James. "The animals get sick" (Johnson 1999, 24).

"Those of us who are dependent on Arctic resources know global warming is occurring," said Caleb Pungowiyi, a Yupik Inuit (Eskimo) elder from Nome, Alaska. "It is affecting us, and the impacts in some cases are quite severe" (Global Warming 1998). Pungowiyi lives in Savoonga, a village of about 500 people, near Nome. The people of Savoonga make their livings mainly from the sea, harvesting seals, walrus, polar bear, whales, salmon, trout, whitefish, and other fish.

Pungowiyi said Alaska's Native peoples are alarmed by what they see. "Real changes are occurring in the Arctic areas," Pungowiyi said. He said that sea-based wildlife now often lack the ice cover needed to support breeding cycles. Because they travel over the ice in sleds, native villagers have shorter trading seasons since ice bridges between villages thaw earlier than previously. "Hunters are saying the walruses are skinny," Pungowiyi added, suggesting Alaskans are beginning to see "cascading effects" of shifting food chains (Global Warming 1998). Pungowiyi works for Nome-based Kawerak Inc., a natural resources program that collects native wildlife harvest data and conducts wildlife research.

GLOBAL WARMING AND NATIVE PEOPLES OF THE ARCTIC

In the Canadian Inuit town of Inuvik, 90 miles south of the Arctic Ocean, the temperature rose to 91 degrees F. on June 18, 1999, a type of weather unknown to living memory in the area. "We were down to our T-shirts and hoping for a breeze," said Richard Binder, 50, a local whaler and hunter. Along the Mackenzie River, according to Binder, "Hillsides have moved even though you've got trees on them. The thaw is going deeper because of the higher temperatures and longer periods of exposure." In some places near Binder's village, the thawing earth has exposed ancestral graves, and remains have been reburied (Sudetic 1999, 106). Some hunters say that seals have moved farther north, killer whales are eating sea otters and beaver are proliferating, something that would not happen if rivers and ponds were freezing to usual depths.

Coastal hunters above the Arctic circle in Alaska say they are definitely seeing a trend: the ice regularly comes a month later than it did 20 years ago, and roughly two months later than 30 years ago. Ice also breaks up earlier than previously, so hunting seasons are becoming shorter.

"Sea ice is a pretty sensitive indicator," said Gunter Weller, a professor of geophysics at the University of Alaska Fairbanks. "It doesn't take much [temperature change] to make a change in the ice" (Kizzia 1998). Weller noted that researchers on the National Science Foundation's ice ship SHEBA, north of Alaska, found the polar ice cap to be considerably thinner than previous studies had indicated. (See also Chapter 4, "Icemelt: Glacial, Arctic, and Antarctic.") In fact, they couldn't find an ice floe thick enough to anchor their icebreaker safely. Lack of sea ice causes a feedback warming effect because open ocean absorbs more of the sun's energy than ice and snow. (Not all icepacks in the Arctic decreased in 1998. Craig Evanego, a researcher at the National Ice Center, agrees that ice north of Alaska was at a record minimum. On the other side of the pole, however, north of Russia, the Barents and Kara seas experienced unusually heavy ice.)

Greenpeace and the Arctic Network, a non-profit conservation group in Alaska, have been interviewing indigenous people of the Arctic about the condition of the ice with which they live on a daily basis. These anecdotal accounts support statistics indicating that Alaska's climate is warming more rapidly than perhaps any other place on Earth due to increased proportions of carbon dioxide and other greenhouse gases in the atmosphere.

Arctic Network reports that weather records from Siberia, Alaska, and Northwest Canada indicate a rise in mean temperature of approximately one degree C. each decade for the last 30 years. The surface area of sea ice has been declining at a rate of 2.8 percent per decade for 20 years, a process that has accelerated since 1987 to a decadal rate of 4.5 percent. Many animals that are dietary staples of indigenous communities live in symbiosis with the sea ice; smaller ice coverage means smaller habitats and smaller sustainable human harvests.

In *Native Americas*, Johnson described the concerns of Siberia-Yupik elder Leonard Apangalook, Sr., who lives on St. Lawrence Island, roughly 200 miles off the coast of mainland Alaska, 38 miles from the easternmost tip of Siberia. According to Johnson, Apangalook, who is retired, "spends most of his time hunting and fishing and maintaining a daily vigil, mentally recording the natural world indicators upon which he depends: temperature, wind, rainfall, currents, ice, clouds, seasonal trends. All are standard components, customary tools of a traditional knowledge system developed in concert with the Yupik geography to sustain human life. When measured over time and compared against memory, these indicators provide information critical to the Yupik people. What Apangalook has witnessed this past decade has him deeply concerned" (Johnson 1999, 8–9).

What Apangalook sees is this: The great winter ice packs that usually flow from the north have disappeared. The seals, which feed on fish at the edges of the ice (providing food for the Yupik people) have declined because the retreating ice streams are their home. They rest on the ice, and give birth there. The ice that brought food to the Yupik did not even reach some of their communities during the El Niño years of the middle and late 1990s. Elsewhere in the Arctic, reports Johnson, "Near Iqaluit, a community on the water's edge of Baffin Island, off Frobisher Bay, where ice has been forming much later in the season, come reports of polar bears pacing up and down the coast waiting for the ice to come inland" (Johnson 1999, 10). The bears are hungry as they pace, and the Inuit are concerned, as they are forced to eat lower-quality, more expensive meats imported great distances from the south.

On Earth Day, 2000, Inuits brought their accounts of dramatic warming to urban audiences in southeastern Canada. Inuit leader Rosemarie Kuptana told a press conference in Ottawa that experienced hunters have fallen through unusually thin ice. Three men had recently died this way. Never-before-seen species (including robins, barn swallows, beetles, and sandflies) have appeared on Banks Island in the Arctic Ocean about 800 miles northwest of Fairbanks, Alaska. Growing numbers of Inuit are suffering allergies from white pine pollen that recently reached Kuptana's home in Sach's Harbour, on Banks Island, for the first time. "If this rate of change continues, our lifestyle may forever change, because our communities are sinking with melting permafrost and our food sources are . . . more difficult to hunt" (Duffy 2000, A-13). Sach's Harbour itself is slowly sinking during the summer into a muddy mass of thawing permafrost. Born in an igloo, Kuptana has been her family's weather watcher for much of her life (she was 46 years of age in 2000). Her job was to scan the morning clouds and test the wind's direction to advise family members on what to wear and to help the hunters decide whether to go out. Now she gathers observations for international weather-monitoring organizations.

Canadian Wildlife Service scientists reported during December 1998 that polar bears around Hudson Bay are 90 to 220 pounds lighter than 30 years ago, apparently because earlier ice melting has given them less time to feed on seal pups. When sea ice fails to reach a particular area, the entire ecological cycle is disrupted. When the ice melts, the polar bears can no longer use it to hunt for ring seals, many of which also have died, having had no ice on which to haul out. Johnson continues his description of the cycle of life that is being broken by retreating ice:

If the polar bears and seals disappear, and if there is no place for the walrus to rest, to give birth and to nurse their pups, and if fisheries change as currents shift because of the redistribution of heat, then Native populations that subsist and maintain their cultures, at least in part, on the polar bear, the seal, the fish, and the walrus, are going to find their way of life in grave danger. (Johnson 1999, 11)

Pungowiyi sees evidence of a warmer climate in the daily life of her village:

Our community has seen real dramatic effects as a result of the warming that is occurring in the Arctic Ocean and the arctic environment. In the springtime we are seeing the ice disappearing faster, which reduces our hunting time for walrus, seals, and whales. The ice freezes later. Ice is a supporter of life. It brings the sea animals from the north into our area and in the fall it also becomes an extension of our land. When it freezes along the shore, we go out on the ice to fish, to hunt marine mammals, and to travel. Ice is a very important element in our lives. . . . When it starts disintegrating and disappearing faster, it affects our lives dramatically. (Moreno 1999, 43)

The offshore ice-based ecosystem is sustained by upwelling nutrients which feed the plankton, shrimp, and other small organisms which feed the fish, which feed the seals, which feed the bears. The Native people of the area also occupy a position in this cycle of life. When the ice is not present, the entire cycle collapses.

[T]he ice melts earlier, in the springtime, in March, when the seals are having their pups, and [as] the ice breaks up, their pups will not be fully weaned so a lot of them will starve and will not be fully developed. Twenty years from now, we'll see a reduction in animals because that generation of pups [did] not reproduce. We'll see a major reduction in seals that we depend upon. So our future generations will feel a major impact as a result of what's happening today. It will be felt in 20 years. (Moreno 1999, 43)

Pungowiyi continued, "[A]s Native peoples, we are the ones that live the closest to the land, to Mother Earth. We live with it, we experience it, with our hearts and souls, and we depend upon it. When this Earth starts to be destroyed, we feel it. We have to do something before it is too late. We can't wait until the economic community of the world is destroyed and we finally come to our senses" (Moreno 1999, 44–45).

NATIVE EXPECTATIONS OF INDUSTRIALISM'S DEMISE

The emerging consensus that Western industrial economic culture is hanging itself from its own fossil-fueled noose is no surprise to many of the indigenous peoples of the world. José Barreiro, who is Taino (from Cuba), and editor of the news magazine *Native Americas*, notes,

For years Native traditionalists have pointed to the growing convergence of scientific prediction and Native prophecy. The intuitive, observational acumen of Native cultural practitioners, particularly when informed by the values and stories that detail prophetic tradition, have upheld certain basic truths. One is that everything in the living world is related. Another is that everything must be in balance; harmony as a positive factor. It has not been lost on elders that new currents of thinking in the academy—ecological, multi-disciplinary, inter-relational—are increasingly working from these premises. (Barreiro 1999, 2)

Barreiro notes that for many years, the Haudenosaunee (Iroquois) chiefs and clanmothers have told the world about their tradition of the Seven Generations, which proposes the concept that the current generation always has a responsibility for the state of the world to the seventh generation yet to come. Many other Native nations have expressions and traditions which articulate long-term, multi-generational visions of human existence on the Earth.

From a Native American perspective, writes Barreiro, contemporary non-Native political and corporate leaders seem exceedingly narrow, short-sighted, and ignorant of the burdens their quest for profits imposes on the world as a whole. Many Native American peoples draw cold comfort from the knowledge that the Earth on which they stand is warming at least as rapidly (and often with direct impacts on them, especially in the Arctic) as the parts of the Earth trod by industrial, fossil-fuel burning capitalism.

Oren Lyons, a faithkeeper of the Onondaga Nation, questions the Western world's "lack of connective knowledge" (Johnson 1999, 18). Lyons believes that Western traditions, which rely on quantification and analytical thought, ignore the spiritual instructions foretelling extraordinary occurrences impending for the Earth. Recently, however, many people, including some in the sciences, have begun to appreciate the relationship between intellectual and spiritual forms of knowledge.

Traditional Native knowledge often cautions that human beings should not meddle with the Earth. The U'wa people of Colombia believe that oil is the Earth's blood, and that its release into the upper world will bring about deleterious environmental changes. The U'wa take this belief seriously enough to threaten collective suicide if Occidental Petroleum drills for oil in their ancestral territories. The Hopi have a similar prophecy which warns of dire changes if people disturb things (such as the coal of the nearby Black Mesa) which lie deep within the Earth. The Code of Handsome Lake, recorded by the Iroquois prophet of the same name, which was formulated two centuries ago, foresees a time when the Earth will be covered with smoke and its water resources ruined.

At a conference held in Albuquerque from October 28 to November 1, 1998, scientists from various institutions, including the National Aeronautics and Space Administration (NASA), met with dozens of Native elders. A year later, NASA and *Native Americas* collaborated on a special issue devoted to the effects of global warming on indigenous Americans (A Call to Action 1999). About 250 people attended the Albuquerque meeting, which was titled "Circles of Wisdom: The Native Peoples/Native Homelands Climate Change Conference." Native scientists, scholars, environmental managers, and spiritual elders convened with a national network of scientists to share their knowledge of history and observations of climate change and to explain Native cultural values and relevant prophecies.

"The Albuquerque Declaration," issued at the conclusion of the Native peoples' meeting with NASA, said,

As indigenous peoples, we are to begin each day with a prayer, bringing our minds together in thanks for every part of the natural world. We are grateful that each part of our natural world continues to fulfill the responsibilities set for it by our Creator, in an unbreakable relationship to each other. As the roles and responsibilities are fulfilled, we are allowed to live our lives in peace. We are grateful for the natural order put in place and regulated by natural laws.

Mother Earth, Father Sky, and all of Creation, from micro-organisms to human, plant, trees, fish, bird, and animal relatives are part of the natural order and regulated by natural laws. Each has a unique role and is a critical part of the whole that is Creation. Each is sacred, respected, and a unique living being with its own right to survive, and each plays an essential role in the survival and health of the natural world.

Because of our relationship with the lands and waters of our natural surroundings, which have sustained us since time immemorial, we carry knowledge and ideas that the world needs today. We know how to live with this land: we have done so for thousands of years.

We express profound concern for the well being of our sacred Mother Earth and Father Sky and the potential consequences of climate imbalance for our indigenous peoples and the significance of these consequences for our communities, our environment, our economies, our cultures and our relationships to . . . natural order and laws. A growing body of Western scientific evidence now suggests what indigenous peoples have expressed for a long time: life as we know it is in danger. We can no longer afford to ignore the consequences of this evidence.

In June 1997, more than 2,000 U.S. scientists [and scientists] from over 150 countries, including Nobel Laureates, signed the Scientists' Statement on Global Climate Disruption which reads, in part, "the accumulation of greenhouse gases commits the sacred Earth irreversibly to further global climate change and consequent ecological, economic, social and spiritual disruption." Climate imbalance will cause the greatest suffering to the indigenous peoples and most pristine ecosystems globally. According to this overwhelming consensus of international scientists, the burning of oil, gas, and coal (fossil fuels) is the primary source of human-induced climate change.

The increasing effects of the indiscriminate use of fossil fuels add to other adverse impacts on natural forests. Natural forests are critical parts of the ecosystems that maintain global climate stability. The mining and drilling for coal, oil, and gas, as well as other mineral extractions, results in substantial local environmental consequences, including severe degradation of air, forests, rivers, oceans and farmlands. Fossil fuel extraction areas are home to some of Mother Earth's last and most vulnerable indigenous populations, resulting in accelerated losses of biodiversity, traditional knowledge, and ultimately in ethnocide and genocide.

For the future of all the children, for the future of Mother Earth and Father Sky, we call upon the leaders of the world, at all levels of governments, to accept responsibility for the welfare of future generations. Their decisions must reflect their consciousness of this responsibility and they must act on it.

We request that the potential consequences of climate imbalance for indigenous peoples and our environments, economies, culture, place and role in the natural order be addressed by:

1. Establishing and funding an Inter-sessional Open-ended Working Group for indigenous peoples within the Conference of the Parties of the UN Framework Convention on Climate Change.

2. Provisions for case studies [should] be established within the framework of that Working Group that would allow for assessing how climate changes affect different regions of indigenous peoples and local communities, assessing climate changes on flora and fauna, freshwater and oceans, forestry, traditional agricultural practices, medicinal plants and other biodiversity that impact subsistence and land-based cultures of indigenous peoples, and other case studies that would provide a clearer understanding of all effects and impacts of climate change and warming upon indigenous peoples and local communities.

3. Indigenous participation. Indigenous peoples of North America were invited by neither the United States nor Canada to participate in the negotiations of the United Nations Convention on Climate Change. We demand a place at the table of this important international discussion.

Indigenous peoples have the right, responsibility and expertise to participate as equal partners at every level of decision-making including needs assessments, case studies, within national and international policy-making activities concerning climate-change impacts, causes and solutions. They need to help establish protocols that would actively promote international energy efficient and sustainable forms of development, including the widespread use of appropriately scaled solar energy and renewable energy technologies as well as sustainable agricultural and forestry practice models; exploration and development in the traditional territories of indigenous peoples of the world must be done with the full consent of indigenous peoples, respecting their right to decline a project that may adversely impact them. Where destruction has already occurred, there should be a legally binding obligation to restore all areas already affected by oil, gas, and coal exploration and exploitation. This restoration must be done such that indigenous peoples can continue traditional uses of their lands. (A Call to Action, 1999)

SACRED AND PROFANE: THE NAVAJOS AND POWER GENERATION

Traditional Dine (Navajo) believe that certain rocks, elements, and sacred places must never be disturbed. If they are disturbed, disease, drought, and devastation will follow. Today, in a homeland that has been extensively polluted by uranium and coal mining, many Navajo traditionalists believe that these forbidden elements include uranium, coal, oil, and natural gas.

Valerie Taliman, a Navajo, sketched the energy geography of her homeland:

More than 100 tons of sulfur and 120 tons of nitrogen oxides are emitted daily by Arizona's 2,040-megawatt Four Corners Power Plant, according to Navajo Environmental Protection Agency officials. Two other regional power plants, the San Juan Generating Station near Farmington, N. Mex., and the Plains Escalante plant near Prewitt, Utah, add several hundred more tons of sulfur and nitrogen oxides each day. Carbon dioxide, methane gas, mercury, selenium and other trace metals also are discharged daily. Coal dust from nearby mines and particulate matter from power plant emissions cause an array of respiratory diseases and damage the region's ecosystem by producing acid rain and smog. (Taliman 1999, 35)

During the last 30 years, energy colonization has stripped trees and shrubs from hillsides now scarred by open-pit coalmines. Electrical transmission lines sometimes cut the sky above Navajo hogans which have no electricity. Some underground water supplies have been poisoned by mining. Precious underground water supplies are being pumped out of the earth to transport coal slurry in pipelines from the Hopis' mother mountain, Black Mesa, to the Mohave Generating Station in Laughlin, Nevada, 273 miles away.

Taliman continues her description of the price the Navajos have paid to furnish other people with electricity generated through the use of fossil fuels and uranium:

On the northern part of the reservation near Cove and Red Valley, more than 1,000 abandoned uranium mines have exposed Navajo families, children and the unborn to dangerous levels of radiation. A 1977 Navajo Division of Health report revealed that Navajo teenagers in that region contracted testicular and ovarian cancers at a rate 17 times higher than the national average. Similarly, a March of Dimes study from the early 1970s found that children in Shiprock, New Mexico, were afflicted by Down's syndrome, cleft palate, and anencephaly—a fatal condition where the brain fails to form—at an alarming rate of almost eight times the national average. (Taliman 1999, 35)

The legacy of uranium in Navajo country includes the largest single release of radioactive material within the borders of the United States. The spill occurred during 1979, after an earthen dam at the United Nuclear Corp. mine in Church Rock, New Mexico, burst, sending 94 million gallons of radioactive waste down the river, through the eastern edge of the reservation. Two decades later, the Rio Puerco area is still polluted with heavy metals and radioactivity. Scientists detect the legacy of the spill in the people living there, as well as in their sheep, cattle, and horses.

When talk turns to global warming among Native American peoples of the Southwest, the major concern is water, the lifeblood of communities that endure prolonged heat and drought even today. Two rivers, the Colorado and the Rio

Grande, are important water sources. These two rivers are used as wastewater for fossil-fueled power plants that help light Los Angeles, San Diego, Las Vegas, and Phoenix. Irrigation from the same two rivers grows a large proportion of North America's winter fruits and vegetables. In a warmer world, water resources which already are taxed to the maximum could become objects of intense conflict.

A 1997 report by the National Center for Atmospheric Research on the climate of the Southwest in 50 years predicted that precipitation is expected to decrease in summer and winter, while average temperatures for the region may increase by five to seven degrees F. (Taliman 1999, 39). Areas already stressed by heat and drought can expect to get more of both.

Taliman sketches a bleak future for her homeland:

Forests, especially in drier areas, are likely to die back as a result of drought, insects and disease, according to assessments by the EPA [Environmental Protection Agency]. Their projections indicate that more frequent summer droughts will increase the risk of fire by creating tinder-dry forests, thereby endangering wildlife habitats, commercial timber production, and recreational activities. (Taliman 1999, 39)

REFERENCES

Barreiro, José. 1999. "A Consciousness of Mother Earth." *Native Americas* 16, no. 3/4(Fall/Winter): 2. [http://nativeamericas.aip.cornell.edu]

"A Call to Action: The Albuquerque Declaration." 1999. *Native Americas* 16, no. 3/4(Fall/Winter): 98. [http://nativeamericas.aip.cornell.edu/fall99/fall99suagee.html]

Crary, David. 1998. "Global Warming a Threat to Canada." Associated Press, 14 December. [http://benetton.dkrz.de:3688/homepages/georg/kimo/0254.html]

Duffy, Andrew. 2000. "Global Warming Why Arctic Town is Sinking: Permafrost is Melting under Sachs: Inuit Leader." *Montreal Gazette*, 18 April, A-13.

"Global Warming Worries Native Americans." 1998. *Deseret News*, 27 November. [http://www.desnews.com/cit/071en14f.htm]

Johnson, Tim. 1999. "World Out of Balance: In a Prescient Time Native Prophecy Meets Scientific Prediction." *Native Americas* 16, no. 3/4(Fall/Winter): 8–25. [http://nativeamericas.aip.cornell.edu]

Kizzia, Tom. 1998. "Seal Hunters Await Late Ice." *Anchorage Daily News*, 28 November. [http://www.adn.com/stories/T98112872.html]

Moreno, Fidel. 1999. "In the Arctic, Ice is Life, and It's Disappearing." *Native Americas* 16, no. 3/4(Fall/Winter): 42–45. [http://nativeamericas.aip.cornell.edu]

Sudetic, Chuck. 1999. "As the World Burns." *Rolling Stone*, 2 September, 97–106, 129.

Taliman, Valerie. 1999. "Reading the Clouds: Native Perspectives on Southwestern Environments." *Native Americas* 16, no. 3/4(Fall/Winter): 34–41. [http://nativeamericas.aip.cornell.edu]

CHAPTER 9

Greenhouse Gases and the Weather: Now, and in the Year 2100

North American prairie weather has always been noted for great daily, seasonal, and interannual variability of temperature and precipitation. The Great Plains and prairies of North America host what must be an Olympics of weather— from the jungle-like heat of summer to the tundra-like cold of Arctic winter. As the only relatively level stretch of continental land in the world which reaches from the Arctic nearly to the tropics, inland North America is a stage which hosts a meteorological tour de force. Omaha, Nebraska, lies near the middle of this land mass.

Even against this background, some of the storminess which visited the area during the middle and late 1990s raised eyebrows. In 1997, a late October snow-thunderstorm lasted eight hours, and poured a foot of half-melted (and almost entirely unforecast) snow the consistency of fresh cement on Omaha. The heavy, sodden snow froze to the stately old maples and elms which frame Omaha's older neighborhoods, stripping many of them like overripe bananas.

At three in the morning, homeowners emerged from their beds to scrape the remains of ice-shattered trees off the power and telephone lines. Loud cracks from splitting trees competed with the sounds of explosions in the distance, some of them from the thunder following blue puffs of snow-shrouded lightning, the others from explosions of electrical substations short-circuited by ice-caked tree branches. It was difficult, during the storm's tumult, to tell the difference. For a night, the world seemed to come apart. The next day, nearly everyone's power was out, and Omaha's trees looked as if they had been to war, as an announcer intoned over a battery-powered radio: "Better take a picture of this. In thirty years, no one will believe you!" Electricity was out across most of the city for several days; because most gas furnaces use electrical ignition, heating also was unavailable for most homes.

Two summers later, during August of 1999, Omaha was doused with 10.5 inches of rain in a similar large thunderstorm. Sydney, Australia, had nearly a foot of rain in two days during the same week; nine months later, on May 7, 2000, the St. Louis area was swamped by as much as 14 inches of rain in one night. Two weeks later, Liberty, Texas, a bedroom community of Houston, was deluged by 19 inches of rain, most of it in seven hours.

During the deluge of August, 1999, streets on many of Omaha's hills became sluiceways; one man stepped out to inspect the flood and was swept to his death. A few weeks later, Omaha began a drought. Between September 4 and November 23, only 0.01 inch of precipitation fell during more than two-and-a-one-half months, a drought that brought soil moisture to levels comparable to the drought years of the 1930s Dust Bowl.

Many greenhouse climate models call for drought in the U.S. Midwest, punctuated by occasional "junk" rainfall. Omaha's 10.5 inches one night during August, followed by 0.01 inches between September 4 and November 23, seemed to be something of a rehearsal for this pattern. The dry winter of 1999–2000 also seemed to be something of a rehearsal for another pattern forecast for the Intergovernmental Panel on Climate Change (IPCC): monthly average temperatures were 5 to 10 degrees F. above average for the entire period, or roughly what some of the IPCC's climate models called for in the area a century from now.

People living on the plains and prairies of North America always have had confidence in the belief that one can hide from a tornado, given a little lead time and a sturdy storm shelter. This old verity still holds true most of the time. However, changes in regional weather during the 1990s included a monster tornado that sucked some people out of their basements in Spencer, South Dakota, as it wiped that small town from the face of the Earth. A summer or two before that monstrous tornado, Nebraska had a 400-mile-wide "derecho" supercell thunderstorm with 110-mile-an-hour winds. The storm formed in the western part of the state on strong downdraft winds and raced eastward at 85 miles an hour. In Omaha, the winds twisted and snapped the tops off 40-foot-high fir trees.

Even though the weather of the prairie seems almost infinitely variable, there are definite seasonal patterns. Some of these patterns were disrupted during the 1990s. One very strong pattern has been the winter freeze-thaw cycle. Ever since weather records have been kept in the area, winter thaws almost always have been followed (often very emphatically) by a return to cold weather. By the winter of 1999–2000, however, the cycle had become one of unseasonable thaw punctuated by brief, snowy freezing periods. This pattern was repeating itself year after year during the late 1990s.

The IPCC forecasts that warmer temperatures will lead to a more vigorous hydrological cycle and that global warming will increase the probability of more severe droughts and floods in many parts of the world. Several climate models utilized by the IPCC indicate an increase in precipitation intensity, suggesting a possibility for flooding rains in one season and withering drought in another in the same locations. Aiguo Dai, Inez Y. Fung, and Anthony D. del Genio, writing in the *Journal of Climate*, analyzed measurements of rainfall rates around the world through the twentieth century to 1988 and found amounts to be increasing on an average of about 2.2 millimeters per decade, a trend that parallels observed rises in temperature over the same period (Dai et al. 1997, 2943).

One may caution against drawing too many inferences from a single annual cycle of weather. Seasonal and annual variability is one thing, but growing mountains of anecdotal evidence that the Earth is warmer and precipitation more variable do mean something. Evidence that the weather is changing in ways forecast by climate modelers is abundant. In Point Barrow, Alaska, for example, the growing season has lengthened by 15 days in two decades (Anderson 1999, 5). Tibet during 1998 experienced its warmest June on record, as temperatures exceeded 77 degrees F. on 23 of 30 days at Lhasa. New York City experienced its warmest month on record during July, 1999. Cairo, Egypt, experienced its warmest August on record in 1998. In England, a third of bird species studied in 1995 laid their eggs an average of 8.8 days earlier than in 1971. Half of the glacial ice in the Caucasus Mountains (southwest of Russia) has melted during the last century.

During 1995, the World Cup skiing competition in Austria was canceled for lack of snow. The same year, the National Oceanic and Atmospheric Administration (NOAA) National Climate Data Center scientists issued a report suggesting that extreme weather events were increasing in the United States, the former Soviet Union, and China. At the same time, saltwater intrusion was killing mangroves in the swamps of Bermuda. Some penguin populations in Antarctica had declined by a third in 25 years because their native sea-ice habitat is shrinking. In Kenya, during 1997, several hundred people died of malaria in the highlands, a region where the disease had not been heretofore reported. In Columbia, mosquitoes capable of carrying dengue and yellow fever appeared at altitudes of 7,200 feet, more than twice as high as their usual range. In Indonesia, also during 1997, malaria was detected at 6,900 feet, for the first time (Greenhouse Effects, 1999, 78–79). The Midwest floods of 1993 surpassed all previously recorded U.S. floods in terms of precipitation amounts, record river stages, flood duration, persons displaced, crop and property damage, and economic impact (Glick 1998). In 1995, the British Meteorological Office reported that Great Britain experienced its warmest summer since 1659 and its driest summer since 1721. During the same summer, almost 600 people died in Chicago alone during the most intense heat wave to settle over the Midwest in many people's lifetimes. In Omaha, the temperature soared to 109 during a week of three-digit highs.

INSURANCE CLAIMS AND REFUGEE NUMBERS AS INDEXES OF STORM SEVERITY

One index of weather severity is the amount of money claimed from insurance companies for natural calamities. The figures are biased upward by increasing populations and land values in areas which are vulnerable to storms, including coastal development (hurricanes) and Midwestern urban expansion (tornadoes). Even with these biases taken into account, the figures are instructive of a trend. Those who believe that warmer weather brings more extreme storminess cited a case in point when insurance losses from 1998 were tallied. Worldwide, during the world's warmest year on record, $90 billion worth of damage was done by the weather, as much, in monetary terms, as during the entire decade of the 1980s.

Insurance claims may be plotted along with temperatures. A National Science Foundation study based on natural indicators such as tree rings, ice cores, and corals, published in 1999, found that the last decade of the millennium was its warmest. Among the years of the decade, 1998 was the warmest (by 0.4 degrees F., an unusually large margin), surpassing the previous record, set in 1997, which eclipsed the 1995 record. Between 1990 and 1995, half a decade, 16 floods, hurricanes, and other storms destroyed more than $130 billion in property, for which insurers paid $57 billion (Gelbspan, A Global Warming, 1997).

"More and more, there's a human fingerprint in natural disasters, in that we're making them more frequent and more intense and we're also . . . making them more destructive," said Seth Dunn, a research associate and climate-change expert at the Worldwatch Institute (Calamities 1998). A high-ranking official of a Swiss reinsurance firm told the World Watch Institute: "There is a significant body of scientific evidence indicating that [the recent] record insured loss from natural catastrophes was not a random occurrence." The Reinsurance Association of America said climate change "could bankrupt the industry" (Gelbspan, A Global Warming, 1997). During September of 1999, insurers watched nervously as Hurricane Floyd inflicted more damage on North Carolina than any event since the Civil War.

The number of federal disaster-area declarations averaged 46 per year in the 1990s, compared to 24 a year during the 1980s. The top 10 most expensive disasters for insurance companies in the United States (eight of them weather events) occurred during the 1990s, even when the expense is adjusted for inflation (Anderson 1999, 1). While some observers attribute these changes to global warming, others argue that the federal government was more likely to declare disasters in the 1990s than previously and that increasing populations, affluence, and migrations to coastal areas are placing many more people in harm's way, thus increasing insurance claims (and premiums).

Another index of climatic severity is the number of people put out of their homes by natural disasters. Once again, the numbers are biased upward by rising populations in vulnerable locations. Even so, the rise in numbers of "environmental refugees" is rather impressive. During 1998, for the first time (according to the International Red Cross) environmental refugees outnumbered people displaced by wars and other human conflicts. More than 25 million people were driven from their homes during that year because of floods, droughts, deforestation, and other environmental problems. The number of people requiring sustained relief from hurricanes, earthquakes, floods, and other natural events jumped from 500,000 in 1992 to more than 5.5 million in 1998 (Ewen 1999, 26).

The number of environmental refugees is growing for two reasons: first, the number of disasters has been increasing; second, an increasing number of people are now living in marginal areas which are prone to disasters. During December, 1999, when an estimated 20,000 people were killed by floods in Venezuela, the toll was inflated by the large number of poor people living on land denuded of its one-time natural forest cover, opening the way to devastating floods. "There are unfortunately thousands of people buried in the mud, and the final count will never be known," said Angel Rangel, Venezuela's director of civil defense (Logan 2000, 28). Some parts of Venezuela received several years' worth of average rainfall during one month.

GREENHOUSE WEATHER

The 1990s witnessed five consecutive El Niño years, leading to speculation that this pattern may become the new norm. According to Kevin Trenberth:

El Niños have occurred more often since the mid-1970s and the duration of the 1990–95 El Niño is unprecedented in the past 120 years. Research shows this to be sufficiently unusual that it suggests that the climate may be changing in such a way as to make this behavior more likely. Because widespread flooding and droughts occur simultaneously in different parts of the world in association with El Niños, this has profound implications for agriculture, water resources, human health, and society in general. (Epstein et al. 1996)

Trenberth adds,

[T]he heating from increased greenhouse gases enhances the hydrological cycle and increases the risk for stronger, longer-lasting or more intense droughts and heavier rainfall events and flooding, even if these phenomena occur for natural reasons. Evidence, although circumstantial, is widespread across the United States. Examples include the intense drought in the central southern U.S in 1996, Midwest flooding in spring of 1995 and extensive flooding throughout the Mississippi Basin in 1993 even as drought occurred in the Carolinas. (Epstein et al. 1996)

(For more information on the relationship of global warming and El Niño climate patterns, see Chapter 5, "Warming Seas.")

A combination of deforestation and climate change may have intensified the effects of 1998's most severe disasters, including Hurricane Mitch, the flooding of China's Yangtze River, and Bangladesh's most extensive flood in a century. An account in *The Age*, a newspaper published in Melbourne, Australia, summarized the situation: "In a sense, we're turning up the faucets . . . and throwing away the sponges [including] . . . the forests and the wetlands" (Calamities 1998).

Hurricane Mitch, the deadliest Atlantic storm in 200 years, caused more than 10,000 deaths in Honduras, Nicaragua, Guatemala, and El Salvador. The flooding was made worse by denuded mountainsides because Central American nations have experienced some of the highest rates of deforestation in the world, losing 2 to 4 percent of their remaining forest cover each year. The costliest disaster of 1998 was the flooding of China's Yangtze River, which killed more than 3,000 people. The Yangtze basin has lost 85 percent of its forest to logging and agriculture in recent decades, as wetlands have been drained (Calamities 1998). (For more information on Hurricane Mitch's ecological effects, see Chapter 6, "Flora and Fauna.")

Two-thirds of Bangladesh, which is located at the mouths of the Ganges and Brahmaputra Rivers, was flooded for several months during 1998, as 30 million people were left temporarily homeless. Logging upriver in the Himalayas of northern India and Nepal exacerbated the disaster, as did the fact that the region's rivers and flood plains have been filled with silt and constricted by human urbanization (Calamities 1998).

Until the 1990s, scientists had believed that climatic changes from ice ages to more moderate climatic periods occurred slowly and relatively evenly over hundreds of thousands of years. Researchers from the Woods Hole Research Center who examined deep-ocean sediment and ice-core samples found that temperature changes of up to seven degrees C. have occurred within 30 to 40 years.

Ross Gelbspan reports,

Over the last 70,000 years, the earth's climate has snapped into radically different temperature regimes. "Our results suggest that the present climate system is very delicately poised," said researcher Scott Lehman. "Shifts could happen very rapidly if conditions are right, and we cannot predict when that will occur." His cautionary tone is underscored by findings that the end of the last ice age, some 8,000 years ago, was preceded by a series of extreme oscillations in which severe regional deep freezes alternated with warming spikes. As the North Atlantic warmed, Arctic snowmelt and increased rainfall diluted the salt content of the ocean, which, in turn, redirected the ocean's warming current from a northeasterly direction to one that ran nearly due east. Should such an episode occur today, say researchers, "the present climate of Britain and Norway would change suddenly to that of Greenland." (Gelbspan, *The Heat is On*, 1997)

A growing number of scientists are coming to believe that the Earth may have entered one of its many quick-changes in climate, with a difference: this one is being initiated, at least in part, by humankind's accelerating consumption of fossil fuels, not solely because of variations in natural phenomena. This human-induced change is faster, and will therefore impose more stress on the flora and fauna of the Earth, than any previous natural climate change. (For more information on the possible speed of climate change, see Chapter 2.)

Climate zones may shift northward (or southward, in the Southern Hemisphere), about 100 miles for each one degree C. of global warming, according to Dean Abrahamson. An increase of atmospheric carbon dioxide at present levels (about 1.5 parts per million per year) places the world's atmosphere at about 40 percent above the levels of the year 2000 by the year 2030 and double present levels by the year 2100. Abrahamson cites a study by National Aeronautic and Space Administration (NASA) climatologist James E. Hansen indicating that a doubling of carbon dioxide levels would produce an average of 12 days at 100 degrees F. or more in Washington, D.C. (compared to one day in the late 1980s), 21 such days in Omaha (compared to three in 1988), and 78 such days in Dallas, compared to 19 in 1988. In other words, in 100 years Omaha may have the present summer temperatures of Dallas, with average temperatures still rising (Abrahamson, Global Warming, 1989, 3, 7). According to Harold W. Bernard, a doubling of carbon dioxide levels would give Boston the present average summer temperatures of northern Florida (Bernard 1993, 131).

Global warming takes place during the winter as well as the summer, of course, although fewer people are apt to object to the amiable mildness of a winter day as compared to an unusually hot, humid summer day. Most climate models indicate that winter temperatures will rise more quickly than summer readings. Winter temperatures (especially nighttime readings) may be among the best indicators that greenhouse weather has arrived. The length of the winter is another factor. The annual springtime rise of water levels in the Great Lakes arrived nearly a month earlier in the late 1990s than it did 139 years earlier, according to climatologist John D. Lenters of the University of Wisconsin, who reported his findings May 24, 2000 at a meeting of the International Association of Great Lakes Research (Maugh 2000).

Bernard suggests the 1930s, the warmest decade in recorded history until the 1980s, as an analog for greenhouse weather to come. He reminds the reader that between 1930 and 1936, nearly 15,000 people in the United States died because of extreme heat, 4,768 of them in 1936 alone (Bernard 1993, 20). Bernard's analog to the 1930s suggests that cold weather will not vanish in a generally warmer world. "Remember," he writes, "the coldest month ever [recorded] in the United States, February, 1936, directly preceded the hottest summer on record" (Bernard 1993, 123). Lincoln, Nebraska recorded 41 days at 100 degrees F. or higher during the summer of 1936; the following winter, the same city recorded 33 days with minimums below zero F. (Anderson 1999, 5). (In Omaha, during the El Niño years of 1982 and 1983, a miserably hot and very dry summer preceded one of the coldest winters in recorded history.)

The U.S. federal government's first study of humidity's relationship with global warming indicates that heat indexes (temperature plus humidity) will rise significantly during the coming century, raising the average annual number of heat-related deaths from about 1,200 to several thousand a year. The report asserts that the heat index on an average summer day will be about 100 degrees F. for much of the country by the middle of the twenty-first century. "Things are going to be hot and sticky, and that's going to be a problem with global warming," said research meteorologist Thomas Knutson, one of the study's co-authors (Many More 1999, 23A). A three-to-eight degree F. increase in temperature, which scientists say is likely during the next century, will translate into a 7 to 10 degree F. rise in heat index, Knutson said (Many More 1999, 23A).

Dian Gaffen and Rebecca Ross of the U.S. National Oceanic and Atmospheric Administration reported in *Nature* that between 1949 and 1995, mean temperatures during July and August in the United States increased by a third of a degree C. Moreover, by combining effects of temperature and humidity (which has risen by several percent each decade) into a measure known as "apparent" temperature (a widely used measure of stress on human health), Gaffen and Ross found that the mean summertime apparent temperature rose by an additional tenth of a degree, for a total of 0.42 degrees C. (Abdulla 1998).

Gaffen and Ross also studied extremes of summer heat. They found that the annual number of exceptionally hot days (defined as days that exceeded location-specific threshold temperatures that are closely correlated with increased morbidity and mortality) increased at most of the 113 weather stations they examined. The number of "high-heat-stress nights" increased by 25 percent in some locations during the last half of the twentieth century. The number of heat waves per year (four or more above-threshold days in a row) went up by 20 percent during the same period. The largest rises in temperature occurred in some of the most heavily populated cities, indicating that urban heat islands are adding additional heating to greenhouse forcing (Abdulla 1998).

Many areas will face increased demands for electricity as summer temperatures rise and demand for air conditioning grows. This increase in energy use will result in numerous costs to society, including increased capital and maintenance costs for utilities, higher energy bills for consumers, and environmental costs such as increased air pollution and carbon emissions (Glick 1998). The U.S. Environmental Protection Agency estimates that by 2010, a global-warming-induced increase in U.S. electricity demand to run air conditioners in buildings could lead to as much as $7.6 billion per year in additional costs for capital, fuel, operation and maintenance each year, increasing to between $42 and $92 billion per year by 2055 (Glick 1998).

Increased costs for air conditioning in the summer probably will outweigh reductions in the use of energy for heating. Air conditioning in buildings is fueled by electricity and is usually run at peak hours, while heating in those regions is generally fueled with natural gas. Because the economic and the environmental costs of electricity are higher than they are for natural gas, the environmental cost of cooling a house one degree is more than the cost of heating it the same amount (Glick 1998).

HEAT, DROUGHT, AND THE GREAT PLAINS OF NORTH AMERICA

Many climate models for global warming foresee the breadbasket of North America becoming hotter and drier than during the Dust Bowl years of the 1930s. C.A. Woodhouse and J.T. Overpeck surveyed tree rings and other natural records and found that the central United States experienced several decades-long droughts before the European-Americans arrived. The authors predict that global warming will intensify heat and drought in the entire area (Woodhouse and Overpeck 1998). Bernard, using his climatic parallels to the 1930s, suggests that the most severe combination of heat and dryness in a greenhouse world will fall on the American Midwest, notably on Kansas and bordering states. "Precipitation decreases for mid-century [about 2050] in Kansas and Nebraska [will be] no greater than those of the worst Dust Bowl years (in 1934 and 1936), [but] temperatures are forecast to be substantially higher than during the Dust Bowl" (Bernard 1993, 124, 125).

During the mid-1980s, the Environmental Protection Agency issued reports projecting what might become of North America's breadbasket under rising greenhouse-gas levels. With a doubling of carbon dioxide levels, for example, the grain-growing breadbasket might migrate northward to the Canadian prairies. "With global temperature rises of only a few degrees, it [the "Heartland"] could become a near-desert, with massive crop failures, complete collapse of major multi-state aquifers . . . a useless expanse of ruined, windblown, overheated land," writes John J. Nance (Nance 1991, 132).

A team of U.S. Geological Survey scientists, led by David Muhs, has made a case that a few years of heat and drought could turn the high plains of the United States into a desert. Recent findings indicate that the climate in this area has been more variable during the last few thousand years than previously thought. Muhs' team found evidence that as late as the eighteenth century, sand dunes migrated across areas that are mainly grazing land today. Early in the history of the United States, the area was sometimes called "The Great American Desert." Slight changes in atmospheric circulation patterns can place this region consistently on the lee side of moisture-bearing winds that dry as they cross the mountains. Some climate models indicate that this sort of circulation pattern will dominate in the area as the earth warms. Such a climate may be punctuated by brief, heavy downpours which erode the parched soil.

A study by Kellogg and Schware (1981) projected that global warming would cause a decrease in rainfall of about 40 percent in the Great Plains of North America (if carbon dioxide levels were doubled over 1980s levels), part of a general trend toward summer drought in the interiors of the world's great land masses that other climate modellers also anticipate. Syukuro Manabe and R.T. Wetherald projected that summer drought under global-warming conditions would affect not only North America's interior, but also southern Europe and Siberia (Manabe and Wetherald 1986). Drought in these areas also would be aggravated by earlier snowmelt and changes in albedo across the affected areas. Manabe and Wetherald's study also indicates that winter precipitation might increase in the same areas.

Roughly 5,000 to 8,000 years ago, when considerable proxy evidence suggests that mean global temperatures were perhaps one to two degrees C. warmer than today, dunes of drifting dust covered portions of the high plains. A portion of north-central Nebraska is appropriately named "the Sand Hills." Under warmer and drier conditions, without irrigation water, parts of this dune-swathed landscape might resemble the Sahara Desert. Anyone who has lived in the area has experienced periodic short droughts, which occur nearly every year.

Regions such as North America's prairies and the Russian steppe are expected to warm more quickly than other areas because the midcontinental areas are relatively dry (allowing carbon dioxide to absorb more heat, as in polar regions) and apt to heat more rapidly than areas closer to large bodies of water. Bernard describes climate models that expect a doubling of carbon dioxide levels to raise temperatures in the American Midwest and on the Russian steppes six degrees C. (or 11 degrees F.) (Bernard 1993, 102–103).

If global warming desiccates the American Midwest, as forecast, the effects of intensifying heat and drought may be compounded by a developing scarcity of groundwater. The Ogallala Aquifer, which lies under the Great Plains from South Dakota to Texas, is the largest body of groundwater in North America. The water in the aquifer was generated during several ice ages over millions of years. The aquifer is being drawn down at several feet per year while rainfall recharges it at inches a year. The Ogallala Aquifer already has begun to go dry in small parts of Texas, and there is a palatable fear among some commentators that it may go dry in a more general fashion just as the Great Plains become hotter and drier because of global warming. Under such conditions the Great Plains could become largely useless for agriculture.

Forecasts of a new dust bowl in North America's agricultural heartland are not unanimous. "A Time to Reap: Global Warming and Iowa," an analysis released by the Environmental Defense Fund, indicates that global warming could increase temperatures in Iowa by 4.5 to 10 degrees F. during the twenty-first century, with as much as 30 percent *more* precipitation than today (Bloomfield 2000). According to the report, these climatic changes will mean "heat waves, floods, and pest infestations could occur more often, last longer and inflict greater damage than today." The report recommends that farmers in Iowa should "adopt money-saving farm-management practices now to reduce greenhouse-gas emissions and avoid some of the potentially damaging impacts of climate change" (Bloomfield 2000). Farm-management techniques that reduce greenhouse gases include conservation tillage, installation of permanently vegetated conservation buffers, and improved manure management, according to the report.

Janine Bloomfield, an Environmental Defense Fund scientist and author of the analysis, said that Iowa's devastating floods of 1993, which caused an estimated $1.15 billion worth of crop losses, may have been the worst natural disaster in Iowa's recorded history. A flood of such magnitude occurs about once every 100 to 200 years, but with current rates of climate change, Iowa could experience a so-called 100-year flood every 10 years by the end of the century, Bloomfield wrote (Bloomfield 2000). Most climate models indicate that occasional floods in this area may alternate with long periods of drought. At present, average precipitation decreases consistently as one moves westward. Such differences may become even greater in a century if temperatures rise as anticipated. Average precipitation, given a doubling of the carbon dioxide level in the atmosphere, may decline on the Great Plains to the leeward of the Rocky Mountains and rise in prairie areas, such as Iowa and Illinois, which get more convective moisture from the Gulf of Mexico.

In Iowa, the number of days over 90 degrees F. is expected to increase with unchecked global warming, bringing the total of 90-plus degree days in Des Moines from about 20 currently to between 35 and 70 by the end of the century, according to computer models cited in Bloomfield's report. "In addition to greater risk of heat-related illness and mortality, high temperatures can significantly decrease corn yield, and very high temperatures can cause severe crop damage," said Bloomfield (Bloomfield 2000). Warmer temperatures and more frequent floods also are expected to increase the range and abundance of weeds and insect pests, according to studies cited in Bloomfield's analysis. Pests also probably will arrive earlier in the season, causing damage to crops during important growing stages.

Global warming's expected effects in the American Midwest could stand as a proxy for many areas in the middle of continental land masses. The spread of deserts is expected to become an acute social and economic problem within a century in many inland areas. About 2,000 delegates from 190 countries and hundreds of environmental organizations gathered in the Senegalese capital, Dakar, late in November 1998, for an international conference on the spread of deserts. Figures released before the conference showed that the spread of deserts was costing about $42 billion a year (Africa 1998).

Desertification reduces the land's resilience to natural climate variability. Soil becomes less productive as topsoil is blown away by the wind or washed away by rainstorms. Vegetation is damaged. Desertification may be compounded by human activities that overtax the land, such as overgrazing. Sometimes the spread of deserts is accelerated by deforestation.

The organizers of the aforementioned conference estimate that the spread of barren land has an impact on 250 million people worldwide and could eventually threaten a billion people. In North America, 74 percent of the dry land is already "seriously or moderately" affected by desertification. Africa is a close second at risk, with 73 percent of its dry land damaged. In the sub-Saharan Sahel region, experts expect 200,000 people to die per year because of desertification (Africa 1998).

ANTICIPATED EFFECTS IN NEW YORK CITY

A report issued by the Environmental Defense Fund (EDF) during June 1999 outlines some of the problems New York City may face as the climate changes (Bloomfield 1999). The report contends that New York City, with its 600 miles of coastline, is very vulnerable to sea-level rises that could be provoked by global warming. "Entry points to many of the tunnels and much of New York City's subway system lie less than or near to ten feet above sea level as do the three major airports serving the New York City Metropolitan Region," the report said (Bloomfield 1999).

A gradual rise in sea level is already eroding New York City's urban shoreline, according to NASA researcher Cynthia Rosenzweig. The rising tide appears poised to bring more salt up the Hudson River in coming years, threatening New York City's freshwater supply (Bloomfield 1999). The EDF report warns that a sea level rise in the New York metropolitan area may be accentuated by slow sinking of some coastlines due to natural processes. Combined with sea level rises caused by global warming, actual sea level in and near New York City will rise from three-quarters of a foot to 3.5 feet during the twenty-first century, the EDF expects.

According to the report,

Sea-level rise will contribute to the temporary flooding or permanent inundation of many of New York City's and the region's coastal areas as increased sea levels accentuate the impact of the storms that already strike the region. Erosion could devastate valuable beaches and wetlands and, as has happened in the past, we can expect that homes will be swept into the ocean by storms. (Bloomfield 1999)

The EDF report examined a powerful nor'easter during 1992 as a possible harbinger of greenhouse weather. That storm shut down New York City's subways, many of which were flooded, along with other commuter rails, some highways, and airports. By 2100, the EDF report expects such a storm to occur on an average of once a year, instead of roughly once a generation, as during the twentieth century. A large part of lower Manhattan Island would be at risk of occasional flooding by the end of the next century, according to the report, even if individual storms do not become more intense or frequent.

The foundations of Battery Park City and the World Trade Center would be flooded regularly in this scenario. The East River would flood Bellevue Medical Center, the FDR Drive and East Harlem between 96th and 114th Streets. Storms would flood much of Coney Island, submerging or creating islands of residential communities there and in Staten Island nearly annually. In New Jersey, storm surges would temporarily transform the Meadowlands into a salty lake. During storm surges, the barrier islands of Long Island—including Jones Beach, Fire Island, and Westhampton Beach—could narrow and fragment into small islets, while highly productive salt marshes could shrink permanently due to higher sea levels. Although marshes adapt to moderate rates of sea-level rise, they may be unable to keep up with the accelerated rates that are expected to occur as a result of human-induced climate change. (Bloomfield 1999)

Between 1880 and 1990, temperatures in New York rose an average of four degrees F., due mainly to the urban heat-island effect, as well as a one degree F. rise in general global readings. Even if temperatures increase no more quickly in New York City than during the last hundred years, the average number of days at over 90 degrees F. is expected to rise to about 27 by the year 2100. The EDF study projects that, by the year 2030, average temperatures in New York City could increase by another one to four degrees F. compared to temperatures in 1990.

Models project that the number of days per year which surpass the temperature threshold (the level at which deaths from heat stress rise rapidly) could increase from an historical average of 13 to between 38 and 80 days per year in New York City by the year 2100. In the best-case scenario sketched by EDF, New York City in the year 2100 would have as many 90-degree days as Miami, twice the present average. In the worst-case scenario, New York City would have the present day summer temperatures of Houston, Texas, six times the present number of 90-degree days. The 90-degree benchmark was chosen for the study because heat-related mortality rises sharply in New York City at that level.

Rising temperatures in major urban centers (New York City included) also could lead to enhanced levels of ground level ozone which causes acute respiratory distress and chronic damage to the lungs. Much of the New York urban area is already listed as being in the "severe" category for ozone pollution by the Environmental Protection Agency. Warming may intensify the effects of air pollution because higher temperatures increase tropospheric ozone production. Ground level ozone in high concentrations causes eye and nose irritation, coughing, and impaired lung function.

WARMING'S POSSIBLE EFFECTS IN WASHINGTON, D.C.

Observers of the Washington, D.C., Cherry Blossom Festival noted during the 1990s that the festival (held at the end of the first week in April) was occurring after the cherry blossoms had fallen from the trees. These observations, among others, compelled botanists working with the Smithsonian Institution's National Museum of Natural History to analyze the first dates of blossoming between the years 1970 and 2000 for 100 species of plants common to the Washington, D.C. area. They found that a rise in the region's average daily minimum temperatures was producing earlier flowering in 89 of the 100 species. Stanwyn Shetler and colleagues Mones Abu-Asab, Paul Peterson, and Sylvia Stone Orli found that, on average, the plants were blossoming 4.5 days earlier in 2000 than in 1970 (Scientists at the Smithsonian 2000).

The Tidal Basin cherry blossoms were arriving an average of a week earlier in the year 2000 than they had in 1970. "This trend of earlier flowering is consistent with what we know about the effects of global warming," said Shetler. "When we compared the records from the Smithsonian study with local, long-term temperature records we discovered statistically significant correlations. The minimum temperature has been going up over these years and the early arrival of the cherry blossoms appears to be one of the results" (Scientists at the Smithsonian 2000). Shetler continued, "Over a long period, the species composition of our local flora could change. Species like the sugar maple that require a long cold period may die out in our region. Invasive alien species, especially from more southern climes, may become more and more of a problem" (Scientists at the Smithsonian 2000).

At about the same time that it profiled the effects of a warmer climate in New York City, the EDF compiled a similar climate-change profile for Washington, D.C. The report asserted that rising sea levels may increase risks of flooding along the Potomac River, threatening such historic landmarks as the Capitol Mall and the aforementioned cherry trees which surround the Tidal Basin near the Jefferson Memorial. Warming of the magnitude predicted would almost double Washington D.C.'s number of days over 90 degrees F.

The District of Columbia is extremely vulnerable to flooding, according to the report, because much of the city has been built on swamps near the Potomac River. The area is so low in elevation that the Potomac sometimes experienced tidal inundation even late in the twentieth century. "Many of the District's most familiar monuments and landmarks, including the Jefferson Memorial, the Mall, the Reflecting Pool, and National Airport, are very low lying, and are therefore susceptible to flooding," the report states (Bloomfield and Showell 1999). With a storm surge on top of a sea level rise of two feet, according to this report, "The Jefferson Memorial would become an island. . . . [A] major storm surge would nearly encircle the Washington Monument and completely surround the Internal Revenue Service, the National Museum of Natural History, the National Gallery of Art and neighboring structures" (Bloomfield and Showell 1999).

The report also warns that the climate which it models for Washington, D.C. in the year 2100 may lead to greater risk of mosquito-borne diseases. The ranges of malaria, dengue or "breakbone" fever, as well as eastern equine encephalitis are expected to expand to include the city. Higher temperatures could increase the number of mosquitoes, make mosquito bites more infectious, and cause mosquitoes to bite more frequently.

Rises in sea level projected by the EDF could threaten beaches and resort areas on the Atlantic coast east of the Washington-Baltimore region, especially those on barrier islands. Recreational beaches, such as Ocean City, Maryland, may be lost as sea levels rise unless vigorous and expensive beach "renourishment projects" are continued and expanded. The shellfish, birds, and fish of the abundant Chesapeake Bay ecosystem may be decimated by a combination of sea level rise and warmer temperatures. Some historic Chesapeake Bay island communities may be completely submerged by the end of the next century (Bloomfield and Showell 1999).

Sea level rise caused by global warming would threaten barrier islands in the area. An Environmental Protection Agency study estimated that a one-foot rise in sea level at Ocean City, Maryland, could double the rate of beach erosion, causing the loss of more than 200 feet of shore. The 1985 EPA study estimated that, for a one-foot rise in sea level, maintaining the beach at Ocean City alone would cost a total of $60–$85 million (U.S. EPA 1997).

Chesapeake Bay's watershed, the largest estuary in the United States, is home to 13 million people, as well as a diverse array of fish, shellfish, mammals, shorebirds, and migratory birds. A combination of subsidence and global warming is expected to produce a sea level rise of two to three feet. Some islands in the Bay, including historic Poplar Island, could become completely submerged, according to the EDF report for the Washington, D.C., area (Bloomfield and Showell 1999).

Higher sea levels also would cause Chesapeake Bay to become saltier,

causing damage to shellfish, such as oysters and soft shell clams. The soft-shell clam, already at the southernmost point of its range, could be eliminated due to a combination of salinity changes and warming waters. Saltwater intrusion could cause large-scale plant die-off, which is already occurring at Chesapeake's Blackwater National Wildlife Refuge as a result of current rates of sea-level rise. Although sea-level rise will convert some low-lying areas to marshes, that process is much slower than the rate at which existing marshes are lost. Onshore human development will constrain the amount of land that can be converted to wetland. (Bloomfield and Showell 1999)

EXPECTED EFFECTS IN CALIFORNIA AND THE PACIFIC NORTHWEST

A team of scientists speculated during the early 1990s regarding the effects of global warming on California, assuming a doubling of atmospheric carbon dioxide. Under such conditions, this group expects winters in California to warm more rapidly (five to six degrees F.) than summers (one to two degrees F.). This state level study regarding the effects of global warming on California also anticipates,

- An increase in annual surface temperatures of between 2 and 4 degrees C. above average levels from 1951 to 1980;
- Precipitation changes of up to 20 percent above or below present amounts, depending on location;
- A rise in sea level between 0.2 and 1 meter;
- A rise in snow level of 100 meters for each degree Celsius of increase in air temperature;

- An increase in urban ozone pollution of up to 20 percent, and an "ultraviolet flux" increase of 50 percent;
- An expectation that storm tracks will move northward as part of a general movement of climate zones. (Knox and Scheuring 1991, 9)

Availability of water may become a major issue in a warmer California. Commercial, public, and private consumers in 1990 used a net 34.2 million acre feet (m.a.f.) of water per year in that state. Groundwater supplied 23 percent of that total, a troublesome proportion because these resources are being depleted. An increasing number of farming areas in California, the largest producer of food in the United States, are suffering from increasing salinity (Lewis et al. 1991, 107, 111). In an average year, California uses about 2.0 m.a.f. of underground water which is not replaced by natural processes, leading to declining water levels and increased salinity. A warmer climate may accelerate this process.

Increased frequency of drought may be caused by a number of factors which could coincide during the twenty-first century in California, as well as along the rest of the U.S. West Coast:

- If global warming causes winter storm tracks to migrate northward, some areas of California (especially parts which already are dry) may receive less precipitation;
- With warmer temperatures and higher snow levels, more snow will melt and run off during winters, depleting reserves needed during the summer-fall dry cycle;
- Problems with overdraft and salinity of groundwater may cause the volume of usable water from this source to decline.

If warming shortens the rainy season along the West Coast, economic activity could be severely curtailed. More than 90 percent of the Pacific Northwest's electricity comes from hydropower, which requires abundant snowfall in mountainous areas during the winter to sustain power generation through usually dry summers. More than 80 percent of California's agricultural land is irrigated, and its massive urban areas are sponges for imported water.

Some areas which receive copious rainfall may get more as the Earth warms. One example could be the Pacific Northwest. Tom Paulson of the *Seattle Post-Intelligencer* reported in late 1999 on projections by a consortium of scientists at the University of Washington (U.W.). "The models say precipitation should increase," Paulson reports Philip Mote, a U.W. professor of atmospheric sciences and head of the consortium, as saying (Paulson 1999). In the Northwest, average temperatures have increased 1.5 degrees C. during the twentieth century, as precipitation has increased 14 percent, Mote said.

Increased rainfall may not translate into deeper mountain snowpacks, however, because freezing levels will rise and more of the precipitation is expected to run off during the winter, sometimes as floods, which the consortium expects to increase as carbon dioxide levels double during the next 50 years. The Northwest's forests also are expected to shrink because higher temperatures reduce trees' resistance to fires and insect infestations. The group expects agriculture in the region to benefit somewhat from slightly higher temperatures, but not from water shortages during the growing season.

WHAT MAY HAPPEN AFTER THE YEAR 2100?

One senses real anxiety in global warming studies when the subject of discussion broaches the possible state of the atmosphere after the year 2100. Beyond that year, it is very possible that humankind's greenhouse-gas emissions may kick start the Earth on a spiral of warming that will, within a few centuries, return the Earth to the climatic days of the dinosaurs. No one knows how long this process may take. In perhaps 500 years, the wink of an eye in geologic time, surviving warm-blooded animals may face a very inhospitable world if humankind burns all the fossil fuels that could be marketed at what the marketplace defines as a salable price.

There is a certain ecological irony to all of this: most of the carbon-based fuels with which we are raising the temperature of the Earth date from the days of the dinosaurs, when the earth experienced the sort of climate that may be our legacy to coming generations. In simplest terms, industrial society is filling the air with carbon which the flora and fauna of earlier ages left behind millions of years ago.

One reason very few scientists venture past the year 2100 is that they find themselves in uncharted territory with no valid models. For example, given trends at the year 2000, the level of carbon dioxide in the atmosphere a century from now will probably have risen to between 560 and 720 parts per million, the highest such level in at least the last 25 million years.

William R. Cline believes that anyone who stops looking at greenhouse-gas effects at the year 2050 or 2100 makes "the same mistake as the man who falls off the twentieth floor [of a building] and concludes, as he passes the sixteenth floor, that all is well" (Cline, Comments, 1991, 223). Cline's *Economics of Global Warming* projects late-twentieth century population and fossil-fuel consumption trends ahead 250 to 300 years. The study projects that global carbon emissions, about six billion gigatons (ggt) in 1990, given present trends, could rise to 20 billion ggt in 2100, and 50 ggt late in the twenty-third century (Cline 1992). According to Cline's study, this eight-fold increase over present day carbon production would imply a global warming of between 6 degrees C. and 18 degrees C., with 10 degrees C. being a median estimate (Cline 1992, 4, 44–46, 289).

Cline faults the Intergovernmental Panel on Climate Change and other bodies for concentrating on the doubling of carbon dioxide levels expected during the twenty-first century. Cline's study looks forward to a world with a fraction of its present permanent ice cover, in which sea levels will submerge the sites of many of today's largest cities.

Cline presents climate models with some very precise temperature forecasts for the end of the twenty-third century, as he cautions, "These estimates should, of course, be interpreted as indicative rather than precise" (Cline 1992, 67). These climate models indicate a wide regional disparity between expected global warming in the temperate zone interiors of the major continents and the Arctic (where warming is projected to be the greatest) and the tropics, or near subtropics, near large bodies of water, where the models anticipate that temperatures will remain nearly stable, compared to late-twentieth century averages.

These models sometimes strain credulity. For example, expected average highs for July range from the upper 80s F. at Galveston, Texas, a lower temperature than the expected high of 94.6 in Seattle or the 103.6 degrees F. in New York City, 108.6 in Omaha, 112.6 in Little Rock, 111.5 in Rapid City, South Dakota, 107.6 in Washington, D.C., and 124.6 in Las Vegas, Nevada (Cline 1992, 68–69). The expected July average high temperature in Helena, Montana is about 108 degrees F., while the expected average highs in Charleston, South Carolina (93.9) and Jacksonville, Florida (95.9) are about the same as for Seattle. The average high projected for Juneau, Alaska (83.6 degrees F.) is not far below that of San Juan, Puerto Rico (88 degrees F.). New Orleans is expected to have a July maximum of 91 degrees F., or 20 degrees below that of nearby Little Rock, Arkansas, a difference based mainly on the fact that New Orleans is near open ocean, the bathwater-warm Gulf of Mexico (Cline 1992, 68–69). It goes without saying that while forecasting temperatures a century ahead is risky business, taking a model out four centuries may be more an exercise in imagination than in climate modeling.

REFERENCES

Abdulla, Sara. 1998. "Hot and Sticky in the States." *Nature*, 17 December. [http://imagine.nature.com/nsu/981217/981217–3.html]

Abrahamson, Dean Edwin. 1989. "Global Warming: The Issue, Impacts, Responses." In Dean Edwin Abrahamson, ed., *The Challenge of Global Warming*. Washington, D.C.: Island Press, 3–34.

"Africa: Fighting Back the Widening Deserts." 1998. BBC News, 30 November. [http://news.bbc.co.uk/hi/english/world/africa/newsid_224000/224597.stm]

Anderson, Julie. 1999. "Weather Runs Hot, Cold in Year of Surprises." *Omaha World-Herald*, 7 December, 1, 5.

Bernard, Harold W., Jr. 1993. *Global Warming: Signs to Watch For*. Bloomington: Indiana University Press.

Bloomfield, Janine. 1999. *Hot Nights in the City: Global Warming, Sea-Level Rise and the New York Metropolitan Region*. Environmental Defense Fund, June. [http://www.edf.org/pubs/Reports/HotNY/index.html]

———. 2000. *Iowans Can Elect To Combat Climate Change Now*. Environmental Defense Fund, 13 January. [http://terra.whrc.org/links/links.htm]

Bloomfield, Janine, and Sherry Showell. 1999. *Global Warming: Our Nation's Capital at Risk*. Environmental Defense Fund. [http://www.edf.org/pubs/Reports/WashingtonGW/index.html]

"Calamities Show a Human Touch." 1998. *The Age* (Melbourne, Australia), 30 November. [http://www.theage.com.au/daily/981130/news/news15.html]

Christianson, Gale E. 1999. *Greenhouse: The 200-Year Story of Global Warming*. New York: Walker and Company.

"Climate Shifts Expected to Alter California Life." 1999. *Los Angeles Times*. In *Omaha World-Herald*, 4 November, 5.

Cline, William R. 1991. "Comments." In Rudiger Dornbusch and James M. Poterba, eds., *Global Warming: Economic Policy Responses*. Cambridge, Mass.: MIT Press, 222–228.

————. 1992. *The Economics of Global Warming*. Washington, D.C.: Institute for International Economics.

Colborn, Theo, Diane Dummanoski, and John Peterson Myers. 1997. *Our Stolen Future: Are We Threatening Our Fertility, Intelligence, and Survival? A Scientific Detective Story*. New York: Plume/Penguin.

Dai, Aiguo, Inez Y. Fung, and Anthony D. del Genio. 1997. "Surface Observed Global Land Precipitation Variations During 1900–1988." *Journal of Climate* 10:2943–2962.

Epstein, Paul, Georg Grabherr, Tom Karl, Ellen Mosley-Thompson, Kevin Trenberth, and George M. Woodwell. 1996. Current Effects: Global Climate Change. An Ozone Action Roundtable, 24 June, Washington, D.C. [http://www.ozone.org/curreff.html]

Ewen, Alexander. 1999. "Consensus Denied: Holy War Over Global Warming." *Native Americas* 16, no. 3/4(Fall/Winter): 26–33. [http://nativeamericas.aip.cornell.edu]

Gaffen, Dian J. 1998. "Falling Satellites, Rising Temperature?" *Nature* 394 (13 August): 615–616.

Gelbspan, Ross. 1997. *The Heat is On: The High Stakes Battle Over Earth's Threatened Climate*. Reading, Mass.: Addison-Wesley Publishing Co.

————. 1997. "A Global Warming." *The American Prospect* 31 (March/April). [http://www.prospect.org/archives/31/31gelbfs.html]

————. 1995. "The Heat is On: The Warming of the World's Climate Sparks a Blaze of Denial." *Harper's Magazine*, December. [http://www.dieoff.com/page82.htm]

Glick, Patricia. 1998. Global Warming: The High Costs of Inaction. San Francisco: Sierra Club. [http://www.sierraclub.org/global-warming/inaction.html]

"Greenhouse Effects: Global Warming is Well Under Way. Here Are Some Telltale Signs." 1999. *Time*, 13 December, 78–79.

Houghton, John. 1997. *Global Warming: The Complete Briefing*. Cambridge, U.K.: Cambridge University Press.

James, P.D. 1994. *The Children of Men*. New York: Warner Books.

Kellogg, W.W., and R. Schware. 1981. *Climate Change and Society*. Boulder, Colo.: Westview Press.

Knox, Joseph P. 1991. "Global Climate Change: Impacts on California. An Introduction and Overview." In Joseph P. Knox and Ann Foley Scheuring, eds., *Global Climate Change and California*. Berkeley: University of California Press, 1–25.

Knox, Joseph P., and Ann Foley Scheuring. 1991. *Global Climate Change and California*. Berkeley: University of California Press.

Knutson, T.R., R.E. Tuleya, and Y. Kurihara. 1999. "Simulated Increase in Hurricane Intensities in a CO2-Warmed Climate." *Science* 279 (13 February): 1018–1020.

Lewis, Lowell, William Rains, and Lynne Kennedy. 1991. "Global Climate Change and

California Agriculture." In Joseph P. Knox and Ann Foley Scheuring, eds., *Global Climate Change and California.* Berkeley: University of California Press, 97–122.

Logan, Leslie. 2000. "Intense Destruction: Storms of the Warming Waters." *Native Americas* 17, no. 1 (Spring): 28–29.

Manabe, S., and R.T. Wetherald. 1986. "Reduction in Summer Soil Wetness Induced by an Increase in Atmospheric Carbon Dioxide." *Science* 232:626–628.

"Many More Hot, Sticky Days to Come, Forecasters Predict." 1999. Knight-Ridder Washington Bureau. In *Omaha World-Herald,* 7 November, 23A.

Maugh, Thomas H. 2000. "Great Lakes Water Levels Show Spring is Arriving Earlier." *Los Angeles Times,* 25 May, B-2.

Nance, John J. 1991. *What Goes Up: the Global Assault on Our Atmosphere.* New York: William Morrow and Co.

Paulson, Tom. 1999. "Global Warming Could Bring Us Wetter Winters." *Seattle Post-Intelligencer,* 9 November. [http://www.post-intelligencer.com/local/clim09.shtml]

Schneider, Stephen H. 1983. "CO2, Climate, and Society: A Brief Overview." In Robert S. Chen, Elise Boulding, and Stephen H. Schneider, eds., *Social Science Research and Climate Change: An Interdisciplinary Appraisal.* Boston: D. Reidel Publishing Co., 9–15.

Scientists at the Smithsonian's National Museum of National History Find Global Warming to be Major Factor in Early Blossoming [of] Flowers in Washington. 2000. Press Release, Smithsonian Institution. March. [http://www.mnh.si.edu/feature.html]

U.S. Environmental Protection Agency (EPA). *The Cost of Holding Back The Sea.* 1997. Washington, D.C.: EPA. [http://users.erds.com/jtitus/holding/NRJ.html#causes]

Woodhouse, C. A., and J.T. Overpeck. 1998. "2000 Years of Drought Variability in the Central United States." *Bulletin of the American Meteorological Society* 79:12 (December): 2693–2714.

CHAPTER 10

Possible Solutions

Addressing the consequences of global warming will demand, on a worldwide scale, the kind of social and economic mobilization experienced in the United States only during its birthing revolution and World War II, and therein lies a problem. The buildup of greenhouse gases in the atmosphere is a nearly invisible, incremental crisis. Carbon dioxide is not going to bomb Pearl Harbor to kick start the mobilization. Author Jonathan Weiner observes, "We do not respond to emergencies that unfold in slow motion. We do not respond adequately to the invisible" (Weiner 1990, 241).

It has been said (not for attribution) that the best thing which could happen to raise worldwide concern about global warming would be a quick collapse of the West Antarctic ice sheet, which would raise worldwide sea level a notable number of feet over a *very* short time. When stock brokers' feet get wet on the ground floor of New York City's World Trade Center, all the world's competing economic interests might mobilize together and provide the sociopolitical responses necessary to address the atmosphere's overload of greenhouse gases before it is too late. The same water that could lap at the ground floor of the Trade Center also would ruin most farmers in Egypt and Bangladesh and slosh in the lobbies of glass towers of Hong Kong and Tokyo. Perhaps, only then, might all of humankind heed the implications of Chief Seah'tl's farewell speech a century and a half ago. We may be brothers (and sisters) after all. We shall see.

So far, at this writing in the year 2000 on the Christian calendar, humankind's collective nervous system—national and international leadership, public opinion, and so forth—hasn't done much about global warming. As of this writing, the flora and fauna of the planet Earth are still in the position of a laboratory frog submerged in steadily warming water.

This is not a secret crisis, just a politically unpalatable one. Al Gore, in *Earth in the Balance: Ecology and the Human Spirit*, raised a sociopolitical call for mobilization against human-induced warming of the Earth: "This point is crucial. A choice to 'do nothing' in response to the mounting evidence is actually a choice to continue and even accelerate the reckless environmental destruction that is creating the catastrophe at hand" (Gore 1992, 37). In his book, Gore, then a U.S. senator, called for a "global Marshall Plan," to include stabilization of world population, the rapid creation and development of environmentally appropriate technologies, and "a comprehensive and ubiquitous change in the economic 'rules of the road' by which we measure the impact of our decisions on the environment" (Gore 1992, 306).

Eight years after Gore issued his manifesto, fossil-fuel emissions had risen in the United States. Gore had captured the Democratic Party's nomination for president of the United States, and global warming had slipped from campaign radar. From this vantage point, one imagines the world lurching through the twenty-first century as global public opinion slowly galvanizes around year after year of high temperature records, and as public policy only slowly begins to catch up with the temperature curve. The temperature (and especially the dewpoint) may wake the global frog before he becomes poached meat. Whatever the outcome of the public policy debate, the odds are extremely high that the weather of the year 2100 will be notably warmer than today, as greenhouse "forcing" exerts an ever-stronger role in the grand dance of the atmosphere which produces climate.

Ross Gelbspan observes,

Global warming need not require a reduction of living standards, but it does demand a rapid shift in patterns of fuel consumption—reduced use of oil, coal, and the lighter-carboned natural gas to an economy more reliant on solar energy, fuel cells, hydrogen gas, wind, biomass, and other renewable energy sources. It is doubtful that capitalistic market forces will bring about this shift on their own, because market prices of fossil fuels do not incorporate their environmental costs. (Gelbspan, A Global Warming, 1997)

George Woodwell has been quoted as saying, "[For] all practical purposes, the era of fossil fuels has passed, and it's time to move on to the new era of renewable sources of energy." The other alternative, says Woodwell, is to accept the fact that "[t]he Earth is not simply moving toward a new equilibrium in temperature. . . . It is entering a period of continuous, progressive, open-ended warming" (Gordon and Suzuki 1991, 219).

In Jeremy Leggett's opinion,

The uniquely frustrating thing about global warming—to the many people who see its dangers—is that the solutions are obvious. There is no denying, however, that creating the necessary changes will require paradigm shifts in human behavior—particularly in the field of cooperation between nation-states—which have literally no precedent in human history. . . . There is no single issue in human affairs that is of greater importance. (Leggett 1990, 457)

According to a Greenpeace Report edited by Leggett, "The main routes to surviving the greenhouse threat are energy efficiency, renewable forms of energy production . . . less greenhouse-gas-intensive agriculture, stopping deforestation, and reforestation" (Leggett 1990, 462). Greenpeace also recommends redirecting spending away from armaments and toward development of a sustainable energy for the future of humankind (Leggett 1990, 470).

Of the broader picture, Michael MacCracken writes,

The underlying challenge is for industrialized society to achieve a balanced and sustainable coexistence with the environment, one that permits use of the environment as a resource, but in a way that preserves its vitality and richness for future generations. . . . The challenge [is] to transform our ways before the world is irrevocably changed . . . toward displacing militarization and the ever-increasing push for greater national consumption as the primary driving forces behind industrial activity. (MacCracken 1991, 35)

According to Donald Goldberg and Stephen Porter of the Center for International Global Law:

The Clinton administration has bungled repeated chances to initiate domestic measures. For example, recent legislation proposed by the White House to restructure the electric utility industry could have been crafted to require utilities to reduce their carbon-dioxide emissions. In fact, the Environmental Protection Agency lobbied hard for the authority to impose a cap-and-trade program on utilities' CO2 emissions, similar to the trading system that has lowered sulfur dioxide (SO2) emissions in a cost-effective way. This was a golden opportunity, as the restructuring bill is projected to save the average consumer roughly $200 a year, which would have more than offset the cost of reducing GHG [greenhouse-gas] emissions. Unfortunately, the White House chose to forgo this opportunity. (Goldberg and Porter 1998)

According to Goldberg and Porter, loopholes in the Kyoto Protocol, adopted at the insistence of the United States, permit richer countries to avoid many of its mandated emission reductions by purchasing allowances from other countries through the protocol's "flexibility mechanisms." The Buenos Aires Climate Conference (1998) negotiated a mechanism allowing trade in greenhouse gas emission rights in two markets. The first market would allow "sellers," nations which exceed greenhouse gas-reduction targets set in the 1997 Kyoto Protocol, to offer their rights for sale to "buyers," countries which have not met their targets. The second market, the Clean Development Mechanism, will allow industrialized countries to meet part of their greenhouse-gas-reduction quotas by transferring clean technology to poorer countries so that antipollution projects can be carried out there.

"If it buys all (or most) of its reductions," Goldberg and Porter write, "the United States will not get its own house in order. In the long run, efficiency and productivity in the U.S. economy will suffer because domestic industry will be shielded from any incentive to adapt" (Goldberg and Porter 1998). Under these provisions, the United States could "purchase" emission reduction credits from nations, such as Russia and Ukraine, which reduced their greenhouse-gas emissions during the 1990s because their economic infrastructure collapsed.

The continuing political wrangling over the Kyoto Protocol illustrates why the world is responding so slowly to the impending crisis of global warming. Climate diplomacy remains an arena dominated by competition of special (mainly national) interests. Meanwhile, a few countries, most of them in Europe, are taking steps to mitigate greenhouse forcing on their own. While British emissions of greenhouse gases by the year 2000 had fallen between five and six percent compared to the Kyoto Protocol 1990 targets, emissions in the United States rose 11 percent between 1990 and 1998. Canada's greenhouse-gas emissions rose 13 percent during the 1990s, while several European countries (including Britain) made substantial progress toward meeting the goals of the Kyoto Protocol by reducing their greenhouse-gas emissions as much as 10 percent compared to 1990 levels.

Denmark (which produces less than one percent of humankind's greenhouse gases) underwent something of a mobilization against global warming during the 1990s. Denmark was planning "farms" of skyscraper-sized windmills in the North and Baltic seas that, if plans materialize, will supply half the nation's electric power within 30 years. The Danish wind-energy manufacturers' association believes that electricity produced through wind power on a large scale will be financially competitive with power from plants burning fossil fuels, which will be phased out if wind power proves itself. Svend Auken, Denmark's environmental and energy minister, said that with half of his country's power coming from Norwegian hydroelectric plants and the other half from wind power, the country is planning to meet its electricity needs within three decades while reducing carbon dioxide production to nearly zero. The wind farms must prove their endurance in winter storms and stand up to the corrosion of seawater, but if they can, Denmark's windmills will prevent the production of 14 million tons of carbon dioxide a year.

While the fossil-fuel economy remained firmly entrenched in most of the world at the turn of the millennium, gains were being achieved in some basic areas of energy conservation. In 1994, for example, the average person in the United States was recycling 380 pounds a year, up from 62 pounds in 1960, a 613 percent increase (Casten 1998, 101–102). Following the passage of the Clean Air Act in 1972, the United States also made a concerted effort to limit the production of nitrous oxides by gas turbine engines. Before regulation, the typical gas turbine engine emitted 200 parts per million (p.p.m.). Since then, several technological innovations have reduced emissions to below 10 p.p.m.. Technology was being developed in the late 1990s which could reduce the rate to two to three p.p.m. (Casten 1998, 117–118).

THE IPCC'S COPING STRATEGIES

The Intergovernmental Panel on Climate Change's (IPCC) *Second Assessment* describes technological coping strategies which may help reduce the rapid rise in atmospheric greenhouse gases. For example, according to the IPCC, gains in energy efficiency of 10 to 30 percent "are feasible at negative to zero cost in each of the sectors in many parts of the world . . . over the next two to three decades" (Bolin et al. 1995). With more attention to technological improvements, according to the IPCC, efficiency gains could reach 50 to 60 percent in some industries. However, according to the IPCC, "Because energy use is growing worldwide, even replacing current technology with more efficient technology could still lead to an absolute increase in greenhouse-gas emissions in the future" (Bolin et al. 1995). The IPCC recommends a wide-ranging, intense examination of transportation, industrial processes, and other human activities which produce greenhouse gases with energy conservation and efficiency in mind. The recommendations extend to alternative, "clean" methods of energy production (biomass, solar, wind, etc.) and to management of the world's forest cover with its ability to absorb carbon dioxide in mind.

A number of measures could conserve and sequester substantial amounts of carbon . . . over the next 50 years. In the forestry sector, measures include sustaining existing forest cover; slowing deforestation; natural forest regeneration; establishment of tree plantations; promoting agroforestry. (Bolin et al. 1995)

The IPCC's *Second Assessment* also examines the use of taxes on carbon emissions, as well as on the production and use of energy derived from fossil fuels, encouragement of research and development into alternative energy technologies, along with "phasing out [of] existing distortionary policies which increase greenhouse-gas emissions, such as some subsidies and regulations" (Bolin et al. 1995).

The world economy and indeed some individual national economies suffer from a number of price distortions which increase greenhouse-gas emissions, such as some agricultural and fuel subsidies and distortions in transport pricing. A number of studies of this issue indicate that global emissions reductions . . . together with increases in real incomes are possible from phasing out fuel subsidies. (Bolin et al. 1995)

The IPCC's estimates of climate change in the next century are based, for the most part, on conservative assumptions that do not include such factors as possible sea level rise from partial disintegration of the West Antarctic Ice Sheet, or major changes in ocean circulation. Even using conservative projections, the tone of many IPCC reports is sobering. The IPCC's outlook for countries such as China and India, with large populations and rapidly growing resource-consuming middle classes, indicates that achievements in energy efficiency and development of fuels which do not consume fossil fuels "are likely to be insufficient to offset rapidly increasing emissions baselines, associated with increased economic growth and overall welfare" (Bolin et al. 1995).

CARBON AS A TAXABLE COMMODITY

As early as the 1980s, Colorado Senator Timothy Wirth introduced legislation in the United States Senate Committee on Energy and Natural Resources (as the National Energy Policy Act of 1988) to place mandatory controls on industries which produce greenhouse gases. The proposal was received by lobbyists (the centurions of the special-interest state) like a proverbial lead balloon, as Senator Wirth's measures languished in committee. Wirth also sought to legislate stringent regulation of energy efficiency, especially in automobiles, which produce 40 percent of human-generated carbon dioxide in the United States. Wirth favored limits on population growth and increased governmental support for alternative energies (wind, solar, hydrogen) which do not burn fossil fuels.

Several economists have suggested that greenhouse-gas emissions be taxed. As of 1990, however, carbon taxes were still a matter for economic theorists in all except a few European nations: Sweden (about $62 a ton), Finland ($6.50 per ton), and the Netherlands ($1.50 a ton) (Poterba 1991, 71, 73).

Chris Flaven of the Worldwatch Institute has proposed a worldwide carbon tax of $50 a ton. He also proposes paying 10 percent of the tax into a fund to subsidize development of new technologies that will reduce emissions of greenhouse gases. A study of greenhouse-gas emission-reduction strategies in Chicago supported a "CO2 Fund . . . which would pool resources from the state and private industry and make funds available in low-income communities in order to encourage emissions reductions" (Energy and Equity N.d.). In *Earth in the Balance*, Gore supports a fossil-fuel tax, part of which would be placed in an Environmental Security Trust Fund, "which would be used to subsidize the purchase by consumers of environmentally benign technologies, such as low-energy light bulbs or high-mileage automobiles" (Gore 1992, 349).

William D. Nordhaus has modeled the economic effects of carbon taxes ranging from $5 to almost $450 per ton (Nordhaus, Economic Approaches, 1991, 56). Nordhaus provides a range of carbon-tax rates with expected reductions in fossil-fuel use and costs to gross national product in the United States: he expects a carbon tax of $13 a ton to reduce carbon emissions six percent (while reducing greenhouse gases 10 percent), at a negligible cost to Gross National Product (GNP). A carbon tax of $98 a ton is expected, by Nordhaus, to reduce greenhouse gases 40 percent, with a half a percentage point decrease in GNP. Nordhaus' highest tax estimate, $448 per ton, would decrease greenhouse gases 90 percent, and lower GNP by more than four percent (Cline 1992, 168).

Nordhaus estimates that a $5 per ton tax would raise the price of coal 10 percent, the price of oil 2.8 percent, and the price of gasoline 1.2 percent. Applied worldwide, the same level of carbon taxation would, according to Nordhaus' models, reduce greenhouse-gas emissions 10 percent, provide $10 billion in tax revenue, and add $4 billion per year to the global economy. A $100 per ton tax on carbon emissions would raise the price of coal 205 percent, the price of oil 55 percent, and the price of gasoline between 23 and 24 percent, according to Nordhaus. The $100 tax would reduce greenhouse-gas emissions by an estimated 43 percent (close to the level recommended by the IPCC to forestall significant global warming). A $100 tax would provide $125 billion in tax revenues and would decrease global "net benefits" $114 billion, according to Nordhaus (1991). (Nordhaus' numbers indicate that achieving the kind of greenhouse-gas reductions sought by the IPCC would impose a high cost in economic dislocation, especially if the changes are undertaken on a crash basis.)

Michael E. Mann and Richard J. Richels contend that a carbon tax of $250 a ton will be necessary to suppress carbon dioxide emissions by 20 percent. Such a tax would add about 75 cents to the cost of a gallon of gasoline, and $30 to the cost of a barrel of oil (Cline 1992, 147).

More than 2,000 economists, including six Nobel Laureates, endorsed a statement on global warming during 1997 which called for the use of market mechanisms, including carbon taxes, to move world economies away from fossil fuels:

As economists we believe that global climate change carries with it significant environmental, economic, social, and geopolitical risks, and that preventive steps are justified. ... Economic studies have found that there are many potential policies to reduce greenhouse-gas emissions for which the total benefits outweigh the total costs. For the United States in particular, sound economic analysis shows that there are policy options that would slow climate change without harming American living standards and these measures may in fact improve U.S productivity in the longer run. (Economists' Letter 1997)

The United States tax code has been suggested as a tool to reduce fossil-fuel emissions, with such proposals as tax credits for energy-efficient appliances, changes in building codes focused on energy efficiency; energy efficiency standards for universities and health centers which receive federal grants, as well as tax incentives for fuel-efficient cars.

Stephen Schneider observes that any significant carbon tax probably would shape the future course of technology:

When the price of conventional energy goes up—and will stay up because everyone knows the carbon tax or quota system will be here to stay—then a host of entrepreneurs and governmentally assisted labs will take up the challenge to develop and test an unimaginable array of more efficient and decarbonized production and end-use alternatives. The higher price of carbon-intensive fuels will spur the R[esearch] and D[evelopment] investments, which economists call induced technological changes. (Schneider, Kyoto Protocol, 2000)

FULL COST-PRICING OF ENERGY

A 1991 National Academy of Sciences report recommended a "study in detail [of] the 'full social cost pricing' of energy, with a goal of gradually introducing such a system" (National Academy 1991, 73). Such a system would price energy to include not only the costs of production and distribution, but environmental costs as well. The study admits that "such a policy would not be easy to design or implement" (National Academy 1991, 73). Such a change would involve a full-scale restructuring of the accounting rules in the capitalistic marketplace, which defines "cost" only in terms of financial (not environmental) assets.

Schneider explains full cost-pricing of energy under the rubric of "integrated assessment."

Integrated assessment is an attempt to merge economy and ecology with the inclusion of externalities and the analysis of end-to-end costs. In other words, the price of a lump of coal isn't simply extraction, storage and transport, but health consequences of mining and burning, as well as the whole range of potential environmental alterations from the production and use of energy. The cost of a car isn't simply the materials, labor and profit, but should also include disposal costs, and tailpipe emissions. . . . Integrated assessment will fall short on several grounds . . . because of the technical difficulty [of] evaluating the whole range of costs and benefits of our activities and because, ultimately, the single unit of comparison . . . is monetary currency. (Schneider 2001)

Thomas R. Casten endorses an end to annual federal government spending of $4 billion on subsidies for producers of fossil fuels in the United States, mainly from tax credits related to depletion allowances. Such a change could take place under present accounting rules and assumptions. In the longer term, several observers, including Al Gore, propose that "the definition of Gross National Product (GNP) should be changed to include environmental costs and benefits," and that "the definition of productivity should be changed to reflect calculations of environmental improvement or decline." Gore calls such definitions "Eco-nomics" (Gore 1992, 346).

The Sierra Club also asserts that subsidies for fossil-fuel exploration should be ended. Instead, according to the Sierra Club, emissions of pollution sould be taxed, as government increases research and development expenditures for solar and other renewable-energy technologies. According to a Sierra Club study, the cost of generating wind energy fell 85 percent between 1981 and the middle 1990s. Improvements in photovoltaic cell technology also have reduced the per-unit cost of generating solar power. Despite the growing competitiveness of alternative fuels, Gelbspan argues that 90 percent of U.S. energy subsidies and incentives still go to fossil fuels, compared to 10 percent for energy alternatives which may help slow the rise in atmospheric greenhouse-gas levels (Gelbspan, *The Heat is On*, 1997, 96–97).

HIGHER RATES OF ENERGY EFFICIENCY

A number of reports have recommended that all energy use be closely scrutinized for technological improvements (Pearman 1991). Electricity-generating industries, automobiles, the home, and so forth have been studied for ways to save energy. Many observers believe that at least 50 percent of present energy use could be saved through changes that are available through existing technology. Only about one-third of the energy consumed by a typical coal-burning power plant in the United States actually drives steam generators which produce electricity. The other two-thirds of the energy consumed by a typical coal- or oil-fired power plant is lost as waste heat through smokestacks and cooling towers. Increasing attention is being paid to "co-generation" strategies that capture more of that waste heat for actual energy production. Industry in the United States made 25 percent more efficient use of energy in 1986 than in 1973.

Existing power plants offer many opportunities for conservation: "For a pulverized coal plant, 90 percent of the carbon dioxide can be recovered using a chemical absorption process to clean up flue gases" (Hendriks et al. 1989, 125). In such a process, according to C.A. Hendriks and colleagues of the Netherlands' University of Utrecht, "[T]he carbon dioxide is recovered by leading the flue gas through a solution containing the absorber" (Hendriks et al. 1989, 127) The cost of removing 90 percent of the carbon dioxide is roughly a 25 percent loss in generation efficiency and roughly a 35 percent increase in the cost of electricity, according to Hendriks.

According to many scenarios, coal- and oil-fired power plants would first be streamlined to generate electricity as efficiently as possible, then phased down, and finally out, in favor of yet-to-be-developed electricity generation capacity derived from biomass, solar photovoltaics, solar thermal, and wind. Carbon dioxide generated by fossil-fuel burning plants would be buried in the deep ocean or under depleted oil and gas fields.

Energy conservation strategies produced some gains in energy efficiency in the United States following the oil-shortage scares of the 1970s. In 1973, roughly 18,000 British Thermal Units (BTUs) of energy was required to produce a dollar of gross domestic product in the United States. By 1999, roughly 11,000 BTUs was required to produce the same amount (in constant-dollar terms), a decline of roughly 30 percent. In the United States, oil expenditures amounted to 8.5 percent of gross domestic product in 1981, and only 3 percent in 1999, according to the U.S. Energy Department (Liesman and Schlesinger 1999, A-1).

BASIC CHANGES WITH EXISTING TECHNOLOGY

Necessary changes could begin incrementally, on an individual basis, with low-technology, close-to-home measures such as painting buildings light colors, replacing standard light bulbs with more energy-efficient types, and requiring significantly higher gas mileage from all forms of motorized transportation.

Eventually, however, human economic activity and energy requirements must be decoupled as much as possible from the fossil fuels that are raising the levels of atmospheric carbon dioxide and other greenhouse gases.

A National Academy of Sciences report, issued in 1991, considers greenhouse-gas mitigation strategies in some detail and concludes that basic energy-conservation strategies (many requiring technological upgrades of infrastructure) could lower greenhouse-gas emissions in the United States by roughly one-third from 1990 levels (National Academy 1991, 63). This report contains a detailed list of such strategies, including many basic conservation measures in existing homes, businesses, cars, and trucks.

The National Academy of Sciences' list of mitigation strategies starts with the most prosaic—a reduction of air-conditioning usage and the urban "heat island" effect through a crash program to paint half the roofs in urban areas white. Next, the report recommends another crash program to replace incandescent lighting with compact fluorescent bulbs. Energy-efficiency technology would then be upgraded in residential and commercial water heaters, commercial lighting, commercial cooking, refrigeration, and appliances. Homes would be heavily insulated to conserve fuel consumed by space-heating and air-conditioning. Industrial energy efficiency would be upgraded through the use of co-generation, fuel switching, and new technology (National Academy 1991, 54–58).

The American Council for an Energy-efficient Economy estimates that 35 to 40 percent of the electricity consumed in the United States could be saved through existing technology and rigorous conservation measures. Electricity consumed by lighting could be cut nearly in half through adoption of compact fluorescent light bulbs. These bulbs consume one-sixth the energy used by "standard" bulbs. Singapore, the Philippines, Indonesia, Malaysia, and Thailand reduced their energy consumption by more than 20 percent in a decade (1985–1995) by implementing more efficient building codes (Gelbspan, *The Heat is On*, 1997, 116). The Marshall Islands began installing solar hot-water heaters to replace oil-fueled machines used by many of its 55,000 residents.

Some proposed changes are more behavioral than technological—such things as building more bicycle paths and urging workers to "telecommute" when possible. Other observers advocate dietary measures, such as avoiding beef in favor of plant-based foods. Cattle are responsible for 72 percent of methane released by livestock in the United States (Lovins and Lovins 1991, 378). Consideration has been given to farming with an eye to reducing greenhouse gases. Under this system, traditional farm machinery is not abandoned, but used less often. Natural sources of fertilization (such as legumes) are substituted when possible for synthetic fertilizers. Such a farm is designed as an ecosystem in which many wastes are recycled. Many of the practices of "sustainable agriculture" (such as spreading manure on fields) are hardly new. Some of them recall farming methods of pre-industrial times.

A.B. Lovins and L.H. Lovins comment,

Organic farming techniques are already rapidly spreading . . . for economic, health, and environmental reasons [which] can simultaneously reduce biotic CO2, N2O [Nitrous Oxide], and CH4 [Methane] emissions, directly from farmland, and indeed can reverse the CO2 emissions. (Lovins and Lovins 1991, 379)

A study by the Illinois Environmental Protection Agency found that a number of basic energy-saving measures (including heating-system retrofits, improved thermostats, better efficiency in hot-water generation, more insulation, installation of storm windows, and lighting retrofits) could reduce energy use in an average home by 22 percent. The same study also found that similar efficiencies in public buildings could reduce their natural-gas consumption by 20 percent, and electricity consumption by 13 percent (Energy and Equity N.d.).

Lovins and Lovins caution that the infrastructure changes required to retool machinery to use energy more efficiently will be expensive: roughly $200 billion a year for the United States, in 1989 dollars (Lovins and Lovins 1991, 432). Lovins and Lovins argue against the claim that mitigating global warming will not drastically curtail American lifestyles.

Nothing could be further from the truth. The fuel-saving technologies that can stabilize global climate while saving money actually provide unchanged services: showers as hot and tingly as now, beer as cold, rooms as brightly lit . . . homes as cozy in the winter and cool in the summer. . . . The quality of these and other services can often be not just sustained but substantially improved by substituting superior engineering for brute force. (Lovins and Lovins 1991, 433)

In Japan, by the beginning of 1999, control of greenhouse-gas emissions had reached the local (or prefecture) level. On April 1, Japan enacted a stringent law to combat global warming. The law is meant to bring Japan into compliance with the 1997 Kyoto Protocol on climate change. Each municipality has been instructed to submit its own plan to bring down greenhouse-gas emissions.

The city of Koga, in Ibaraki Prefecture, during July 1998, began to actively promote the bicycle as a means of daily transport. Two months earlier, the city hosted a conference among local governments from throughout Japan. Participants traveled between conference sites by bicycle. The city published a pamphlet offering hints on how to best use bicycles, and purchased twenty bikes for official use. Cyclists have been given the right-of-way on some roads in the city.

The San Francisco Board of Supervisors passed a resolution in 1998 supporting the Kyoto Protocol on climate change. The city of Oakland quickly followed (Eco Bridge N.d.).

Whereas, the City of San Francisco has begun to address its local contribution to global climate change through:
  1. Setting a long-term goal of eliminating climate-changing and ozone-depleting emis-

sions and toxics associated with energy production and use, as set forth in the Sustainability Plan adopted in 1997,

2. Passage of Resolution #227–95, supporting San Francisco's Participation in the Cities for Climate Protection Campaign, in which it pledged to take a leading role in reducing energy consumption, establishing a greenhouse gas reduction goal, and developing a local action plan to reduce local greenhouse gas emissions. . . . Therefore, be it resolved that the City and County of San Francisco supports the Kyoto Protocol on Climate Change as a small but significant means to reduce greenhouse gas emissions and stabilize the global atmosphere and as a necessary first step toward maintaining the health and quality of life for future generations. (Eco Bridge N.d.)

During 2000, Oakland, California, approved a plan to buy alternative power for all of its municipal needs. "It leads us in the direction of reducing global warming, stimulating new industry, and sets the pace for the national government," said Oakland Mayor Jerry Brown, a former California governor and presidential candidate (Boxall 2000, 3). Santa Monica last year became the first city in California to purchase green power for all its municipal needs during 1999. Palmdale soon followed and several other communities were preparing to do the same during 2000, including Santa Barbara, San Jose, and Santa Cruz. By the year 2000, one-eighth of the power consumed in California came from nonfossil-fueled sources, which include wind, geothermal, solar, methane, and hydropower from small dams. Large hydroelectric plants are not considered environmentally friendly.

URBAN ZONING CHANGES

Even as climate scientists rang alarm bells regarding global warming during the 1990s, the suburban rings of most major urban areas in the United States continued to grow, intensifying the reliance of many residents on their automobiles, which often were being driven longer distances to and from work and other engagements, raising greenhouse-gas emissions. A number of observers have recommended denser urban development that is more amenable to mass transit and walking. A group of scientists led by S.T. Boyle graphed gasoline consumption against urban density. Not surprisingly, the greatest energy consumption occurred in the least dense, newer urban areas in the United States (Houston, Denver, Los Angeles, etc.) which expanded rapidly after the automobile diffused early in the twentieth century. The densest (low-consumption) urban areas were in Europe and Asia, with large areas of urban infrastructure predating the automobile, such as London, Hamburg, and Tokyo (Boyle et al. 1991, 240).

"A LAW OF THE AIR"

The phrase "law of the air" was first used by Margaret Mead during a meeting in the early 1970s and repeated in "The Atmosphere: Endangered and Endangering," a Fogarty Conference report that she and and William Kellogg edited (Mead and Kellogg 1980). Stephen Schneider picked it up in his works during the 1980s. "Of course, we now have a 'law of the air' known as the Kyoto Protocol," commented Schneider (Schneider, Personal communication, 2000). Thus, we now have a *design* for a "Law of the Air." We will have a *real* Law of the Air when countries and companies can be taken to court, convicted, and penalized for illegal emission of greenhouse gases.

CHANGE THE AUTOMOBILE

In a laboratory near Stuttgart, Germany, engineers from DaimlerChrysler have been developing an experimental motor vehicle (called the NECAR4) which will run on a hydrogen fuel cell, emitting only water vapor from its exhaust pipe. Fuel cells of this type have been used by NASA to generate power aboard space vehicles. Only during the 1990s, however, did hydrogen fuel cells become small enough to use as automobile engines. The prototype NECAR4 is modeled on a Mercedes sedan and is capable of carrying five people and luggage 280 miles between fueling stops at speeds of up to 90 miles (145 kilometers) per hour. Similar vehicles may be on the market by the year 2005. In any fuel cell, electricity is required to separate water into hydrogen and oxygen atoms. Someday, this electricity may be produced by solar and wind power, but with existing technology, such generation would be too expensive to make hydrogen a competitive fuel. The first hydrogen-fueled cars probably will be charged with electricity from existing sources, including fossil fuels.

Less than 20 percent of the energy consumed by a typical Ford Escort (an "economy car") is used to move the NECAR4. The rest is discharged, largely as waste heat, through the car's radiator and tailpipe (Oppenheimer and Boyle 1990, 125). Late in the twentieth century, one-fourth of the world's greenhouse gas total was being produced by more than 600 million automobiles which are, for the most part, generally as inefficient as that Ford Escort.

Proposals have been aired to steadily change the technology and fuel sources of automobiles to reduce greenhouse emissions. Under most of these plans, automakers would be required to produce more efficient internal combustion engines. As technology becomes available, automobiles would be adapted to operate on "hydrogen, electricity, and carbon fuels from biomass sources" (Jenney 1991, 300). By the year 2100, the gasoline-burning automobile of the twentieth century may be a museum piece. Combustion of fossil fuels may seem as old-fashioned to people of that time as a horse and buggy appears today.

According to a study by the National Academy of Sciences (NAS), released in 1991, greenhouse-gas emissions could be reduced markedly through increasingly higher mileage requirements for private automobiles and heavy trucks. The NAS study proposes that mileage standards should rise to about 48 miles per gallon (m.p.g.) for private vehicles and 40 m.p.g. for heavy trucks. People who commute in an automobile alone also would be severely taxed to encourage what the report calls "transportation-demand management" (National Academy 1991, 56). The Sierra Club recommends raising Corporate Average Fuel Economy (CAFE) standards in the United States to 45 m.p.g. for cars and 34 m.p.g. for light trucks by 2005 as "the biggest single step the U.S. can take to curb global warming and reduce our dependence on oil" (Glick 1998). Patricia Glick comments, "By simply adding existing technology to their vehicles, automobile manufacturers can slash global warming pollution and save consumers money at the same time" (Glick 1998). The Sierra Club points out that an automobile which meets its mileage standards (the Honda Civic VX) was being manufactured during the 1990s.

During 1997, Toyota introduced the world's first hybrid gas-electric car, the Prius, with much lower greenhouse-gas emissions than conventional automobiles. Marketed as a "green" sedan, the Prius has since sold so quickly in Japan that Toyota has opened a second production plant. By the year 2000, Honda was selling the Insight, a hybrid gasoline-electric car that gets 61 miles to a gallon of gasoline in start-and-stop city driving and 70 m.p.g. on the open road. Such cars were capturing only a small fraction of the market in the United States, however. Honda's initial shipment of Insights to the United States totaled only 4,000 cars (Welsh 2000, W-15C).

In Detroit, General Motors President John Smith said in 1998, "No car company will be able to survive in the twenty-first century by relying on the internal-combustion engine alone" (Worldwatch 1998). Early in 2000, General Motors unveiled a five-passenger, 54-horsepower automobile that it said will have a fuel efficiency of 80 miles to a gallon on the open road. The Precept uses a small diesel engine with an electric motor. One problem cited by engineers on these cars is that the diesel engine produces more nitrous oxides than air-pollution standards allow. The Precept's engine is in the rear of the car (to improve airflow). The car cuts weight by using aluminum and titanium rather than steel for some parts. This car is similar in size to a Chevrolet Malibu, but weighs 460 pounds less.

One problem with technological fixes is that humankind's mechanical infrastructure does not refashion itself instantly, and the poorer a region, the longer the transition. Automobiles are a very good example of this delay. Fuel standards mandated today will not apply to the entire fleet until older vehicles are retired from the road. A report on reducing greenhouse-gas emissions in Chicago endorsed "clunker" trade-in programs, which help to reduce greenhouse-gas emissions by paying nominal amounts of cash to retire "high emitters" from the road. Several clunker-purchase programs have been implemented in California, Delaware, and Illinois. The Illinois Environmental Protection Agency implemented a clunker trade-in program that targeted vehicles built before 1980. The cost per vehicle ranged from $647 to $950 depending on its age. The average cost was $860 per vehicle (Energy and Equity N.d.).

Improved fuel efficiencies may reduce greenhouse-gas emissions for the average individual automobile, but these savings are being more than outweighed by increasing numbers of cars on the road. The world fleet of automobiles and light trucks was 53 million in 1950 and 400 million by 1990. Annual production was 10 million in 1950 and 50 million in 1990. A report by the World Resources Institute, published in 1990 (MacKenzie and Walsh 1990), asserts that fundamental changes in transportation technology will be required to stabilize or reduce greenhouse gases. Hydrogen-powered and electric cars are mentioned, but the report does not speculate on how the electricity that would power these vehicles should be generated.

WIND ENERGY AND OTHER ALTERNATIVE SOURCES

Some major energy and natural resource companies were developing large-scale wind-power projects by the late 1990s. One example was Northern Alternative Energy's $32 million project, which entails construction of two wind-energy projects with Northern States Power (NSP) Company. The two projects will add a combined installed capacity of 23 megawatts to NSP's wind-energy resources. The new projects will be located on Buffalo Ridge near Hendricks, Minnesota. Northern Alternative Energy's partners in the Minnesota project include Edison Capital, a subsidiary of Southern California Edison.

The world's largest single wind-power generation facility was dedicated September 26, 1998, near Lake Benton, Minn. Constructed, owned, and operated by Enron Wind Corp., the wind turbines of this facility are based on an earlier machine developed in cooperation with the U.S. Department of Energy. Electricity generated by this wind facility is sufficient to power 43,000 homes and will displace greenhouse gases equivalent to removing 50,000 new cars and light trucks from the road. Power is being purchased by the Northern States Power Company, with a service territory which includes much of Minnesota, as well as parts of Wisconsin, North and South Dakota, and Michigan.

Wind-power investment in the United States increased from $600 million in 1993 to $2 billion in 1996. Some third world countries are beginning use wind power to reduce a small measure of their reliance on fossil fuels. India, by the middle 1990s, had installed 900 megawatts of wind-powered electrical energy and had plans to develop 500 megawatts of solar (photovoltaic) energy. India also concluded an agreement for a joint venture with Enron and Amoco to build a 50 megawatt solar-energy plant that would provide electricity to about 200,000 homes.

BIOMASS FUEL

The manufacture of fuel from biomass has been proposed, mostly for ground transportation. In some corn-producing areas of the United States (one example is Nebraska), ethanol has been mixed with gasoline for several years. Such fuels have not done well in the competitive marketplace because their manufacture is from two to four times as expensive as oil-derived fuels. A fuel-cost comparison indicates that while gasoline could be refined for 15 to 16 cents per liter (in the late 1980s), the cost of biofuels ranged from an average of about 30 cents per liter (for methanol derived from biomass) to 63 cents per liter (for ethanol derived from beets in the United Kingdom) (Barbier et al. 1991, 142).

As an attempt to reduce greenhouse gases, the growth and manufacture of biomass fuels has experienced problems other than cost of manufacture. Under some circumstances (if a biomass field replaced a forest, for example) this type of fuel might actually produce a net increase in emissions of greenhouse gases. Densely populated nations also would be hard pressed to spare from food production the large amounts of land required to raise vegetable matter destined for the world's gas tanks. When Brazil subsidized biomass production, for example, some farmers abandoned food because the government guarantees for biomass paid them more than they could earn for food on the open market (Barbier et al. 1991).

REFORESTATION

Proposals have been made to ameliorate increases in greenhouse-gas emissions through reforestation, the purposeful planting of large forests to absorb some of humankind's surplus carbon dioxide. The 1997 Kyoto Protocol contains mechanisms whereby governments of countries such as the United States, which produce more greenhouse gases per capita than average, may earn credit toward meeting their emissions goals by subsidizing the preservation of forests in poorer nations.

Reforestation could help slow greenhouse warming, but only if trees are planted on a very large scale. For example, if an area of 465 million hectares was planted, the trees on this land area, once mature, would remove almost 3 billion tons of carbon dioxide from the atmosphere per year, or about 40 percent of the carbon that human beings add to the air. The creation of such a carbon sink would require a land area roughly half the size of the United States (Silver and DeFries 1990, 122–123).

In England, some private firms have been planting trees to offset their contributions to global warming. The *London Sunday Independent* conducted a campaign during which readers bought more than 7,000 trees to offset the amount of carbon dioxide created by the manufacture of the newspaper over a year's time. The newspaper itself contributed 750 trees. The Glastonbury arts festival sold 1,333 trees to offset the equivalent amount of carbon dioxide produced by all the emissions created in the set-up, running, and dismantling of the show (Rowe 2000, 5). The trade in trees is coordinated by an organization called Future Forests. Some musical groups, such as The Pet Shop Boys and Neneh Cherry, have produced 1.5 million "carbon-neutral" compact discs, meaning they have bought enough trees to offset the carbon emitted by production of their recording as well as movement of their stage materials around the country.

Ecologist George Woodwell estimates that one to two million square kilometers of newly planted trees would remove one billion tons of CO_2 (1 ggt) annually, of the roughly seven million ggt being placed in circulation at the turn of the century. The problem would be finding large tracts of land fertile enough to support trees that isn't being used by human beings for other purposes. William R. Cline points out that reforestation is "a temporary remedy because a forest stores additional carbon only when it is expanding; once it reaches a steady state, the carbon released by dying trees often offsets that sequestered by new and growing trees" (Cline 1992, 216–217).

A study of greenhouse-gas reduction potential in Chicago endorsed urban tree-planting projects as a way to reduce air pollution in cities, where reductions are needed the most.

A strategy focused on tree-planting would have a number of benefits [which] include CO_2 absorption, removal of air pollution, and cooling and sheltering effects. Trees in Chicago have removed significant amounts of pollutants. In 1991, trees in Chicago removed an estimated. . . . 15 metric tons of carbon monoxide, 84 tons of sulfur dioxide, 89 tons of nitrogen dioxide, and 191 tons of ozone. Heating and cooling costs for buildings can also be reduced by strategically planting trees to shield buildings from wind and to provide shade in the summer months. (Energy and Equity N.d.)

PHOTOVOLTAICS AND OTHER SOLAR ENERGY SOURCES

Photovoltaic cells require no fuel (other than that provided by the sun) and little maintenance. The conversion of solar energy to electricity takes place with no moving parts and causes very little environmental disturbance. Because the sun must be shining to produce energy, however, it is likely that photovoltaic solar will develop, at least initially, with backup power from existing power-generation networks.

By the 1990s, photovoltaic cells already were cheaper than conventional, centralized power distribution to some remote locations, such as weather stations. This vast improvement in their efficiency probably will continue through the twenty-first century. By the end of the century, some present-day observers expect that many urban homes will produce their own electricity through photovoltaic solar panels mounted on rooftops. Centralized, fossil-fueled electricity generation may become obsolete, or one of several choices for electric-power generation.

During the 1970s, photovoltaic cells were manufactured by sawing large crystals of silicon into thin slices, an expensive and inefficient process. The cost of solar photovoltaic energy at the time was about $20 per peak watt. ("Peak," in solar energy terminology, means the cost of energy generated when the sun is at its peak elevation in the sky. The same word at a centralized, fossil-fueled or nuclear power plant means the cost of energy produced when customer demand is highest.) By the 1990s, however, work was underway to produce photovoltaics from semiconducter (computer) chips, potentially at a much lower cost once the technology is refined.

A study cited by Barry Commoner in his book *Making Peace With the Planet* indicates that aggressive expansion of photovoltaic technology could bring its cost down to $2-$3 per peak watt in one year, $1 per peak watt in three years, and 50 cents per watt in five years. At $1, photovoltaics would be cost-competitive with centralized electricity for roadside lighting. At 50 cents per peak watt, decentralized solar power could compete on a price basis with household power generated by fossil fuels in many U.S. markets (Commoner 1990, 207). Gelbspan wrote in 1997 that by the middle 1990s, photovoltaic electricity was being produced at 3.2 cents per kilowatt hour (k.w.h.), and wind power at 3.0 cents per k.w.h., rates which could compete with coal or oil-fired electricity in some situations (Gelbspan, *The Heat is On*, 1997, 96–97).

By the late 1980s, the Luz company of Southern California was selling enough solar-generated energy to Southern California Edison to meet the electricity demands of 270,000 homes. Technology also is being developed to produce "solar shingles," photovoltaic cells that can be positioned on the roofs of homes and businesses to generate electricity on-site. These are, perhaps, a harbinger of a new energy-delivery system, as well as a new source of power.

Solar power creates opportunities for decentralized "appropriate-scale" technology, especially in countries with large rural populations. One example is India, which averages 210 days a year of nearly direct sunlight, a large rural population, and a tradition of local, basic, small-scale problem-solving that stems from Mahatma Gandhi, who turned homespun cloth from a small spinning machine into a powerful political symbol vis-à-vis a centralized weaving industry controlled by the British.

By 1995, six thousand villages in India that had no access to conventional power grids were drawing electricity from banks of photovoltaic solar cells. Using the same model of small-scale, locally controlled technology, photovoltaic modules and solar cooking stoves are being used increasingly in India's villages. Many villagers also use biogas digesters which convert the dung of cows and other animals to energy. The resulting methane is burned as energy before it bubbles into the atmosphere as a greenhouse gas. New technology also allows dung to be turned to an energy-rich sludge without smoke and fire. A million digesters were operating in India by 1990, despite the fact that one of them costs about $50 (with half the amount paid by a government subsidy), or almost one-fifth of the average rural Indian's annual cash income (Oppenheimer and Boyle 1990, 137, 139). The digester-financing program is administered by the Indian federal government's Department of Non-conventional Energy.

These programs should not leave an impression that India, as a whole, is reducing its greenhouse-gas emissions. In India's cities, a growing middle class is creating demands for more energy, most of it generated from fossil fuels, especially coal. India has only one percent of the world's coal reserves, but it is fourth among the world's nations in coal combustion. During the late 1980s,

India had only 55,000 megawatts of electrical-generating capacity, twice the capacity of New York State, serving a population of more than 800 million (Oppenheimer and Boyle 1990, 137). These figures suggest that electricity generation via fossil fuels is still in its infancy on the Indian subcontinent.

INJECTION OF CARBON DIOXIDE INTO THE DEEP OCEANS

The oceans are well known to scientists as a major "sink," or repository, for atmospheric carbon dioxide and methane. This fact has led to various proposals to remove the nettlesome oversupply of these gases from the atmosphere and inject at least some of it into the depths of the oceans.

Peter G. Brewer, Gernot Friederich, Edward T. Peltzer, and Franklin M. Orr, Jr. have demonstrated that deep-ocean disposal of carbon dioxide is technologically feasible. They described a series of experiments in *Science* during which carbon dioxide was lowered into several hundred meters of seawater, at different depths. The carbon dioxide, which is a gas at the surface, formed solid hydrates that were expected to remain in the ocean depths for "quite long residence times" (Brewer et al. 1999).

This idea has been dissected and dismissed by Hein J.W. de Baar of the Netherlands Institute for Ocean Sciences:

The crucial problem with fossil fuel CO2 is its very rapid introduction within 100 to 200 years into the atmosphere, as opposed to the very slow response of many thousands to millions of years of the deep ocean in absorbing such CO2. Eventually, the capacity for storage of CO2 in the deep ocean is very large. Yet, in the meantime, we will witness a transient peak of atmospheric CO2 which may yield catastrophic changes in the climate. Only after several thousands to millions of years most, but not all, of the fossil fuel CO2 will be taken up by the oceans. (Baar and Stoll 1992, 143)

In addition, according to Baar and Stoll, deep-sea carbon dioxide injection is suitable only for large stationary energy plants (30 percent of total human emissions) and would raise the cost of generation 30 to 45 percent, while decreasing efficiency by a similar percentage. Much of the carbon dioxide injected into the deep oceans would eventually return to the surface, doubling seawater's acidity, which would be toxic for fish, plankton, and other life in the oceans. "Deep-sea injection is at best a partial, expensive, and temporal [temporary] remedy to the CO2 problem," write Baar and Stoll (1992, 144).

ATMOSPHERIC MODIFICATION

The mitigation strategies described in the 1991 NAS report conclude with consideration of atmospheric-modification projects. These include the placing of mirrored platforms in orbit to reflect sunlight, the use of guns or balloons to add dust to the stratosphere (to reflect sunlight), and the placing of billions of aluminized, hydrogen-filled balloons in the stratosphere, also to reflect incoming solar radiation. Additional strategies include the use of aircraft to maintain a dust cloud between earth and sun by making their engines less efficient—in other words, intentional air pollution (National Academy 1991).

Another proposed solution along the same lines involves the burning of sulfur in ships and power plants to form sulfate aerosols, or particles. It is believed that the sulfur particles will stimulate the formation of sunlight-reflecting clouds over the oceans. Sulfur also has a well-known cooling effect in the atmosphere. Mikhail Budyko, a Russian climatologist, has proposed a massive atmospheric infusion of sulfur that would form enough sulfur dioxide to wrap the earth in radiation-deflecting thin, white clouds within months. The net effect, according to Budyko, would be to cool the earth in a fashion similar to the massive eruption of the volcano Tambora in 1815. The eruption of Tambora ejected enough sulfur into the air to produce, in 1816, an annual cycle known to climate historians as "the year without a summer." Crops across New England and upstate New York were devastated by frosts which continued into the summer. Mohawk Indians at Akwesasne, in far northern New York State, reported frosts into June.

Such atmospheric modification probably will remain an intellectual parlor game because the environmental costs of filling the stratosphere with sulfur dioxide (or other pollutants) far outweigh the benefits, even in a warmer and more humid world. The sulfur dioxide would have to be refreshed at least twice a month, as the previous load washes onto the earth, planet-wide, as acid rain. Sulfuric acid also tends to attract chlorine atoms, creating a chemical combination that could assist chloroflourocarbons (CFCs) in devouring stratospheric ozone.

Another instance of proposed intentional pollution involves fertilizing colder ocean waters with iron to stimulate the reproduction of oxygen-generating photoplankton. Tsung-Hung Peng believes that iron fertilization could lower carbon dioxide levels, but only by about 10 percent, under perfect conditions (Peng 1993). J.H. Martin and R.M. Gordon also have been proponents of the iron-fertilization idea (Martin and Gordon 1988).

Yet another proposal involves the use of "lasers to break up CFCs in the atmosphere" (National Academy 1991, 58), an idea which evokes images of the Star Wars movies. Another proposal would create fields of photovoltaic solar cells on stations in orbit around the Earth or from bases on the moon. The tenuous nature of this technology is illustrated by a proposal that energy be transmitted from stations on the Moon to Earth via "some sort of energy beam" (Jenkins 1993, 107). Little thought is given to the energy costs of transporting materials and human workers who would construct such stations, nor the energy cost of maintaining bases in space or on the Moon, much less the method by which the energy would be transmitted to consumers on Earth. At present, these proposals remain within the pie-in-the-sky, when-pigs-fly range of possibilities.

THE ERADICATION OF INDUSTRIAL CIVILIZATION?

The German philosopher Friedrich Nietzsche once said, "The Earth is a beautiful place, but it has a pox called man" (Browne 1998). Bill McKibben's thesis, in *The End of Nature*, is similar: "We have built a greenhouse, a human creation, where once bloomed a sweet and wild garden" (McKibben 1989, 91). "If industrial civilization is ending nature, it is not utter silliness to talk about ending—or, at least, transforming—industrial civilization," writes McKibben (1989, 186). The "inertia of affluence, the push of poverty, [and] the soaring population" make McKibben pessimistic about humankind's chances of averting a general ruination of the Earth under greenhouse conditions (McKibben 1989, 204).

John Gribben's *Hothouse Earth*, one of several popular treatments of global warming which appeared shortly after the notably hot summer of 1988, begins with a tribute to the philosophy of the Earth as *Gaia*, a living organism that will restore its ecological balance, even if restoring that balance requires extinguishing the human race. At the beginning of the book, Gribben quotes Jim Lovecock: "People sometimes have the attitude that 'Gaia will look after us.' But that's wrong. If the concept means anything at all, Gaia will look after *herself*. And the best way for her to do that might well be to get rid of us" (Gribben 1990, frontispiece).

Gribben believes that what humankind is doing to the Earth's atmosphere will not be undone by human hands. His book ends with a profession that humankind will be unable to forestall the greenhouse effect by its own devices: "There is no prospect at all of bringing a halt to the release of carbon dioxide and other greenhouse gases, and thereby allowing the carbon cycles and the temperature of Gaia to return to normal" (Gribben 1990, 263).

Given rising populations, rising affluence, and rising levels of most greenhouse gases in the atmosphere, it is difficult, early in the twenty-first century, to conceive how the entire planet will be able to effectively achieve the technological and ideological paradigm shifts necessary to decouple human prosperity from the burning of fossil fuels. It is more difficult to imagine how our squabbling collection of nations and peoples will be able to respond as required to stop (much less reverse) an accelerating toward a warmer, more humid, and more miserable world. That said, a lot of authors have made retrospective fools of themselves by trying to forecast the future, so I can trust in my own fallibility, and hope, for the sake of the seventh generation, that my pessimism is at least partially in error.

REFERENCES

Baar, Hein J.W. de, and Michel H.C. Stoll. 1992. "Storage of Carbon Dioxide in the Oceans." In F. Stuart Chapin III, Robert L. Jefferies, James F. Reynolds, Gaius

R. Shaver, and Josef Svoboda, eds., *Arctic Ecosystems in a Changing Climate: An Ecophysiological Perspective*. San Diego: Academic Press, 143–177.

Barbier, Edward B., Joanne C. Burgess, and David W. Pearce. 1991. "Technological Substitution Options for Controlling Greenhouse-gas Emissions." In Rutiger Dornbusch and James M. Poterba, eds., *Global Warming: Economic Policy Reponses*. Cambridge, Mass.: MIT Press, 109–161.

Bolin, Bert, John T. Houghton, Gylvan Meira Filho, Robert T. Watson, M.C. Zinyowera, James Bruce, Hoesung Lee, Bruce Callander, Richard Moss, Erik Haites, Roberto Acosta Moreno, Tariq Banuri, Zhou Dadi, Bronson Gardner, J. Goldenberg, Jean-Charles Hourcade, Michael Jefferson, Jerry Melillo, Irving Mintzer, Richard Odingo, Martin Parry, Martha Perdomo, Cornelia Quennet-Thielen, Pier Vellinga, and Narasimhan Sundararaman. 1995. *Intergovernmental Panel on Climate Change. Second Assessment Synthesis of Scientific-Technical Information Relevant to Interpreting Article 2 of the United Nations Framework Convention on Climate Change*. Approved by the IPCC at its Eleventh Session, 11–15 December, Rome. [http://www.unep.ch/ipcc/pub/sarsyn.htm]

Boxall, Bettina. 2000. "Oakland Switches to 'Green' Power." *Los Angeles Times*, 29 June 3.

Boyle, S.T., W. Fulkerson, R. Klingholz, I.M. Mintzer, G.I. Pearman, G. Oinchera, J. Reilly, F. Staib, R.J. Swart, and C.-J. Winter. 1991. "Group Report: What Are the Economic Costs, Benefits, and Technical Feasibility of Various Options Available to Reduce Greenhouse Potential per Unit of Energy Service?" In G.I. Pearman, ed., *Limiting Greenhouse Effects: Controlling Carbon-dioxide Emissions*. Report of the Dahlem Workshop on Limiting the Greenhouse Effect, Berlin, December 9–14, 1990. New York: John Wiley & Sons, 229–260.

Brewer, Peter G., Gernot Friederich, Edward T. Peltzer, and Franklin M. Orr, Jr. 1999. "Direct Experiments on the Ocean Disposal of Fossil Fuel CO2." *Science* 284 (May 7): 943–945.

Browne, Malcolm W. 1998. "Will Humans Overwhelm the Earth? The Debate Goes On." *New York Times*, 8 December. [http://benetton.dkrz.de:3688/homepages/georg/kimo/0254.html]

Casten, Thomas R. 1998. *Turning Off the Heat: Why America Must Double Energy Efficiency to Save Money and Reduce Global Warming*. Amherst, N.Y.: Prometheus Books.

Cline, William R. 1992. *The Economics of Global Warming*. Washington, D.C.: Institute for International Economics.

Commoner, Barry. 1990. *Making Peace With the Planet*. New York: Pantheon.

Eco Bridge. N.d. "What Can We Do About Global Warming?" [http://www.ecobridge.org/content/g_wdo.htm]

Economists' Letter. 1997. Economists' Letter on Global Warming, 23 June. [http://uneco.org/Global_Warming.html]

"Energy & Equity: The Full Report." N.d. Illinois Environmental Protection Agency. [http://www.cnt.org/ce/energy&equity.htm]

Gelbspan, Ross. 1997. *The Heat is On: The High Stakes Battle Over Earth's Threatened Climate*. Reading, Mass.: Addison-Wesley Publishing Co.

———. 1997. "A Global Warming." *The American Prospect* 31 (March/April). [http://www.prospect.org/archives/31/31gelbfs.html]

Glick, Patricia. 1998. "Global Warming: The High Costs of Inaction." San Francisco: Sierra Club. [http://www.sierraclub.org/global-warming/inaction.html]

Goldberg, Donald, and Stephen Porter. 1998. "In Focus: Global Climate Change." *Center for International Environmental Law* 3, no. 12 (May). [http://www.foreignpolicy-infocus.org/briefs/vol3/v3n12cli.html]

Gordon, Anita, and David Suzuki. 1991. *It's a Matter of Survival.* Cambridge, Mass.: Harvard University Press.

Gore, Albert, Jr. 1992. *Earth in the Balance: Ecology and the Human Spirit.* Boston: Houghton Mifflin Co.

Gribben, John. 1990. *Hothouse Earth: The Greenhouse Effect and Gaia.* London: Bantam Press.

Hendriks, C.A., K. Blok, and W.C. Turkenburg. 1989. "The Recovery of Carbon Dioxide from Power Plants." In P.A. Okken, R. J. Swart, and S. Zwerver, eds., *Climate and Energy: The Feasibility of Controlling CO2 Emissions.* Dordrecht, Germany: Kluwer Academic Pubishers, 125–142.

Jenkins, Lyle M. 1993. "Space-based Geoengineering Options for Dealing with Global Change." In Richard A. Geyer, ed., *A Global Warming Forum: Scientific, Economic, and Legal Overview.* Boca Raton, Fla.: CRC Press, 101–109.

Jenney, L.L. 1991. "Reducing Greenhouse-gas Emissions From the Transportation Sector." In G.I. Pearman., ed., *Limiting Greenhouse Effects: Controlling Carbon-dioxide Emissions.* Report of the Dahlem Workshop on Limiting the Greenhouse Effect, Berlin, December 9–14, 1990. New York: John Wiley & Sons, 283–302.

Leggett, Jeremy, ed. 1990. *Global Warming: The Greenpeace Report.* New York: Oxford University Press, 149–162.

Liesman, Steve, and Jacob M. Schlesinger. 1999. "The Price of Oil Has Doubled This Year; So, Where's the Recession?" *Wall Street Journal,* 15 December, A-1, A-10.

Lovins, A.B., and L.H. Lovins. 1991. "Least-cost Climatic Stabilization." In G.I. Pearman, ed., *Limiting Greenhouse Effects: Controlling Carbon-dioxide Emissions.* Report of the Dahlem Workshop on Limiting the Greenhouse Effect, Berlin, December 9–14, 1990. New York: John Wiley & Sons, 351–442.

MacCracken, Michael. 1991. "Greenhouse Gases: Changing the Global Climate." In Joseph P. Knox and Ann Foley Scheuring, eds., *Global Climate Change and California.* Berkeley: University of California Press, 26–39.

MacKenzie, James J., and Michael P. Walsh. 1990. *Driving Forces: Motor Vehicle Trends and Their Implications for Global Warming, Energy Strategies, and Transportation Planning.* Washington, D.C.: World Resources Institute.

Martin, J.H., and R. M. Gordon. 1988. "Northeast Pacific Iron Distributions in Relation to Phytoplankton Productivity." *Deep-sea Resources* 35:177–196.

McKibben, Bill. 1989. *The End of Nature.* New York: Random House.

Mead, Margaret, and William Kellogg, eds. 1980. *The Atmosphere: Endangered and Endangering.* Tunbridge Wells, England: Castle House.

National Academy of Sciences. 1991. *Policy Implications of Greenhouse Warming.* Washington, D.C.: National Academy Press.

Nordhaus, William D. 1991. "Economic Approaches to Greenhouse Warming." In Rudiger Dornbusch and James M. Poterba, eds., *Global Warming: Economic Policy Responses.* Cambridge, Mass.: MIT Press, 33–66.

Oppenheimer, Michael, and Robert H. Boyle. 1990. *Dead Heat: The Race Against the Greenhouse Effect.* New York: Basic Books.

Pearman, G.I., ed. 1991. *Limiting Greenhouse Effects: Controlling Carbon-Dioxide Emissions.* Report of the Dahlem Workshop on Limiting the Greenhouse Effect, Berlin, December 9–14, 1990. New York: John Wiley & Sons.

Peng, Tsung-Hung. 1993. "Possible Reduction of Atmopsheric CO2 by Iron Fertilization in the Antarctic Ocean." In Richard A. Geyer, ed., *A Global Warming Forum: Scientific, Economic, and Legal Overview.* Boca Raton, Fla.: CRC Press, 263–285.

Poterba, James. 1991. "Tax Policy to Combat Global Warming." In Rutiger Dornbusch and James M. Poterba, eds., *Global Warming: Economic Policy Responses.* Cambridge, Mass.: MIT Press, 71–98.

Rowe, Mark. 2000. "When the Music's Over . . . a Forest will Rise." *London Independent,* 25 June, 5.

Schneider, Stephen. H. 1989. *Global Warming: Are We Entering the Greenhouse Century?* San Francisco: Sierra Club Books.

———. 2000. Kyoto Protocol: The Unfinished Agenda: An Editorial Essay. Unpublished mss. supplied by the author, 18 March.

———. 2001. "No Therapy for the Earth: When Personal Denial Goes Global." In Michael Aleksiuk and Thomas Nelson, eds., *Nature, Environment & Me: Explorations of Self In A Deteriorating World.* Montreal: McGill-Queens University Press.

———. 2000. Personal communication, 18 March.

Silver, Cheryl Simon, and Ruth S. DeFries. 1990. *One Earth, One Future: Our Changing Global Environment.* Washington, D.C.: National Academy Press.

Turkenberg, W.C., E.A. Alsema, and K. Blok. 1989. "The Prospects of Photovoltaic Solar Energy Conversion." In P.A. Okken, R. J. Swart, and S. Zwerver, eds., *Climate and Energy: The Feasibility of Controlling CO2 Emissions.* Dordrecht, Germany: Kluwer Academic Pubishers, 95–106.

Weiner, Jonathan. 1990. *The Next One Hundred Years: Shaping the Fate of Our Living Earth.* New York: Bantam Books.

Welsh, Jonathan. 2000. "Drive Buys: Honda Insight. A Car With an Extra Charge." *Wall Street Journal,* 10 March, W-15C.

"Wind Facility Provides Clean, Renewable Energy to Thousands in Midwest." 1998. EarthVision Reports, 25 September. [http://climatechangedebate.com/archive/09–18_10–27_1998.txt]

Worldwatch (Institute) Pre-Buenos Aires Briefing. 1998. Press Release: Kyoto Protocol Faces Crucial Test. [http://bonanza.lter.uaf.edu/~davev/nrm304/glbxnews.htm]

Yellen, Janet, Chair, White House Council of Economic Advisers. 1998. Statement on the Economics of the Kyoto Protocol before the Committee on Agriculture, Nutrition, and Forestry, U.S. Senate, March. Washington, D.C. [http://www.fetc.doe.gov/products/gcc/research/economic.html]

Postscript

As *The Global Warming Desk Reference* was going to press, research was published indicating that old, wild forests are far better than plantations of young trees at ridding the air of carbon dioxide. The analysis, published in the journal *Science*, was authored by Dr. Ernst-Detlef Schulze, the director of the Max Planck Institute for Biogeochemistry in Jena, Germany; and two other scientists at the Institute. The study provides an important new argument for protecting old-growth woods. The German study, together with other similar research, has produced a picture of mature forests that differs sharply from long-held notions of forestry science. This is important because the Kyoto Protocol is based on the older notions, and favors new-growth forests with credits.

This research raised some diplomatic eyebrows as representatives of 175 countries gathered at The Hague in the Netherlands to refine the enforcement mechanisms of the Kyoto Protocol. Several thousand protestors built a wall of sandbags around the site of the talks to dramatize the effects that flooding could have on the low-lying Netherlands during the coming century. Netherlands farmers, suffering a third straight year of flooding rains, had another global-warming worry: salt inundation of their "polder," fields reclaimed from a now-rising North Sea. Nearly a third of the Netherlands lies below sea level. Storm surges are a particular worry along the dikes. In 1953, 2,000 Dutch died in a storm surge that covered one-sixth of the Netherlands' territory (Hale 2000, 27-A).

Dr. Andrew Dlugolecki, a climate-change specialist with CGNU, the world's sixth-largest insurance company, told climate talks at The Hague that with world economic growth (GDP) averaging 3 percent a year and losses from climate disasters rising 10 percent a year, the two curves will cross in the year 2065, portending bankruptcy for the world. The cost of damage will exceed the value of the world's resources (Brown, Islands, 2000, 21; McCarthy, Climate Change, 2000, 6). The same day (November 23, Thanksgiving in the United States), island nations pleaded for relief from rising sea levels, and Frank Loy, U.S.

chief negotiator at the climate talks, was hit in the face by protestors with a raspberry, chocolate, and cream pie. The pie was a symbol of frustration at the U.S. role in the collapse of the climate talks.

At the same climate talks, the DuPont Company, once the world's largest producer of ozone-consuming CFCs, announced that an ambitious corporate program had eliminated half of the company's greenhouse-gas emissions since 1990, nearly all of it without losing sales or profits.

During the first week of October 2000, the chief executive officer of Ford Motor Company, speaking before a Greenpeace business conference in London, predicted the demise of the internal-combustion engine. Bill Ford, 43, great-grandson of Henry Ford, described an all-out race by auto makers to design the first mass-produced hydrogen fuel-cell vehicles, anticipating that major Japanese and American car makers would be using fuel-cell vehicles to supersede gasoline engines by the year 2003 or 2004. According to one newspaper account, Ford called global warming "the most challenging issue facing the world." Anyone who denies this fact, according to Ford, was said to be "in denial." "I believe fuel cells will finally end the 100-year reign of the internal-combustion engine," said Ford (McCarthy, Ford Predicts, 2000, 10).

At about the same time, the International Panel on Climate Change's 2000 assessment, a summary of which was distributed to government officials world-wide during late October of that year, said that while uncertainties remain, studies during the last five years, including more sophisticated computer modeling, show that "there is now stronger evidence for a human influence" on the climate and more certainty that man-made greenhouse gases have contributed substantially to observed warming during the last 50 years. The new assessment forecasts that if greenhouse emissions are not curtailed the earth's average surface temperatures may increase substantially more than previous IPCC estimates. The panel concluded that average global temperature increases ranging from 2.7 to as much as 11 degrees F. can be expected by the end of this century if current trends in emissions of heat-trapping gases continue. In its 1995 assessment, the panel had placed the projected increases in a range 1.8 to 6.3 degrees F. The new assessment said higher temperature forecasts stem mainly from more sophisticated computer modeling and an expected decline in sulfate releases into the atmosphere, especially from power plants for environmental reasons.

Also during October, researchers from the British Meteorological Office's Hadley Center for Climate Prediction said that land temperatures may rise by 6 degrees C. during the twenty-first century, 2 degrees C. more than the same body's previous estimate. This projection resulted from a computer model forecasting that global warming will accelerate as warmer conditions reduce the amount of carbon dioxide absorbed by soil and vegetation. Michael Meacher, British environment minister, said the implications of the research, published in the journal *Nature*, were "mind-blowing" (Houlder 2000, 2).

The oceans, soil and vegetation now absorb more than half of the carbon dioxide emitted into the atmosphere. This study projects that soil and vegetation will stop absorbing the gas and start emitting it, on balance, by 2050. The main causes will be greater respiration by plants in warmer soils and damage to the Amazonian rainforest caused by drier conditions. The new study indicates that within a century human society will experience a temperature rise unprecedented in its history; by the twenty-second century, if fossil-fuel use is not sharply curtailed, the Earth may be returning to the climatic days of the dinosaurs. "We have had warmer periods in the past," said Geoff Jenkins of the Hadley Centre, "but the rate of warming is quite unprecedented" (Radford 2000, 10).

The worst storms experienced by England in several decades caused road and rail chaos across the country during October and November. Torrential rain and winds up to 90 m.p.h. uprooted trees, blocked roads, and cut electricity supplies across southern England and Wales. According to newspaper reports, a tornado ripped through a trailer park in Selsey in West Sussex less than 48 hours after a similar twister devastated parts of Bognor Regis.

Marilyn McKenzie Hedger, head of the United Kingdom Climate Impacts Program based at Oxford University, said: "These events should be a wake-up call to everyone to discover how we are going to cope with climate change" (Brown October 31, 2000, 1). Michael Meacher, U.K. environment minister, said that while it would be foolish to blame global warming every time extreme weather conditions occur, "The increasing frequency and intensity of extreme climate phenomena suggest that although global warming is certainly not the sole cause, it is very likely to be a major contributory factor" (Brown October 31, 2000, 1). The storm included the lowest barometric pressure on record (951 millibars) during October in the United Kingdom.

As Great Britain and the rest of Western Europe continued to be pummeled by repeated storms in early November, thirty climate scientists coordinated by Martin Parry produced projections of twenty-first century weather for Europe in a study for the European Union by the Jackson Environment Institute of the University of East Anglia. The study projects that northern Europe will be stormier and warmer, with truly cold winters becoming rare by century's end. It foresees southern Europe becoming hotter and drier—so hot, in some cases, that many vacation destinations will be too hot for tourism in mid-summer. The study also projects sea-level rises that will eliminate all of the Mediterranean Sea's marshland, endangering several species of birds and other animals with habitats in the inter-tidal zones. Agriculture and water resources will be stressed in southern Europe, while forest growth rates are expected to increase in the north, which also will experience longer growing seasons, more flooding, and less snow pack. Between 50 and 90 percent of Europe's glaciers are expected to melt by the end of the century. The report was released during early November, 2000, as record floods (and even a few tornadoes) wracked Great Britain, following a summer of intense heat in southeastern Europe.

As I read these dire forecasts of future warming, during November, Omaha experienced a sharp, early winter. November, 2000, was refreshingly cool in Omaha, at 6 degrees F. below average—13.5 degrees below the notably mild

November of 1999, described in the Preface, a reminder that climate is still very variable, even in a globally warmed world. On the first anniversary of our 82-degree F. record high in November, 1999, the high temperature was 30 degrees F., with snow on the ground.

By mid-December, Omaha was experiencing its most intense spell of traditional Nebraska winter nastiness in almost two decades. During the month, in fact, Omaha was treated to a pair of an old-fashioned Midwestern blizzards, with the wind chill pushing minus 40 degrees F. on several days. In the meantime, my e-mail correspondent Lars Olle Engaa reported that he had swapped his cross-country skis for a bicycle as his home town, Olso, Norway, experienced a record spell without snow. As snow and cold swept the heartland of North America, in Britain, much higher-than-average temperatures prompted a pair of black swans to breed five months out of season. Elsewhere in England, rhododendrons flowered in December. The World Meteorological Organization said that the year 2000 was the fifth warmest year on record worldwide, even with interior North America's brutal December factored in.

In point of meteorological fact, having experienced its warmest January through October on record in the year 2000, the United States' November-December was the *coldest* on record. By the first day of calendar winter (December 21), Omaha had its snowiest December, at 20.4 inches, surpassing the old record, set in 1969, of 19.9. Chicago got twice as much the same month. The temperature for the month, in Omaha, averaged 9.5 degrees F. below average. Throughout the winter, this paradoxical season continued to build its resume. By February 1, 2001, Omaha had its average annual snowfall of 32.5 inches, to which was added another eight in a blizzard eight days later—another day, another snow storm and cold wave, yet another emphatic reminder that we are not traveling a straight line back to the climatic days of the dinosaurs—not just yet.

As I picked my way over old ice in Omaha, reports arrived that Mount Kilimanjaro had lost 90 percent of its glacial mass in the last ninety years, and could be expected to lose the rest of it by roughly 2020. The same study said that a third of the ice cap that Ernest Hemingway celebrated had melted during the last twelve years. Glacier National Park (in northern Montana) lost two-thirds of its ice during roughly the same period, raising the possibility that it could be glacier-free within a half century.

The world as a whole warmed up considerably in January, 2001. According to the Goddard Institute for Space Studies' temperature reports, the month tied for the third-warmest January in the historical record. In Washington, D.C., during late March, the annual Cherry Blossom Festival Parade was moved ahead a week to catch up with earlier blooms. Omaha, meanwhile, was enduring a plague of axle-eating potholes, the end result of a streak of freezing nights that lasted from late November to early March, the longest such streak in the city's recorded history—a winter worth describing to wide-eyed great-grandchildren, perhaps, in warmer times.

REFERENCES

Brown, Paul. October 31, 2000. "Global Warming: It's with Us Now: Six Dead As Storms Bring Chaos Throughout The Country." *The Guardian* (London), 1.
———. November 24, 2000. "Islands in Peril Plead for Deal." *The Guardian* (London), 21.
Drozdiak, William. November 24, 2000. "U.S. Firms Become 'Green' Advocates; Global Warming Talks Near End." *Washington Post*, E-1, 2.
Hale, Ellen. November 22, 2000. "Climate Changes Could Devastate the Netherlands." *USA Today*, 27-A.
Houlder, Vanessa. November 9, 2000. "Faster Global Warming Predicted; Met Office Research has 'Mind-blowing' Implications." *The Financial Times* (London), 2.
McCarthy, Michael. November 24, 2000. "Climate Change Will Bankrupt the World." *The Independent* (London), 6.
———. October 6, 2000. "Ford Predicts End of Car Pollution; Boss Predicts the End of Petrol." *The Independent* (London), 10.
Radford, Tim. November 9, 2000. "World May be Warming up Even Faster; Climate Scientists Warn New Forests Would Make Effects Worse." *The Guardian* (London), 10.

Bibliography

Abdulla, Sara. 1998. "Hot and Sticky in the States." *Nature*, 17 December. [http://imagine. nature.com/nsu/981217/981217-3.html]

Abrahamson, Dean Edwin. 1989. "Global Warming: The Issue, Impacts, Responses." In Dean Edwin Abrahmson, ed., *The Challenge of Global Warming*. Washington, D.C.: Island Press, 3–34.

———, ed. 1989. *The Challenge of Global Warming*. Washington, D.C.: Island Press, 1989.

Abramovotiz, J.N., and S. Dunn. "Record Year for Weather-Related Disasters." Vital Signs Brief 98–5. Washington, D.C.: Worldwatch Institute. [www.worldwatch. org/alerts/981127.html]

Abrupt Climate Change During Last Glacial Period Could be Tied to Dust-Induced Global Warming. 1996. Press Release NOAA 96–78, 4 December. [http://www. noaa.gov/public-affairs/pr96/dec96/noaa96–78.html]

Ackerman, A.S., O.B. Toon, D.E. Stevens, A.J. Heymsfield, V. Ramanathan, and E.J. Welton. 2000. "Reduction of Tropical Cloudiness by Soot." *Science* 288 (12 May): 1042–1047.

Ackerson, M.D., E.C. Clausen, and J.L. Gaddy. 1993. "The Use of Biofuels to Mitigate Global Warming." In Richard A. Geyer, ed., *A Global Warming Forum: Scientific, Economic, and Legal Overview*. Boca Raton, Fla.: CRC Press, 475–485.

Adams, Jonathan, Mark Masli, and Ellen Thomas. N.d. "Sudden Climate Transitions During the Quaternary." Article in press in *Progress in Physical Geography*. [http: //www.esd.ornl.gov/projects/qen/transit.html]

Adams, Richard M., Ronald A. Fleming, Ching-Chang Chang, Bruce A. McCarl, and Cynthia Rosenzweig. 1995. "A Reassessment of the Economic Effects of Global Climate Change on U.S. Agriculture." *Climatic Change* 30, no. 2: 147–167.

Adger, W.N., and K. Brown. 1994. *Land Use and the Causes of Global Warming*. Chicester, U.K.: Wiley.

Adkins, J.F., E.A. Boyle, L. Keigwin, and E. Cortijo. 1997. "Variability of the North Atlantic Thermohaline Circulation During the Last Interglacial Period." *Nature* 390:154–156.

AEI Study Calls for Delayed Action on Global Warming. 1999. 28 July. [http://
www.weathervane.rff.org/negtable/AEIStudy.html]

"Africa: Fighting Back the Widening Deserts." 1998. BBC News, 30 November. [http://
news.bbc.co.uk/hi/english/world/africa/newsid_224000/224597.stm]

Agarwal, Anil, and Sunita Narain. 1991. *Global Warming in an Unequal World: A Case
of Environmental Colonialism*. New Delhi, India: Centre for Science and Environment.

Agostini, P., M. Botteon, and C. Carraro. 1992. "A Carbon Tax to Reduce CO2 Emissions in Europe." *Energy Economics* 14, no. 4: 279–290.

"Aircraft Pollution Linked to Global Warming; Himalayan Glaciers are Melting, with
Possibly Disastrous Consequences." 1999. Reuters in *Baltimore Sun*, 13 June,
13A.

Aldhous, Peter. 2000. "Global Warming Could be Bad News for Arctic Ozone Layer."
Nature 404 (6 April): 531.

Allen, L.H. 1990. "Plant Responses to Rising Carbon Dioxide and Potential Interactions
with Air Pollutants." *Journal of Environmental Quality* 19:15–34.

Alley, Richard B. 2000. "Ice-core Evidence of Abrupt Climate Changes." *Proceedings
of the National Academy of Sciences of the United States of America* 97:4 (15
February): 1331–1334.

Almendares, J., and M. Sierra. 1993. "Critical Conditions: A Profile of Honduras." *Lancet*
342 (4 December): 1400–1403.

Alward, Richard D., James K. Detling, and Daniel G. Milchunas. 1999. "Grassland Vegetation Changes and Nocturnal Global Warming." *Science* 283 (8 January): 229–
231.

Amazon Destruction More Rapid Than Expected. 1999. GS Report, 10 April. [http://
www.gsreport.com/articles/art000099.html]

"Analysis: Climate Warming at Steep Rate." 2000. *Los Angeles Times*. In *Omaha World-
Herald*, 23 February, 12.

Anderson, Christopher. 1993. "Tales of the Coming Mega-greenhouse." *Science* 261:553.

Anderson, Glen D. 1995. "Joint Implementation Projects to Reduce Greenhouse Gas
Emissions: Prospects for Poland." *Geographia Polonica* 65, no. 2: 111–122.

Anderson, Julie. 1999. "Weather Runs Hot, Cold in Year of Surprises." *Omaha World-
Herald*, 7 December, 1, 5.

———. 2000. "UNL Student Helps Shed New Light on El Nino." *Omaha World-Herald*,
14 May, 12B.

Anderson, J.W. 1999. "The History of Climate Change as a Political Issue." The Weathervane: A Global Forum on Climate Policy Presented in August by Resources
for the Future. [http://www.weathervane.rff.org/features/feature005.html]

Anisimov, O.A. 1989. "Changing Climate and Permafrost Distribution in the Soviet
Arctic." *Physical Geography* 10:285–293.

"Antarctic Meltdown." 1998. *World Press Review*, November, 36.

Appenzeller, C., T.F. Stocker, and M. Anklin. 1998. "North Atlantic Oscillation Dynamics Recorded in Greenland Ice Cores." *Science* 282 (16 October): 446–449.

"Arctic Ice Melting, With Help." 1999. Associated Press. In *Omaha World-Herald*, 3
December, 8.

"Arctic Region Quickly Losing Ozone Layer." 2000. Knight-Ridder News Service. In
Omaha World-Herald, 6 April, 4.

Arkin, William M., Damian Durrant, and Marianne Cherni. 1991. *On Impact: Modern*

Warfare and the Environment, A Case Study of the Gulf War. Washington, D.C.: Greenpeace.

Aronson, Richard B., William F. Precht, Ian G. MacIntyre, and Thaddeus J.T. Murdoch. 2000. "Coral Bleach-out in Belize." *Nature* 405 (4 May): 36.

Arrhenius, E., and T. Waltz. N.d. "The Greenhouse Effect: Implications for Economic Development." Discussion Paper 78. Washington D.C.: World Bank.

Arrhenius, Svante. 1896. "On the Influence of Carbonic Acid in the Air Upon the Temperature of the Ground." *The London, Edinburgh, and Dublin Philosophical Magazine and Journal of Science*, 5th ser. (April): 237–276.

———. 1912. "Electrolytic Dissociation." *Journal of the American Chemical Society* 34 (April): 353–365.

———. 1908. *Worlds in the Making: The Evolution of the Universe.* New York: Harper & Bros.

Augenbraun, Harvey, Elaine Matthews, and David Sarma. N.d. The Greenhouse Effect, Greenhouse Gases, and Global Warming. [http://icp.giss.nasa.gov/research/methane/greenhouse.html]

Avok, Michael. 2000. "Where's My Sunblock? Midlanders Relish Taste of Summer in March." *Omaha World-Herald*, 6 March, 11.

Ayres, R., and J. Walter. 1991. "The Greenhouse Effect: Damages, Costs, and Abatement." *Environmental and Resource Economics* 1, no. 1: 237–270.

Azar, C., and H. Rodhe. 1997. "Targets for Stabilization of Atmospheric Carbon Dioxide." *Science* 276:1818–1819.

Baar, Hein J.W. de, and Michel H.C. Stoll. 1992. "Storage of Carbon Dioxide in the Oceans." In F. Stuart Chapin III, Robert L. Jefferies, James F. Reynolds, Gaius R. Shaver, and Josef Svoboda. *Arctic Ecosystems in a Changing Climate: An Ecophysiological Perspective.* San Diego: Academic Press, 143–177.

Bach, Wilfrid. 1984. *Our Threatened Climate: Ways of Averting the CO2 Problem Through Rational Energy Use.* Trans. Jill Jager. Dordrecht, Germany: D. Reidel.

Balling, Robert C., Jr. 1992. *The Heated Debate: Greenhouse Predictions Versus Climate Reality.* San Francisco: Pacific Research Institute for Public Policy.

———. 1993. "The Global Temperature Data." *Research and Exploration* 9, no. 2: 201–207.

———. N.d. The Spin on Greenhouse Hurricanes. Fraser Institute. [http://www.fraserinstitute.ca/publications/books/g_warming/hurricanes.html]

Bamber, Jonathan L., David G. Vaughan, and Ian Joughin. 2000. "Widespread Complex Flow in the Interior of the Antarctic Ice Sheet." *Science* 287 (18 February): 1248–1250.

Barber, D.C., A. Dyke, C. Hillaire-Marcel, A.E. Jennings, J.T. Andrews, M.W. Kerwin, G. Bilodeau, R. McNeely, J. Southon, M.D. Morehead, and J.-M. Gagnon. 1999. "Forcing of the Cold Event of 8,200 Years Ago by Catastrophic Drainage of Laurentide Lakes." *Nature* 400 (22 July): 344–351.

Barber, Timothy R., and William M. Sackett. 1993. "Anthropogenic Fossil Carbon Sources of Atmospheric Methane." In Richard A. Geyer, ed., *A Global Warming Forum: Scientific, Economic, and Legal Overview.* Boca Raton, Fla.: CRC Press, 209–223.

Barber, Valerie A., Glen Patrick Juday, and Bruce P. Finney. 2000. "Reduced Growth of Alaskan White Spruce in the Twentieth Century from Temperature-induced Drought Stress." *Nature* 405 (8 June): 668–673.

Barbier, Edward B., Joanne C. Burgess, and David W. Pearce. 1991. "Technological

Substitution Options for Controlling Greenhouse-gas Emissions." In Rutiger Dornbusch and James M. Poterba, eds., *Global Warming: Economic Policy Reponses*. Cambridge, Mass.: MIT Press, 109–161.

Barnett, T.P. 1984. "The Estimation of 'Global' Sea Level Change: A Problem of Uniqueness." *Journal of Geophysical Research* 89:7980–7988.

Barnola, J.M., D. Raynaud, H. Oeschger, and A. Neftel. 1983. "Comparison of CO2 Measurements By Two Laboratories on Air from Bubbles in Polar Ice." *Nature* 303:410.

Barnola, J.M., D. Raynaud, Y.S. Korotkevich, and C. Lorius. 1987. "Vostok Ice Core Provides 160,000-year Atmospheric Record of CO2." *Nature* 329:408–414.

Barras, J.A., P.E. Bourgeois, and L.R. Handley. 1994. "Land Loss in Coastal Louisiana: 1956–1990." National Biological Survey, Fish and Wildlife Service, U.S. Dept. of Interior, National Wetland Center Open File Report 94–01.

Barrel, J., C. Schuchert, C. Woodruff, R. Lull, and E. Huntington. 1919. *The Earth and Its Inhabitants*. New Haven, Conn.: Yale University Press.

Barry, R.G. 1978. "Cryospheric Responses to a Global Temperature Increase." In Jill Williams, ed., *Carbon Dioxide, Climate, and Society: Proceedings of a IIASA Workshop Co-sponsored by WMO, UNEP, and SCOPE, February 21–24, 1978*. Oxford, UK: Pergamon Press, 169–180.

Barth, M.C., and J.G. Titus, eds. *Greenhouse Effect and Sea-Level Rise: A Challenge for This Generation*. New York: Van Nostrand Reinhold.

Basu, Janet. 1997. Ecologists' Statement on the Consequences of Rapid Climatic Change: May 20. [http://www.dieoff.com/page104.htm]

Bates, Albert K., and Project Plenty. 1990. *Climate in Crisis: The Greenhouse Effect and What We Can Do*. Summertown, Tenn.: The Book Publishing Co.

Bazzaz, Fakhri A., and Eric D. Fajer. 1992. "Plant Life in a CO2-rich World." *Scientific American* 266, no. 1 (January): 68–74.

Beal, Carole N., William Wagner, and James C. White. 1991. *Global Climate Change: The Economic Costs of Mitigation and Adaptation: Proceedings of a Conference*. New York: Elsevier.

Beardsley, Tim. 1992. "Night Heat." *Scientific American* 266, no. 2 (February): 21–22.

Bell, Art, and Whitley Strieber. 1999. *The Coming Global Superstorm*. New York: Pocket Books.

Benarde, Melvin A. 1992. *Global Warming—Global Warning*. New York: Wiley.

Bender, David, and Bruno Leone, eds. 1997. *Global Warming: Opposing Viewpoints*. San Diego: Greenhaven Press.

Benedick, Richard Elliott. 1991. *Greenhouse Warming: Negotiating a Global Regime*. Washington, D.C.: World Resources Institute.

Bengtsson, L., M. Botzet, and M. Esch. 1996. "Will Greenhouse Gas-induced Warming over the Next 50 Years Lead to a Higher Frequency and Greater Intensity of Hurricanes?" *Tellus* 48A:57–73.

Bengtsson, L., E. Roeckner, and M. Stendel. 1999. "Why is the Global Warming Proceeding Much Slower Than Expected?" *Journal of Geophysical Research* 104: 3865–3876.

Benjamin, Craig. 1999. "The Machu Picchu Model: Climate Change and Agricultural Diversity." *Native Americas* 16, no. 3/4(Fall/Winter): 76–81.

Bentley, Charles R. 1983. *West Antarctic Ice Sheet: Diagnosis and Prognosis*. Washington, D.C.: U.S. Department of Energy.

———. 1980. "Response of the West Antarctic Ice Sheet to CO2-Induced Global Warming." In *Environmental and Societal Consequences of a Possible CO2-Induced Climate Change*, vol. 2. Washington, D.C.: Department of Energy.

Bentley, C.R., and M.B. Giovinetto. 1990. "Mass Balance of Antarctica and Sea-level Change." In *International Conference on the Role of Polar Regions in Global Change*. Fairbanks: University of Alaska, 481–488.

Bering Sea Task Force. 1999. *Status of Alaska's Oceans and Marine Resources: Bering Sea Task Force Report to Governor Tony Knowles*. Presented in March. Juneau: State Government of Alaska.

Berk, Richard A., and Daniel Schulman. 1995. "Public Perceptions of Global Warming." *Climatic Change* 29, no. 1: 1–33.

Bernard, Harold W., Jr. 1993. *Global Warming: Signs to Watch For*. Bloomington: Indiana University Press.

Berner, Robert A., and Antonio C. Lasaga. 1989. "Modeling the Geochemical Carbon Cycle." *Scientific American* 260, no. 3: 74.

Bernstein, Mark, Scott Hassell, and Robert Lempert. 1999–2000. "May Cooler Tempers Prevail: Let Technology Reduce Hot Air over Global Warming." *RAND Review* 23, no. 3 (Winter): 12–17.

Bhaskaran, B., J.F.B. Mitchell, J.R. Lavery, and M. Lal. 1995. "Climatic Response of the Indian Subcontinent to Doubled CO2 Concentrations." *International Journal of Climatology* 15:873–892.

Bhattacharya, N. C. 1993. "Prospects of Agriculture in a Carbon-Dioxide-Enriched Environment." In Richard A. Geyer, ed., *A Global Warming Forum: Scientific, Economic, and Legal Overview*. Boca Raton, Fla.: CRC Press, 487–505.

Bianchi, Giancarlo, and I. Nicoholas McCave. 1999. "Holocene Periodicity in North Atlantic Climate and Deep-ocean Flow South of Iceland." *Nature* 397 (11 February): 515–517.

Bilger, Burkhard. 1992. *Global Warming*. New York: Chelsea House Publishers.

Billet, J. D., 1974. "Direct and Indirect Influences of Temperature on the Transmission of Parasites from Insects to Man." In A.E.R. Taylor and R. Muller, eds., *The Effects of Meteorological Factors Upon Parasites*. Oxford: Blackwell Scientific Publications, 79–95.

Billings, W.D. 1992. "Phytogeographic and Evolutionary Potential of the Arctic Flora and Vegetation in a Changing Climate." In F. Stuart Chapin III, Robert L. Jefferies, James F. Reynolds, Gaius R. Shaver, and Josef Svoboda, eds., *Arctic Ecosystems in a Changing Climate: An Ecophysiological Perspective*. San Diego: Academic Press, 91–109.

———. 1995. "What We Need to Know: Some Priorities for Research on Biotic Feedbacks in a Changing Biosphere." In G. Woodwell and F.T. Mackenzie, eds., *Biotic Feedbacks in the Global Climate System*. New York: Oxford University Press, 377–392.

Billings, W.D., J.O. Luken, D.A. Mortensen, and K.M. Peterson. 1982. "Arctic Tundra: A Source or Sink for Atmospheric Carbon Dioxide in a Changing Environment?" *Oecologia* 53:7–11.

Billings, W.D., K.M. Peterson, J.O. Luken, and D.A. Mortensen. 1984. "Interaction of Increasing Atmospheric Carbon Dioxide and Soil Nitrogen on the Carbon Balance of Tundra Microcosms." *Oecologia* 65:26–29.

Bindoff, N. L., and J. A. Church. 1992. "Warming of the Water Column in the Southwest Pacific." *Nature* 357:59–62.

Birks, Hilary H., and Brigitta Amman. 2000. "Two Terrestrial Records of Rapid Climatic Change During the Glacial-Holocene Transition (14,000–9,000 Calendar Years BP [Before Present]) from Europe." *Proceedings of the National Academy of Sciences of the United States of America* 97, no. 4 (15 February): 1390–1394.

Blake, D.R. 1984. Increasing Concentrations of Atmospheric Methane. Ph.D. dissertation, University of California at Irvine.

Bloomfield, Janine. 1999. *Hot Nights in the City: Global Warming, Sea-Level Rise and the New York Metropolitan Region.* Environmental Defense Fund, June. [http://www.edf.org/pubs/Reports/HotNY/index.html]

———. 2000. Iowans Can Elect To Combat Climate Change Now. Environmental Defense Fund. January 13. [http://terra.whrc.org/links/links.htm]

Bloomfield, Janine, and Sherry Showell. 1997. *Global Warming: Our Nation's Capital at Risk.* Washington, D.C.: Environmental Defense Fund. [http://www.edf.org/pubs/Reports/WashingtonGW/index.html]

Blunier, Thomas. 2000. " 'Frozen' Methane Escapes from the Sea Floor." *Science* 288 (7 April): 68–69.

Boer, M., and E. Koster, eds. 1992. *Greenhouse-Impact on Cold-Climate Ecosystems and Landscapes.* Selected papers of a European Conference on Landscape Ecological Impact: Impact of Climatic Change, Lunteren, the Netherlands, December 3–7, 1989. CARTENA Supplement 22. Cremlingen, Germany: Cartena Verlag.

Bolin, Bert, Bo Doos, Jill Jager, and Richard Warrick, eds. 1986. *The Greenhouse Effect: Climatic Change and Ecosystems.* Chichester, England, and New York: Wiley.

Bolin, Bert, John T. Houghton, Gylvan Meira Filho, Robert T. Watson, M.C. Zinyowera, James Bruce, Hoesung Lee, Bruce Callander, Richard Moss, Erik Haites, Roberto Acosta Moreno, Tariq Banuri, Zhou Dadi, Bronson Gardner, J. Goldenberg, Jean-Charles Hourcade, Michael Jefferson, Jerry Melillo, Irving Mintzer, Richard Odingo, Martin Parry, Martha Perdomo, Cornelia Quennet-Thielen, Pier Vellinga, and Narasimhan Sundararaman. 1995. *Intergovernmental Panel on Climate Change. Second Assessment Synthesis of Scientific-Technical Information Relevant to Interpreting Article 2 of the United Nations Framework Convention on Climate Change.* Approved by the IPCC at its eleventh session, 11–15 December, Rome. [http://www.unep.ch/ipcc/pub/sarsyn.htm]

Bonan, G.B., D.D. Pollard, and S.L. Thompson. 1992. "Effects of Boreal Forest Vegetation on Global Climate." *Nature* 359:716–718.

Bond, G., H. Heinrich, W. Broecker, L. Labeyrie, J. McManus, J. Andrews, S. Huon, R. Jantschik, S. Clasen, C. Simet, K. Tedesco, M. Klas, G. Bonani, and S. Ivy. 1992. "Evidence for Massive Discharges of Icebergs Into the North Atlantic Ocean During the Last Glacial Period." *Nature* 360:245–249.

Bond, G.C., and R. Lotti. 1995. "Iceberg Discharges into the North Atlantic on Millennial Time Scales During the Last Deglaciation." *Science* 267:1005–1010.

Bongaarts, John. 1992. "Population Growth and Global Warming." *Population and Development Review* 18, no. 2: 299–319.

Bonnie, Robert, Stephan Schwartzman, Michael Oppenheimer, and Janine Bloomfield. 2000. "Counting the Cost of Deforestation." *Science* 288 (9 June): 1763–1764.

"Borehole Temperatures Confirm Global Warming." 2000. *Nature,* 17 February. [http://www.cnn.com/2000/NATURE/02/17/boreholes.enn/]

Borenstein, Seth. 2000. "Arctic Lost 60 Per Cent of Ozone Layer; Global Warming Suspected." Knight-Ridder News Service, 6 April, in LEXIS.

Botkin, D.B., and R.A. Nisbet. 1992. "Projecting the Effects of Climate Change on Biological Diversity in Forests." In R.L. Peters and T. Lovejoy, eds., *Global Warming and Biological Diversity*. New Haven, Conn.: Yale University Press, 277–293.

Bouma, M.J., H.E. Sondorp, and J.H. van der Kaay. 1994. "Health and Climate Change." *Lancet* 343:302.

Bouma, M.J., H.E. Sondorp, and J.H. van der Kaay. 1994. "Climate Change and Periodic Epidemic Malaria." *Lancet* 343:1440.

Bowes, Michael D., and Pierre R. Crosson. 1993. "Consequences of Climate Change for the MINK [Missouri-Iowa-Nebraska-Kansas] Economy: Impacts and Responses." *Climatic Change* 24, nos. 1–2: 131–158.

Bowles, Scott. 1999. "National Gridlock: Traffic Really *is* Worse Than Ever—Here's Why." *USA Today*, 23 November, 1A–2A.

Boyle, Robert H. 1999. "Global Warming: You're Getting Warmer." *Audubon*, November–December, 80–87.

Boyle, S.T., W. Fulkerson, R. Klingholz, I.M. Mintzer, G.I. Pearman, G. Oinchera, J. Reilly, F. Staib, R.J. Swart, and C.-J. Winter. 1991. "Group Report: What are the Economic Costs, Benefits, and Technical Feasibility of Various Options Available to Reduce Greenhouse Potential per Unit of Energy Service?" In G.I. Pearman, ed., *Limiting Greenhouse Effects: Controlling Carbon-dioxide Emissions*. Report of the Dahlem Workshop on Limiting the Greenhouse Effect, Berlin, December 9–14, 1990. New York: John Wiley and Sons, 229–260.

Bradley, Ray. 2000. "1000 Years of Climate Change." *Science* 288 (26 May): 1353–1354.

Bradsher, Keith. 2000. "Ford is Conceding S.U.V. Drawbacks." *New York Times*, 12 May, Business Section, 1.

Brandenburg, John E., and Monica Rix Paxson. 1999. *Dead Mars, Dying Earth*. Freedom, Calif.: The Crossing Press.

Brasseur, Guy P., John J. Orlando, and Geoffrey S. Tyndall, eds. 1999. *Atmospheric Chemistry and Global Change*. New York: Oxford University Press.

Bremner, Charles, Richard Owen, and Mark Henderson. 2000. "Heat Wave Uncovers the Grim Secrets of the Snows." *Times of London*, 26 August, in LEXIS.

Brewer, Peter G., Gernot Friederich, Edward T. Peltzer, and Franklin M. Orr, Jr. 1999. "Direct Experiments on the Ocean Disposal of Fossil Fuel CO2." *Science* 284 (7 May): 943–945.

"Britain Urges U.S. to Get Tough on Global Warming." 1997. British Broadcasting Corp. On-line, 11 June. [http://benetton.dkrz.de:3688/homepages/georg/kimo/0254.html]

British Atmospheric Data Centre. N.d. [http://tornado.badc.rl.ac.uk/who/nasa_aircraft/SteveHipskind.html]

Britt, Robert Roy. 1999. "Antarctic Ice Shelves Falling Apart." Explorezone.com, 9 April. [http://www.explorezone.com/archives/99_04/09_antarctic_ice.htm]

Broccoli, A.J., and S. Manabe. 1990. "Can Existing Climate Models be Used to Study Anthropogenic Changes in Tropical Cyclone Intensity?" *Geophysical Research Letters* 17:1917–1920.

Broecker, Wallace. S. 1985. *How to Build a Habitable Planet*. Palisades, N.Y.: Eldigio Press.

———. 1987. "Unpleasant Surprises in the Greenhouse?" *Nature* 328:123–126.

———. 1979. "Fate of Fossil Fuel Carbon Dioxide and the Global Carbon Budget." *Science* 206:409–418.

———. 1997. "Thermohaline Circulation, the Achilles Heel of our Climate System: Will Man-made CO2 Upset the Current Balance?" *Science* 278:1582–1588.

———. 2000. "Was a Change in Thermohaline Circulation Responsible for the Little Ice Age?" *Proceedings of the National Academy of Sciences of the United States of America* 97, no. 4 (15 February): 1339–1342.

Broecker, Wallace S., Stewart Sutherland, and Tsung-Hung Peng. 1999. "A Possible 20th-Century Slowdown of Southern Ocean Deep Water Formation." *Science* 286 (5 November): 1132–1135.

Brook, E.J., T. Sowers, and J. Orchardo. 1996. "Rapid Variations in Atmospheric Methane Concentration During the Past 110,000 Years." *Science* 273:1087–1091.

Broom, J. 1992. *Counting the Cost of Global Warming*. Cambridge: White Horse Press.

Brown, B.E. 1987. "Worldwide Death of Corals: Natural Cyclical Events or Man-made Pollution?" *Marine Pollution Bulletin* 18, no. 1: 9–13.

Brown, Kathryn. 1999. "Climate Anthropology: Taking Global Warming to the People." *Science* 283 (5 March): 1440–1441.

Brown, Paul. 1999. "Global Warming: Worse Than We Thought." *World Press Review*, February, 44.

———. 1998. "World's Biggest Super-computer Predicts Runaway Greenhouse Effect that will Bring Drought, Deserts, and Disease in its Wake." *The Guardian* (London), 3 November. [http://bonanza.lter.uaf.edu/~davev/nrm304/glbxnews.htm]

———. 2000. "Over-fishing and Global Warming Land Cod on Endangered List." *The Guardian* (London), 20 July, 3.

Browne, Malcolm W. 1998. "Will Humans Overwhelm the Earth? The Debate Goes On." *New York Times*, 8 December. [http://benetton.dkrz.de:3688/homepages/georg/kimo/0254.html]

———. 1999. "Under Antarctica, Clues to an Icecap's Fate: Radar Uncovers a Network of Ice Streams Larger and Faster Than Expected, and More Ominous." *New York Times*, 26 October, F-1, F-6.

———. 1991. "War and the Environment." *Audubon* 93 (September): 89–99.

Bryant, D., L. Burke, J. McManus, and M. Spaulding. 1998. *Reefs at Risk: A Map-based Indicator of Threats to the World's Coral Reefs*. Washington, D.C.: World Resources Institute.

Bryson, Reid A. 1994. "On Integrating Climate Change and Culture Change Studies." *Human Ecology* 22, no. 1: 115–128.

Budyko, M.I., and Yu A. Izrael. 1991. *Anthropogenic Climate Change*. Tucson: University of Arizona Press.

Burroughs, William J. 1999. *The Climate Revealed*. Cambridge, U.K.: Cambridge University Press.

Burton, John F. 1995. *Birds and Climate Change*. London: A & C Black.

Buxton, James. 2000. "Suspects in the Mystery of Scotland's Vanishing Salmon: Fish Farms, Seals and Global Warming are All Blamed for What Some See as a Crisis." *Financial Times* (London), 13 June, 11.

"Calamities Show a Human Touch." 1998. *The Age* (Melbourne, Australia), 30 November. [http://www.theage.com.au/daily/981130/news/news15.html]

Caldeira, Ken, and Philip B. Duffy. 2000. "The Role of the Southern Ocean in the Uptake and Storage of Anthropogenic Carbon Dioxide." *Science* 287 (28 January): 620–622.

"A Call To Action: The Albuquerque Declaration." 1999. *Native Americas* 16, no. 3/4(Fall/Winter): 98. [http://nativeamericas.aip.cornell.edu/fall99/fall99suagee.html]

Callendar, G.D. 1938. "The Artificial Production of Carbon Dioxide and its Influence on Temperature." *Quarterly Journal of the Royal Meteorological Society* 64:223–237.

Callendar, G.S. 1949. "Can Carbon Dioxide Influence Climate?" *Weather* 4:310–314.

———. 1958. "On the Amount of Carbon Dioxide in the Atmosphere." *Tellus* 10:243–248.

Canadian Temperatures Obliterate Records. 1998. *Toronto Star*, 15 December. [http://benetton.dkrz.de:3688/homepages/georg/kimo/0254.html]

Cannell, M.G.R., J. Grace, and A. Booth. 1989. "Possible Impacts of Climatic Warming on Trees and Forests in the United Kingdom: A Review." *Forestry* 62:337–364.

"Cape Hatteras, N.C. Lighthouse Lights Up Sky From New Perch." 1999. Associated Press. In *Omaha World-Herald*, 14 November, A-16.

Carnell, R.E., C.A. Senior, and J.F.B. Mitchell. 1996. "An Assessment of Measures of Storminess: Simulated Changes in Northern Hemisphere Winter Due to Increasing CO2." *Climate Dynamics* 12:467–476.

Carrel, Chris. 1998. "Boeing Joins Fight Against Global Warming." *Seattle Weekly*, 17 September. [http://climatechangedebate.com/archive/09–18_10-27_1998.txt]

Cassidy, Martin M. 1998. "Global Climate Change: Panel Agrees, 'In 10 Years We Will Know.' " [http://hgs.org/artcpics/climat99.htm]

Casten, Thomas R. 1998. *Turning Off the Heat: Why America Must Double Energy Efficiency to Save Money and Reduce Global Warming.* Amherst, N.Y.: Prometheus Books.

Cerveny, R.S., and R.C. Balling, Jr. 1998. "Weekly Cycles of Air Pollutants, Precipitation and Tropical Cyclones in the Coastal NW Atlantic Region." *Nature* 394:561–563.

Cess, R.D., M.H. Zhang, P. Minnis, L. Corsetti, E.G. Dutton, B.W. Forgan, D.P. Garber, W.L. Gates, J.J. Hack, E.F. Harrison, X. Jing, J.T. Kiehl, C.N. Long, J.J. Morcrette, G.L. Potter, V. Ramanathan, B. Subasilar, C.H. Whitlock, D.F. Young, and Y. Zhou. 1995. "Absorption of Solar Radiation by Clouds: Observations Versus Models." *Science* 267:496–499.

Changnon, Stanley A., ed. 2000. *El Niño 1997–1998: The Climate Event of the Century.* New York: Oxford University Press.

Chao, Hung-po. 1995. "Managing the Risk of Global Climate Catastrophe: An Uncertainty Analysis." *Risk Analysis* 15, no. 1: 69–78.

Chapin, F. Stuart III, Robert L. Jefferies, James F. Reynolds, Gaius R. Shaver, and Josef Svoboda. 1992. *Arctic Ecosystems in a Changing Climate: An Ecophysiological Perspective.* San Diego: Academic Press.

Chapman, W.L., and J.E. Walsh. 1993. "Recent Variations of Sea Ice and Air Temperatures in High Latitudes." *Bulletin of the American Meteorological Society* 74:33–47.

Charlson, Robert J. 1999. "Giants' Footprints in the Greenhouse: The Seeds of Our Understanding of Global Warming Were Sown by Early Heroes." *Nature* 401 (21 October): 741–742.

Charlson, R.J., S.E. Schwartz, J.M. Hales, R.D. Cess, J.A. Coakley, Jr., J.E. Hansen, D.J. Hofmann. 1992. "Climate Forcing by Anthropogenic Aerosols." *Science* 255 (24 January): 423.

Charlson, Robert J., and Tom M.L. Wigley. 1994. "Sulfate Aerosol and Climatic Change." *Scientific American* 270:48–55.

Charney, Jule. 1979. *Carbon Dioxide and Climate: A Scientific Assessment.* Washington, D.C.: National Academy Press.

Chen, Robert S., Elise Boulding, and Stephen H. Schneider, eds. 1983. *Social Science Research and Climate Change: An Interdisciplinary Appraisal.* Boston: D. Reidel Publishing Co.

Chevron Stockholder Vote on Global Warming Resolution Surprises Annual Meeting. 1999. Environmental Media Services/Ozone Action, 29 April. [http://www.corpwatch. org/trac/corner/worldnews/other/355.html]

Choudhury, B., and G. Kukla. 1979. "Impact of CO2 on Cooling of Snow and Water Surfaces." *Nature* 280:668–671.

Christianson, Gale E. 1999. *Greenhouse: The 200-Year Story of Global Warming.* New York: Walker and Company.

Ciborowski, Peter. 1989. "Sources, Sinks, Trends, and Opportunities." In Edwin Abrahamson, ed., *The Challenge of Global Warming.* Washington, D.C.: Island Press, 213–230.

Clark, J.A., and C.S. Lingle. 1977. "Future Sea-level Changes Due to West Antarctic Ice Sheet Fluctuation." *Nature* 269:206–209.

Clark, Peter U., and Alan C. Mix. 2000. "Global Change: Ice Sheets by Volume." *Nature* 406 (17 August): 689–690.

Clark, Peter U., Robert S. Webb, and Lloyd D. Keigwin. 1999. *Mechanisms of Global Climate Change at Millennial Time Scales.* Washington, D.C.: American Geophysical Union.

Climate Change Impacts on the United States. 2000. Report of the Government of the United States. June. [http://www.usgcrp.gov]

"Climate Shifts Expected to Alter California Life." 1999. *Los Angeles Times.* In *Omaha World-Herald,* 4 November, 5.

Cline, William. R. 1995. *Pricing Carbon Dioxide Pollution.* Victoria, British Columbia: Ministry of Environment, Lands and Parks.

———. 1992. *The Economics of Global Warming.* Washington, D.C.: Institute for International Economics.

———. 1991. "Scientific Basis for the Greenhouse Effect." *Economic Journal* 100 (September): 904–919.

———. 1991. "Comments." In Rudiger Dornbusch and James M. Poterba, eds., *Global Warming: Economic Policy Reponses.* Cambridge, Mass.: MIT Press, 222–228.

Clinton, Bill. 1999. "President Clinton's State of the Union Address." *New York Times,* 20 January. [http://geography.rutgers.edu/courses/99spring/370sp99/news01_20_ 99.html#anchor33828]

———. 1997. Remarks by the President on Global Warming and Climate Change. Presented October 22 at the National Geographic Society, Washington, D.C. [http:// uneco.org/Global_Warming.html]

Clover, Charles. 1999. "Air Travel is a Threat to Climate." *London Daily Telegraph,* 5 June.

————. 2000. "Thousands of Species 'Threatened by Warming.' " *London Daily Telegraph*, 31 August, 9.

CO2/Climate Panel. 1982. *Carbon Dioxide and Climate: A Second Assessment*. Report of the CO2/Climate Panel to the Climate Research Committee on Atmospheric Sciences and the Carbon Dioxide Assessment Committee of the Climate Board, Commission on Physical Sciences, Mathematics, and Resources, National Research Council. Washington, D.C.: National Academy Press.

"The CO2 Fertilization Factor and the 'Missing' Carbon Sink: An Editorial Comment." 1996. *Climatic Change* 33:63–68. [http://www.bren.ucsb.edu/~keller/climch33.html]

Cogan, Douglas. 1992. *The Greenhouse Gambit: Business and Investment Reponses to Climate Change*. Washington, D.C.: Investor Responsibility Research Center.

Colborn, Theo, Diane Dummanoski, and John Peterson Myers. 1997. *Our Stolen Future: Are We Threatening Our Fertility, Intelligence, and Survival?* A Scientific Detective Story. New York: Plume/Penguin.

Cole, J. J., N. Caraco, G.W. Kling, and T. Kratz. 1994. "Carbon Dioxide Supersaturation in the Surface Waters of Lakes." *Science* 265:1568–1570.

Commoner, Barry. 1990. *Making Peace With the Planet*. New York: Pantheon.

————. 1991. "Rapid Population Growth and Environmental Stress." *International Journal of Health Services* 21, no. 2: 199–227.

Congressional Budget Office. 1990. *Carbon Charge as a Response to Global Warming: The Effects of Taxing Fossil Fuels*. Washington, D.C.: Congressional Budget Office.

Connor, Steve. 2000. "Ozone Layer Over Northern Hemisphere is Being Destroyed at 'Unprecedented Rate.' " *The Independent* (London), 5 March, 5.

————. 2000. "Global Warming is Blamed for First Collapse of a Caribbean Coral Reef." *The Independent* (London), 4 May, 12.

————. 2000. "Ice Cores From a Himalayan Glacier Confirm Global Warming." *The Independent* (London), 15 September, 9.

Cook, E.R., A.H. Johnson, and T.J. Blasing. 1987. "Forest Decline: Modeling the Effect of Climate in Tree Rings." *Tree Physiology* 3:27–40.

Cooper, R.N. 1998. "Toward a Real Global Warming Treaty." *Foreign Affairs* 77:66–79.

Corson, Walter H., ed. 1990. *The Global Ecology Handbook: What You Can Do About the Environmental Crisis*. Washington, D.C.: The Global Tomorrow Coalition.

Cotton, William R., and Roger A. Pielke. 1995. *Human Impacts on Weather and Climate*. New York: Cambridge University Press.

Countdown to Kyoto: The Consequences of Mandatory Global CO2 Emission Reductions. 1997. Remarks by United States Senator Chuck Hagel, 21 August, Canberra, Australia. Office of Senator Chuck Hagel. [http://www.altgreen.com.au/gge/Ctk_Hagel.html]

Couzin, Jennifer. 1999. "Landscape Changes Make Regional Climate Run Hot and Cold." *Science* 283 (15 January): 317–318.

Cowen, Robert C. 1998. "New Research Shows That Hurricanes Pump More CO2 into the Air by Roiling Oceans." *Christian Science Monitor*, 3 September. [http://www.csmonitor.com/durable/1998/09/03/fp4s1-csm.htm]

Crary, David. 1998. "Global Warming a Threat to Canada." Associated Press, 14 December. [http://benetton.dkrz.de:3688/homepages/georg/kimo/0254.html]

Crawford, Elisabeth. 1996. *Arrhenius: From Ionic Theory to the Greenhouse Effect.*
Canton, Mass.: Science History Publications.

Criswell, David R., and Robert D. Waldron. 1993. "Results of Analyses of a Lunar-based
Power System to Supply Earth with 20,000 GW [Gigawatts] of Electric Power."
In Richard A. Geyer, ed., *A Global Warming Forum: Scientific, Economic, and
Legal Overview.* Boca Raton, Fla.: CRC Press, 111–124.

Crowley, Thomas J. 2000. "Causes of Climate Change Over the Past 1000 Years." *Science* 289 (July 14): 270–277.

Cuo, C., C. Lindberg, and D.J. Thomson. 1990. "Coherence Established Between Atmospheric Carbon Dioxide and Global Temperature." *Nature* 343:709–714.

Cure, J.D., and B. Acock. 1986. "Crop Responses to Carbon Dioxide Doubling: A Literature Survey." *Agricultural and Forest Meteorology* 38:127–145.

Curren, Thomas. 1991. *Forests and Global Warming.* Ottawa: Research Branch, Library
of Parliament.

———. 1979. *The Greenhouse Theory and Climate Change.* Ottawa: Research Branch,
Library of Parliament.

Cushman, John H., Jr. "Pressure Points in Global Warming." 1997. *New York Times*, 7
December, A-23.

———. 1998. "Industrial Group Plans to Battle Climate Treaty." *New York Times*, 26
April, A-1, A-24.

Dahl-Jensen, Dorthe. 2000. "The Greenland Ice Sheet Reacts." *Science* 289 (July 21):
404–405.

Dai, Aiguo, Inez Y. Fung, and Anthony D. del Genio. 1997. "Surface Observed Global
Land Precipitation Variations During 1900–88." *Journal of Climate* 10:2943–
2962.

Daley, Suzanne. 1999. "Battered by Fierce Weekend Storm, Western Europe Begins an
Enormous Cleanup Job." *New York Times* (International Edition), 28 December,
A-10.

Dansgaard, W., S.J. Johnson, H.B. Clausen, D. Dahl-Jensen, S. Gundenstrup, C.U. Hammer, C.S. Hvidberg, C.S. Steffensen, and J.P. Sveinbjrnsdottir. 1993. "Evidence
for General Instability of Past Climate from a 250-kyr [thousand-year] Ice-core
Record." *Nature* 364:218–220.

Dansgaard, W., J.W.C. White, and S.J. Johnsen. 1989. "The Abrupt Termination of the
Younger Dryas Climate Event." *Nature* 339:532–534.

d'Arge, Ralph C., William D. Schulze, and David S. Brookshire. 1982. "Carbon Dioxide
and Intergenerational Choice." *American Economic Review* 72, no. 2: 251–256.

Davidson, Keay. 1999. "Jet-bred Cirrus Clouds Could Pose Threat; Research Suggests
Increase in Air Travel May Speed Global Warming." *Milwaukee Journal-Sentinel*,
16 March.

Davis, M.B. 1989. "Lags in Vegetation Response to Greenhouse Warming." *Climatic
Change* 15:75–81.

December 1997 is Coldest Month on Record in the Stratosphere. 1998. NASA, January
20. [http://science.msfc.nasa.gov/newhome/headlines/essd20jan98_1.htm]

DeCicco, John, James Cook, Dorene Bolze, and Jan Beyea. 1990. *CO2 Diet for a Greenhouse Planet: A Citizen's Guide to Slowing Global Warming.* Washington, D.C.:
National Audubon Society.

Del Genio, Anthony P., Andrew A. Lacis, and Reto A. Ruedy. 1991. "Simulations of the Effect of a Warmer Climate on Atmospheric Humidity." *Nature* 351 (May 30): 382–385.

Delworth, T.D., S. Manabe, and R.J. Stouffer. 1993. "Interdecadal Variations of the Thermohaline Circulation, a Coupled Ocean-Atmosphere Model." *Journal of Climate* 6:1993–2011.

Delworth, Thomas L., and Thomas R. Knutson. 2000. "Simulation of Early 20th Century Global Warming." *Science* 287:2246–2250.

Dennis, Roger L.H. 1993. *Butterflies and Climate Change.* New York: St. Martin's Press.

Devitt, Terry. 1998. "Landscape Changes Seen as Bad as Greenhouse Gases." *Uniscience Research News,* 9 December. [http://benetton.dkrz.de:3688/homepages/georg/kimo/0254.html]

Dewey, K.F., and R. Heim, Jr. 1981. "Satellite Observations of Variations in Northern Hemisphere Seasonal Snow Cover." NOAA Technical Report NESS 87. Washington, D.C.: U.S. Department of Commerce.

Diaz, H.F., M. Beniston, and R.S. Bradley. 1997. *Climatic Change at High-elevation Sites.* New York: Kluwer Academic.

Diaz, Henry F., and Raymond S. Bradley. 1997. "Temperature Variations During the Last Century at High-elevation Sites." *Climatic Change* 36:253–279.

Diaz, H.F., and N.E. Graham. 1996. "Recent Changes in Tropical Freezing Heights and the Role of Sea Surface Temperature." *Nature* 383:152–155.

Diaz, S., J.P. Grime, J. Harris, and E. McPherson. 1993. "Evidence of a Feedback Mechanism Limiting Plant Response to Elevated Carbon Dioxide." *Nature* 364:616–617.

Dickinson, R.E., and R.J. Cicerone. 1986. "Future Global Warming From Atmospheric Trace Gases." *Nature* (London) 319:109–115.

Dickson, R.R., and J. Brown. 1994. "The Production of North Atlantic Deep Water: Sources, Rates, and Pathways." *Journal of Geophysical Research* 99:12319–12341.

Dietz, T., and E. Rosa. 1997. "Effects of Population and Affluence on CO2 Emissions." *Proceedings of the National Academy of Sciences of the United States of America* 94:175–179.

Dixon, R.K., S. Brown, R.A. Houghton, A.M. Solomon, M.C. Trexler, and J. Wisniewski. 1994. "Carbon Pools and Flux of Global Forest Ecosystems." *Science* 263:185–190.

Dixon, R.K., and O.N. Krankina. 1993. "Forest Fires in Russia: Carbon Dioxide Emissions to the Atmosphere." *Canadian Journal of Forest Research* 23:700–705.

Dlugokencky, E.J., K.A. Masrie, P.M. Lang, and P.P. Tans. 1999. "Continuing Decline in the Growth Rate of the Atmospheric Methane Burden." *Nature* 398 (4 June): 447–450.

Doake, C.S.M., H.F.J. Corr, H. Rott, P. Skvarca, and N.W. Young. 1998. "Breakup and Conditions for Stability of the Northern Larsen Ice Shelf, Antarctica." *Nature* 391:778–780.

Dobson, A., and R. Carper. 1992. "Global Warming and Potential Changes in Host-parasite and Disease-vector Relationships." In R.L. Peters and T. Lovejoy, eds., *Global Warming and Biological Diversity.* New Haven, Conn.: Yale University Press, 201–217.

Dobson, Andy, Alison Jolly, and Dan Rubenstein. 1989. "The Greenhouse Effect and Biological Diversity." *Trends in Ecology and Evolution* 4, no. 3 (March): 64–68.

Donn, Farrand, and Ewing. 1962. "Pleistocene Ice Volumes and Sea-level Lowering." *Journal of Geology* 70:206–214.

Donner, L., and V. Ramanathan. 1980. "Methane and Nitrous Oxide: Their Effects on the Terrestrial Climate." *Journal of Atmospheric Sciences* 37:119–124.

Dornbusch, Rudiger, and James M. Poterba, eds. 1991. *Global Warming: Economic Policy Reponses.* Cambridge, Mass.: MIT Press.

Duffy, Andrew. 2000. "Global Warming Why Arctic Town is Sinking: Permafrost is Melting under Sachs: Inuit Leader." *Montreal Gazette,* 18 April, A-13.

Durant, John. "Everybody Talks About It, But Nobody Does Anything About the Weather, Except the Wrong Thing." 2000. Review of William K. Stevens, *The Change in the Weather: People, Weather, and the Science of Climate* (1999). *New York Times Book Review,* 31 January, 20.

Dye, Chris, and Paul Reiter. 2000. "Climate Change and Malaria: Temperatures Without Fevers?" *Science* 289 (8 September): 1697–1698.

Dyurgerow, Mark B., and Mark F. Meier. 2000. "Twentieth-century Climate Change: Evidence from Small Glaciers." *Proceedings of the National Academy of Sciences of the United States of America* 97, no. 4 (15 February): 1351–1354.

Eamus, D., and P.G. Jarvis. 1989. "The Direct Effects of Increase in the Global Atmospheric CO_2 Concentration on Natural and Commercial Temperate Trees and Forests." *Advances in Ecological Research* 19:1–55.

"Earth Out of Balance." 2000. *Indian Country Today,* 9 February, editorial, A-4.

Easterling, David R., Briony Horton, Phillip D. Jones, Thomas C. Peterson, Thomas R. Karl, David E. Parker, M. James Salinger, Vyacheslav Razuvayev, Neil Plummer, Paul Jamason, and Christopher K. Folland. 1997. "Maximum and Minimum Temperature Trends for the Globe." *Science* 277:363–367.

Easterling, William E. III, Pierre R. Crosson, Norman J. Rosenberg, Mary S. McKenney, Laura A. Katz, and Kathleen M. Lemon. 1993. "Agricultural Impacts of and Responses to Climate Change in the Missouri-Iowa-Nebraska-Kansas (MINK) Region." *Climatic Change* 24, nos. 1–2: 23–61.

Eckaus, R. 1992. "Comparing the Effects of Greenhouse-gas Emissions on Global Warming." *The Energy Journal* 13, no. 1: 25–35.

Eco Bridge. N.d. "What Can We Do About Global Warming?" [http://www.ecobridge.org/content/g_wdo.htm]

Economists' Letter. 1997. Economists' Letter on Global Warming, 23 June. [http://uneco.org/Global_Warming.html]

Edgerton, Lynne T., and the Natural Resources Defense Council. 1991. *The Rising Tide: Global Warming and World Sea Levels.* Washington, D.C.: Island Press.

Eisentadt, M., and J. Sorenson. 1989. "World-wide Solar Power Satellite System to Reduce CO_2: So You Have a Better Way?" Center for National Security Studies Paper No. 22, November.

Ekins, P. 1995. "Rethinking the Costs Related to Global Warming." *Environmental and Resource Economics* 6, no. 1: 231–277.

Elderfield, H., and R.E.M. Rickaby. 2000. "Oceanic Cd/P Ratio and Nutrient Utilizatation in the Glacial Southern Ocean." *Nature* 405 (18 May): 305–310.

"Electrical System Hard Hit by Storm." 1999. *Omaha World-Herald,* 6 January, 4.

Emanuel, Kerry A. 1987. "The Dependence of Hurricane Intensity on Climate." *Nature* 326, no. 2 (April): 483–485.

———. 1986. "An air-sea Interaction Theory for Tropical Cyclones. Part I: Steady-state Maintenance." *Journal of the Atmospheric Sciences* 43:585–604.

———. 1988. "The Maximum Intensity of Hurricanes." *Journal of the Atmospheric Sciences* 45:1143–1156.

———. 1988. "Toward a General Theory of Hurricanes." *American Scientist* 76:370–379.

———. 1995. "Comments on 'Global Climate Change and Tropical Cyclone': Part I." *Bulletin of the American Meteorological Society* 76:2241–2243.

———. 1999. "Thermodynamic Control of Hurricane Intensity." *Nature* 401 (14 October): 665–669.

Emanuel, W.R., H.H. Shugart, and M.P. Stevenson. 1985. "Climatic Change and the Broad-scale Distribution of Terrestrial Eco-system Complexes." *Climatic Change* 7:29–43.

"Energy & Equity: The Full Report." N.d. Illinois Environmental Protection Agency. [http://www.cnt.org/ce/energy&equity.htm]

Environment Canada. 1994. *Potential Impacts of Global Warming on Salmon Production in the Fraser River Watershed*. Ottawa: Environment Canada.

Environmental Media Services. 1997. *Understanding the Science of Global Climate Change*. Washington, D.C.: Environmental Media Services.

Epstein, Joshua W., and Raj Gupta. 1990. *Controlling the Greenhouse Effect: Five Global Regimes Compared*. Washington, D.C.: The Brookings Institution.

Epstein, Paul R. 1998. "Climate, Ecology, and Human Health." 18 December. [http://www.iitap.iastate.edu/gccourse/issues/health/health.html]

———. 1999. "Profound Consequences: Climate Disruption, Contagious Disease and Public Health." *Native Americas* 16, no. 3/4(Fall/Winter): 64–67. [http://nativeamericas.aip.cornell.edu/fall99/fall99epstein.html]

———. 2000. "Is Global Warming Harmful to Health?" *Scientific American*, August, 50–57.

———, ed. 1999. *Extreme Weather Events: The Health and Economic Consequences of the 1991/1998 El Niño and La Niña*. Boston: Center for Health and the Global Environment, Harvard Medical School.

Epstein, Paul, Georg Grabherr, Tom Karl, Ellen Mosley-Thompson, Kevin Trenberth, and George M. Woodwell. 1996. Current Effects: Global Climate Change. An Ozone Action Roundtable, 24 June, Washington, D.C. [http://www.ozone.org/curreff.html]

Epstein, P.R., H.F. Diaz, S. Elias, G. Grabherr, N.E. Graham, W.J.M. Martens, E. Mosley-Thompson, and J. Susskind. 1998. "Biological and Physical Signs of Climate Change: Focus on Mosquito-borne Diseases." *Bulletin of the American Meteorological Society* 79, no. 3 (March): 409–417.

Epstein, P.R., O.C. Pena, and J.B. Racedo. 1995. "Climate and Disease in Colombia." *Lancet* 346:1243.

"Europe Faces an Ice Age as the World Warms." 1999. *Times of London*, 17 January. [http://geography.rutgers.edu/courses/99spring/370sp99/news01_20_99.html#anchor33828]

Everts, C.H. 1985. "Effects of Sea Level Rise and Net Sand Volume Change on Shoreline Position at Ocean City, Maryland." In J.G. Titus, ed., *Potential Impacts of Sea*

Level Rise on the Beach at Ocean City, Maryland. Washington, D.C.: U.S. Environmental Protection Agency.

Ewen, Alexander. 1999. "Consensus Denied: Holy War Over Global Warming." *Native Americas* 16, no. 3/4(Fall/Winter): 26–33. [http://nativeamericas.aip.cornell.edu]

Fallows, James. 1999. "Turn Left at Cloud 109." *New York Times Sunday Magazine*, 21 November, 84–89.

Fan, S., M. Gloor, J. Mahlman, S. Pacala, J. Sarmiento, T. Takahashi, and P. Tans. 1998. "A Large Terrestrial Carbon Sink in North America Implied by Atmospheric and Oceanic Carbon Dioxide Data and Models." *Science* 282 (16 October): 442–446.

Fankhauser, Samuel. 1995. *Valuing Climate Change: The Economics of the Greenhouse.* London: Earthscan Publications, Ltd.

Fauber, John, and Tom Vanden Brook. 2000. "A Change in the Seasons: Location Makes Wisconsin Vulnerable to Climate Shift." *Milwaukee Journal-Sentinel*, 28 May, 1A.

————. 2000. "Global Warming May Take Great Lakes Gulp; Plunge in Coming Century Would Have Significant Ripple Effect, Reports Say." *Milwaukee Journal-Sentinel*, 14 June, 1A.

Fearnside, P.M. 1997. "Greenhouse Gases from Deforestation in Brazilian Amazonia: Net Committed Emissions." *Climate Change* 35:321–360.

Feely, Richard A., Rik Wanninkhof, Taro Takahashi, and Peter Tans. 1999. "Influence of El Niño on the Equatorial Pacific Contribution to Atmospheric CO2 Accumulation." *Nature* 398 (15 April): 597–601.

Fedorow, Alexey V., and S. George Philander. 2000. "Is El Niño Changing?" *Science* 288 (16 June): 1997–2002.

Ferguson, H.L. 1989. "The Changing Atmosphere: Implications for Global Security." In Dean Edwin Abrahamson, ed., *The Challenge of Global Warming.* Washington, D.C.: Island Press, 48–62.

Fialka, John J. 2000. "U.S. Study on Global Warming May Overplay Dire Side." *Wall Street Journal*, 26 May, A-24.

Field, Christopher, and Inez Y. Fung. 1999. "The Not-so-Big U.S. Carbon Sink." *Science* 285 (23 July): 544–547.

Firor, John. 1990. *The Changing Atmosphere: A Global Challenge.* New Haven: Yale University Press.

Firth, Penelope, and Stuart G. Fisher, eds. 1992. *Global Climate Change and Freshwater Ecosystems.* Berlin: Springer-Verlag.

Fisher, David E. 1990. *Fire & Ice: The Greenhouse Effect, Ozone Depletion, and Nuclear Winter.* New York: Harper & Row.

Fishman, J., V. Ramanathan, P.J. Crutzen and S.C. Liu. 1980. "Tropospheric Ozone and Climate." *Nature* 282:818–820.

Flavin, Christopher. 1994. "Storm Warnings: Climate Change Hits the Insurance Industry." *World Watch* 7, no. 6: 10–20.

Fleming, James Rodger. 1998. *Historical Perspectives on Climate Change.* New York: Oxford University Press.

Flesher, John. 2000. "The Great Loss: Lakes' Water Drop Incites Debate on Cause, Concern about Impact." *Toledo Blade*, 21 May, B-1, B-3.

Flower, Benjamin P. 1999. "Warming Without High CO2?" *Nature* 399 (27 May): 313–314.

Fourier, Jean-Baptiste. 1824. "Les Temperatures du Globe Terrestre et des espaces planetaires." *Memoires de L'Academe Royale des Sciences de L'Institut de France* 7:569–604.

Frank, K.T., R.I. Perry, and K.F. Drinkwater. 1990. "Predicted Response of Northwest Atlantic Invertebrate and Fish Stocks to CO2-induced Climate Change." *Transactions of the American Fisheries Society* 119:353–365.

Frank, Robert. 1999. "How Thailand Became the 'Detroit of the East,' " *Wall Street Journal*, 8 December, B-1, B-5.

French, H.M., ed. 1986. *Impact of Climatic Change on the Canadian Arctic*. Proceedings of a Conference Held 3–5 March at Orilla, Ontario. Ottawa: Environment Canada.

Friedl, Randall R. 1999. "Perspectives: Atmospheric Chemistry; Unraveling Aircraft Impacts." *Science* 286 (1 October): 57–58.

Friis-Christensen, E., and K. Lassen. 1991. "Length of the Solar Cycle: An Indicator of Solar Activity Closely Associated with Climate." *Science* 254:698–700.

"From the Desk of the Editor: Warm Enough for Ya? Global Warming is More Than A State of Mind." 1999. *Indian Country Today*, 22 December, A-4.

Fung, Inez. 1995. "Perturbations to the Biospheric Carbon Cycle: Uncertainties in the Estimates." In George M. Woodwell and Fred T. MacKenzie, eds., *Biotic Feedbacks in the Global Climate System: Will the Warming Feed the Warming?* New York: Oxford University Press, 367–392.

Gaffen, Dian J. 1998. "Falling Satellites, Rising Temperature?" *Nature* 394 (13 August): 615–616.

Garrett, Laurie. 1994. *The Coming Plague: Emerging Diseases in a World Out of Balance*. New York: Farrar, Straus.

Gash, J.H.C., and W.J. Shuttleworth. 1991. "Tropical Deforestation: Albedo and the Surface-energy Balance. *Climatic Change* 19:123–133.

Gates, D.M. 1993. *Climate Change and Its Biological Consequences*. Sunderland, Md.: Sunderland Associates Inc.

Gay, Kathlyn. 1986. *The Greenhouse Effect*. New York: F. Watts.

Gelbspan, Ross. 1995. "The Heat is On: The Warming of the World's Climate Sparks a Blaze of Denial." *Harper's Magazine*, December. [http://www.dieoff.com/page82.htm]

———. 1997. "Hot Air on Global Warming; Science and Academia in the Service of the Fossil Fuel Industry." *Multinational Monitor* 18, no. 11 (November): 14–17.

———. 1997. *The Heat is On: The High Stakes Battle Over Earth's Threatened Climate*. Reading, Mass.: Addison-Wesley Publishing Co.

———. 1997. "A Global Warming." *American Prospect* 31 (March/April). [http://www.prospect.org/archives/31/31gelbfs.html]

———. 1998. "Beyond Kyoto." *Amicus Journal*, Winter. [http://www.nrdc.org/nrdc/nrdc/nrdc/eamicus/98win/toc.html]

Geyer, Richard A., ed. 1993. *A Global Warming Forum: Scientific, Economic, and Legal Overview*. Boca Raton, Fla.: CRC Press.

Giardina, Christian P., and Michael G. Ryan. 2000. "Evidence that Decomposition Rates of Organic Carbon in Mineral Soil Do Not Vary with Temperature." *Nature* 404 (20 April): 858–861.

Gibbs, Mark T., Karen L. Bice, Eric J. Barron, and Lee R. Kump. 2000. "Glaciation in the Early Paleozoic 'Greenhouse': The Roles of Paleo-geography and Atmospheric CO2." in Brian T. Huber, Kenneth G. MacLeod, and Scott L. Wing. eds.,

Warm Climates in Earth History. Cambridge, U.K.: Cambridge University Press, 386–422.

Gifford, Roger M., Damian J. Barrett, Jason L. Lutze, and Ananda B. Samarakoon. 2000. "The CO2 Fertilizing Effect: Relevance to the Global Carbon Cycle." In T.M.L. Wigley and D. S. Schimel, eds., *The Carbon Cycle.* Cambridge, U.K.: Cambridge University Press, 77–92.

Gill, Richardson Benedict. 2000. *The Great Maya Droughts: Water, Life, and Death.* Albuquerque: University of New Mexico Press.

Gillon, Jim. 2000. "The Water Cooler." *Nature* 404 (6 April): 555.

Glantz, Michael H. 1991. "The Use of Analogies in Forecasting Ecological and Societal Responses to Global Warming." *Environment* 33, no. 5: 10–32.

———, ed. 1992. *Climate Variability, Climate Change, and Fisheries.* Cambridge, U.K.: Cambridge University Press.

Gleick, Peter H. 1987. "Regional Hydrological Consequences of Increases in Atmospheric CO2 and Other Trace Gases." *Climatic Change* 10:137–160.

———. 1989. "The Implications of Global Climate Changes for International Security." *Climate Change* 15 (October): 303–325.

Glick, Patricia. 1998. "Global Warming: The High Costs of Inaction." San Francisco: Sierra Club. [http://www.sierraclub.org/global-warming/inaction.html]

"Global Warming." 1997. *The Financial Times,* 11 March, editorial. [http://benetton. dkrz.de:3688/homepages/georg/kimo/0254.html]

"Global Warming Pests and Pestilence." 1995–2000. In *What's Hot: World Climate Report Archives: 1995–2000.* Vol. 2 no. 2. [http://www.nhes.com/back_issues/ Volland2/WH/hot.html]

"Global Warming Seen as Cause of Antarctic Melting." 1999. Associated Press, 8 April. [http://www.gsreport.com/articles/art000101.html]

"Global Warming Worries Native Americans." 1998. *Deseret News,* 27 November. [http: //www.desnews.com/cit/07len14f.htm]

Glynn, P.W. 1991. "Coral Reef Bleaching in the 1980s and Possible Connections with Global Warming." *Trends in Ecology and Evolution* 6:175–179.

Goddard Institute for Space Studies. 1999. "Global Temperature Trends: 1998 Global Surface Temperature Smashes Record." [http://www.giss.nasa.gov/research/ observe/surftemp]

Goldberg, Donald, and Stephen Porter. 1998. "In Focus: Global Climate Change." Center for International Environmental Law, May. [http://www.foreignpolicy-infocus. org/briefs/vol3/v3n12cli.html]

Gordon, Anita, and David Suzuki. 1991. *It's a Matter of Survival.* Cambridge: Harvard University Press.

Gore, Albert. 1992. *Earth in the Balance: Ecology and the Human Spirit.* Boston: Houghton-Mifflin.

Goreau, T.J., R. Hayes, A. Strong, E. Williams, G. Smith, J. Cervino, and M. Goreau. 1998. "Coral Reefs and Global Change: Impacts of Temperature, Bleaching, and Emerging Diseases." *Sea Wind* 12, no. 3: 2–6.

Gorham, Eville. 1995. "The Biochemistry of Northern Peatlands and Its Possible Responses to Global Warming." In George M. Woodwell and Fred T. MacKenzie, eds., *Biotic Feedbacks in the Global Climate System: Will the Warming Feed the Warming?* New York: Oxford University Press, 169–187.

Gornitz, V. 1994. "Sea-level Rise: A Review of Recent Past and Near-future Trends." *Earth Surface Processes and Trends* 4:316–318.

Gornitz, V., S. Lebedeff, and J. Hansen. 1982. "Global Sea Level Trend in the Past Century." *Science* 215:1611–1614.

Gough, Robert. 1999. "Stress on Stress: Global Warming and Aquatic Resource Depletion." *Native Americas* 16, no. 3/4(Fall/Winter): 46–48. [http://nativeamericas. aip.cornell.edu]

Grabherr, G., N. Gottfried, and H. Pauli. 1994. "Climate Effects on Mountain Plants." *Nature* 369:447.

Grace, John, and Mark Rayment. 2000. "Respiration in the Balance." *Nature* 404 (20 April): 819–820.

Graedel, T.E., and P.J. Crutzen. 1989. "The Changing Atmosphere." *Scientific American* 261, no. 3: 58–68.

Gray, Vincent R. 1998. "The IPCC Future Projections: Are They Plausible?" *Climate Research* 10:155–162.

———. 1999. "Hansen's Reappraisal." Greenhouse Bulletin No. 118 (January). [http:// www.microtech.com.au/daly/bull-118.htm]

"Greenhouse Effects: Global Warming is Well Under Way. Here Are Some Telltale Signs." 1999. *Time*, 13 December, 78–79.

Greening the Planet. 1998. Climate Change Debate, 8 December. [http:// climatechangedebate.com/archive/12–08_12–15_1998.txt]

Greenpeace. N.d. "Countering The Sceptics." [http://greenpeace.org/~climate/kindustry/ culprits/sceptics.html]

Gregory, S. 1988. *Recent Climatic Change: A Regional Report*. London: Bellhaven Press.

Gribben, John. 1971. "Weather Warning: You are Now Experiencing a Climatic Change." *Nature* 252 (November 15): 182.

———. 1981. "The Politics of Carbon Dioxide." *New Scientist* 90:82–84.

———. 1990. *Hothouse Earth: The Greenhouse Effect and Gaia*. London: Bantam Press.

———. 1990. "An Assault on the Climate Consensus." *New Scientist* 128:26.

———. 1989. "The End of the Ice Ages?"*New Scientist* 122:48.

———. 1982. *Future Weather and the Greenhouse Effect*. New York: Delacorte Press/ Eleanor Friede.

Gribben, John, and Mick Kelly. 1989. *Winds of Change: Living in the Global Greenhouse*. London: Hodden and Stoughton.

Grubb, M. 1990. *Energy Policies and the Greenhouse Effect*. Vol. 1, *Policy Appraisal*. London: Royal Institute of International Affairs.

Grubb, Michael, Christian Vrolijk, and Duncan Brack. 1999. *The Kyoto Protocol: A Guide and Assessment*. London: Earthscan.

Gupta, S., and R.K. Pachauri, eds. 1990. *Global Warming and Climate Change: Perspectives From Developing Countries*. New Delhi: Tata Energy Research Institute.

Gurney, R.J., J.L. Foster, and C.L. Parkinson. 1993. *Atlas of Satellite Observations Related to Global Change*. Cambridge, U.K.: Cambridge University Press.

Guzzo, Louis R., and Dixy Lee Ray. 1993 *Environmental Overkill: Whatever Happened to Common Sense?* Washington, D.C.: Regnery Gateway.

Haarsma, R.J., J.F.B. Mitchell, and C.A. Senior. 1993. "Tropical Disturbances in a GMC [Global Climate Model]." *Climate Dynamics* 8:247–257.

Haeberli, W. 1992. "Possible Effects of Climatic Change on the Evolution of Alpine Permafrost." In M. Boer and E. Koster, eds., *Greenhouse-Impact on Cold-Climate*

Ecosystems and Landscapes. Selected papers of a European Conference on Landscape Ecological Impact: Impact of Climatic Change, Lunteren, the Netherlands, December 3–7, 1989. CARTENA Supplement 22. Cremlingen, Germany: Cartena Verlag, 23–35.

Hagel, Chuck, Senator. N.d. Interview on "Global View." [http://www.oche.de/~norb/chuck.html]

Haimson, Leonie, Michael Oppenheimer, and David Wilcove. 1995. *The Way Things Really Are: Debunking Rush Limbaugh on the Environment*. Washington, D.C.: Environmental Defense Fund.

Haines, Andrew. 1990. "The Implications for Health." In Jeremy Leggett, ed., *Global Warming: The Greenpeace Report*. New York: Oxford University Press, 149–162.

Hall, Carl T. 2000. "Spring Scorches the Record Books; It was the Hottest in U.S. History. Study Rekindles Global Warming Debate." *San Francisco Chronicle*, 17 June, A-1.

Hallegraeff, G.M. 1993. "A Review of Harmful Algal Blooms and Their Apparent Global Increase." *Phycologia* 32, no. 2: 79–99.

Hallermeir, R.J. 1981. "A Profile Zonation for Seasonal Sand Beaches from Wave Climate." *Coastal Engineering* 4:253.

Hamilton, Martha M. 1998. "British Petroleum Sets Goal of 10 Per Cent Cut in 'Greenhouse' Gases." *Washington Post*, 18 September, A-6. [http://climatechangedebate.com/archive/09–18_10–27_1998.txt]

Hamilton, W. 1980. *Electric Automobiles*. New York: McGraw-Hill.

Hammel, Paul. 2000. "Warmth Lures Texas Critters." *Omaha World-Herald*, 16 January, 2A.

Hansen, James E. 1989. The Greenhouse, the White House, and Our House. Typescript of a speech delivered at the International Platform Association, 3 August, Washington, D.C.

Hansen, J.E. 1997. "Public Understanding of Global Climate Change." In Y. Terzian and E. Bilson, eds., *Carl Sagan's Universe*. Cambridge, U.K.: Cambridge University Press.

Hansen J.E. 1998. GISS Website. [http://www.giss.nasa.gov]

Hansen, James E. 1999. "Global Warming, a Time for Action." WHOH-TV (Boston, Mass.). [http://www.7almanac.com/1999/pg10.htm]

Hansen, J.E., and S. Lebedeff. 1987. "Global Trends of Measured Surface Temperature." *Journal of Geophysical Research* 92, no. D11: 13345–13372.

Hansen, J.E., and A. Lacis. 1990. "Sun and Dust Versus Greenhouse Gases: An Assessment of Their Relative Roles in Global Climate Change." *Nature* 346:713–719.

Hansen, J.E., D. Johnson, A. Lacis, S. Lebedeff, P. Lee, D. Rind, and G. Russell. 1981. "Climatic Impact of Increasing Atmospheric Carbon Dioxide." *Science* 213:957–966.

Hansen, J.E., G. Russell, D. Rind, P. Stone, A. Lacis, S. Lebedeff, R. Ruedy, and L. Travis. 1983. "Efficient Three-dimensional Global Models for Climate Studies: Models I and II." *Monthly Weather Review* 111:609–662.

Hansen, J.E., A. Lacis, D. Rind, G. Russell, P. Stone, I. Fung, R. Ruedy, and J. Lerner. 1984. "Climate Sensitivity: Analysis of Feedback Mechanisms." *Geophysical Monograph 29 Maurice Ewing*. Vol. 5. American Geophysical Union, 130–163.

Hansen, J.E., G. Russell, A. Lacis, I. Fung, D. Rind, and P. Stone. 1985. "Climate Response Times: Dependence on Climate Sensitivity and Ocean Mixing." *Science* 229 (30 August 30): 857–859.

Hansen, J.E., I. Fung, A. Lacis, S. Lebedeff, D. Rind, R. Ruedy, G. Russell, and P. Stone. 1987. "Prediction of Near-term Climate Evolution: What Can We Tell Decision-makers Now?" In *Preparing for Climate Change: Proceedings of the First North American Conference on Preparing for Climate Change: A Cooperative Approach.* Proceedings of a Conference Held 27–29 October in Washington, D.C.

Hansen, J.E., I. Fung, A. Lacis, D. Rind, S. Lebedeff, R. Ruedy, and G. Russel. 1988. "Global Climate Changes as Forecast by GISS's Three-dimensional Model." *Journal of Geophysical Research* 93:9341–9364.

Hansen, J.E, D. Rind, A. Del Genio, A. Lacis, S. Lebedeff, M. Prather, R. Ruedy, and T. Karl. 1989. "Regional Greenhouse Climate Effects." In *Coping With Climate Change: Proceedings of the Second North American Conference on Coping With Climate Change,* 6–8 December. Washington, D.C.: Climate Institute.

Hansen, J.E., A. Lacis, R. Ruedy, and M. Sato. 1992. "Potential Climate Impact of the Mount Pinatubo Eruption." *Geophysical Research Letters* 19, no. 2 (24 January): 215–218.

Hansen, J.E., M. Sato, and R. Ruedy. 1995. "Long-term Changes of the Diurnal Temperature Cycle: Implications about Mechanisms of Global Climate Change." *Atmospheric Research* 37:175–209.

Hansen, J.E., W. Rossow, B. Carlson, A. Lacis, L. Travis, A. Del Genio, I. Fung, B. Cairns, M. Mishchenko, and M. Sato. 1995. "Low-cost, Long-term Monitoring of Global Climate Forcings and Feedbacks." *Climatic Change* 31:117–141.

Hansen, J.E., R. Ruedy, M. Sato, and R. Reynolds. 1996. "Global Surface Air Temperatures in 1995: Return to Pre-Pinatubo Levels." *Geophysical Research Letters* 23: 1665–1668.

Hansen, J.E., M. Sato, and R. Ruedy. 1997. "The Missing Climate Forcing." *Philosophical Transactions of the Royal Society of London* B352:231–240.

Hansen, J.E., M. Sato, and R. Ruedy. 1997. "Radiative Forcing and Climate Response." *Journal of Geophysical Research* 101:6831–6864.

Hansen, J.E., M. Sato, R. Ruedy, A. Lacis, K. Asamoah, K. Beckford, S. Borenstein, E. Brown, B. Cairns, B. Carlson, B. Curran, S. de Castro, L. Druyan, P. Etwarrow, T. Ferede, M. Fox, D. Gaffen, J. Glascoe, H. Gordon, S. Hollandsworth, X. Jiang, C. Johnson, N. Lawrence, J. Lean, J. Lerner, K. Lo, J. Logan, A. Luckett, M.P. McCormick, R. McPeters, R. Miller, P. Minnis, I. Ramberran, G. Russell, P. Russell, P. Stone, I. Tegen, S. Thomas, L. Thomason, A. Thompson, J. Wilder, R. Wilson, and J. Zawodny. 1997. "Forcings and Chaos in Interannual and Decadal Climate Change." *Journal of Geophysical Research* 102 (27 November): 25,679–25,720.

Hansen, J.E., M. Sato, A. Lacis, R. Ruedy, I. Tegen, and E. Matthews. 1998. "Climate Forcings in the Industrial Era." *Proceedings of the National Academy of Sciences of the United States of America* 95 (October): 12753–12758.

Hansen, J.E., M. Sato, J. Glascoe, and R. Ruedy. 1998. "A Common-sense Climate Index: Is Climate Changing Noticeably?" *Proceedings of the National Academy of Sciences of the United States of America* 95:4113–4120.

Hansen, J.E., R. Ruedy, A. Lacis, M. Sato, L. Nazarenko, N. Tausnev, I. Tegen, and D.

Koch. 1999. "Climate Modeling in the Global Warming Debate." In D. Randall, ed., *Climate Modeling: Past, Present, and Future.* San Diego: Academic Press, 127–164.

Hansen J.E., R. Ruedy, J. Glascoe, and M. Sato. 1999. "GISS Analysis of Surface Temperature Change." *Journal of Geophysical Research* 104 (27 December): 30,997–31,022.

Hansen, J.E., M. Sato, R. Ruedy, A. Lacis, and V. Oinas. 2000. "Global Warming in the Twenty-first Century: An Alternative Scenario." *Proceedings of the National Academy of Sciences of the United States of America* 97, no. 18 (29 August): 9875–9880.

Hardin, Garrett, and John Baden, eds. 1977. *Managing the Commons.* San Francisco: W.H. Freeman.

Hare, Tony, and Aziz Khan. 1990. *The Greenhouse Effect.* New York: Gloucester Press.

Hariss, Robert C., Terry Bensel, and Denise Blaha. 1993. "Methane Emissions to the Global Atmosphere From Coal Mining." In Richard A. Geyer, ed., *A Global Warming Forum: Scientific, Economic, and Legal Overview.* Boca Raton, Fla.: CRC Press, 339–346.

Hartmann, Dennis L., John M. Wallace, Varavut Limpasuvan, David W.J. Thompson, and James R. Holton. 2000. "Can Ozone Depletion and Global Warming Interact to Produce Rapid Climate Change?" *Proceedings of the National Academy of Sciences of the United States of America* 97, no. 4 (15 February): 1412–1417.

Harvell, C.D., K. Kim, J.M. Buckholder, R.R. Colwell, P.R. Epstein, D.J. Grimes, E.E. Hofmann, E.K. Lipp, A.D.M.E. Osterhaus, R.M. Overstreet, J.W. Porter, G.W. Smith, and G.R. Vasta. 1999. "Emerging Marine Diseases—Climate Links and Anthropogenic Factors." *Science* 285, no. 3 (3 September): 1505–1510.

Hashimoto, Ryutaro, Prime Minister of Japan. 1997. Excerpts from Statements on Climate Change by Foreign Leaders at Earth Summit, 23 June. [http://uneco.org/Global_Warming.html]

Hastenrath, S., and P.D. Kruss. 1992. "Greenhouse Indicators in Kenya." *Nature* 355: 503–504.

Hawkes, Nigel. 2000. "Giant Cities are Creating Their Own Weather." *The Times of London*, 23 February, in LEXIS.

Haywood, J.M., V. Ramaswamy, and B.J. Soden. 1999. "Tropospheric Aerosol Climate Forcing in Clear-sky Satellite Observations Over the Oceans." *Science* 283 (26 February): 1299–1303.

Hegerl, Gabriele. 1998. "Climate Change: The Past as Guide to the Future." *Nature* 392 (23 April): 758–759.

Helfferich, Carla. 1990. "Alaska Science Forum: Cloudy Picture of a Changing Globe." Geophysical Institute, University of Alaska Fairbanks. 6 June. [http://www.gi.alaska.edu/ScienceForum/ASF9/981.html]

Helm, Dieter. 1990. "Who Should Pay for Global Warming?" *New Scientist* 128:36.

Hendee, David. 2000. "Mild Conditions Carry a Price." *Omaha World-Herald*, 16 January, 1A, 2A.

Henderson-Sellers, A., and R. Blong. 1989. *The Greenhouse Effect: Living in a Warmer Australia.* Kensington, Australia: New South Wales University Press.

Hendriks, C.A., K. Blok, and W.C. Turkenburg. 1989. "The Recovery of Carbon Dioxide from Power Plants." In P.A. Okken, R.J. Swart, and S. Zwerver eds., *Climate and Energy: The Feasibility of Controlling CO2 Emissions.* Dordrecht, Germany: Kluwer Academic Publishers, 125–142.

Hengeveld, Henry. 1994. "The Global Warming Challenge: Understanding and Coping with Climate Change in Canada." *Environmental Science and Technology* 28, no. 12: 519–523.

Herbert, H. Josef. 2000. "Study: World's Oceans Warming." Associated Press, 24 March, in LEXIS.

Hernandez, M., I. Robinson, A. Aguilar, L.M. Gonzalez, L.F. Lopez-Jurado, M.I. Reyero, E. Cacho, and J. Franco. 1998. "Did Algal Toxins Cause Monk Seal Mortality?" *Nature* 393:28–29.

Hertsgaard, Mark. 1999. "Will We Run Out of Gas? No, We'll Have Plenty of Carbon-based Fuel to See Us Through the Next Century. That's the Problem." *Time*, 5 November, 110–111.

Hesselbo, Stephen P., Darren R. Grocke, Hugh C. Jenkyns, Christian J. Bjerrum, Paul Farrimond, Helen S. Morgans Bell, and Owen R. Green. 2000. "Massive Dissociation of Gas Hydrate during a Jurassic Oceanic Anoxic Event." *Nature* 406 (27 July): 392–395.

Hesshaimer V., M. Heimann, and I. Levin. 1994. "Radiocarbon Evidence for a Smaller Oceanic Carbon Dioxide Sink than Previously Believed." *Nature* 370:201–203.

Hewitt, C.N., and W.T. Sturges. 1993. *Global Atmospheric Chemical Change.* New York: Elsevier Applied Science.

Hileman, Bette. 1995. "Climate Observations Substantiate Global Warming Models." *Chemical & Engineering News*, 27 November. [http://pubs.acs.org/hotartcl/cenear/951127/pg1.html]

Hobbs, P.V., J.S. Reid, and R.A. Kotchenruther. 1997. "Direct Radiative Forcing by Smoke from Biomass Burning." *Science* 275:1776–1778.

Hodell, D.A., J.H. Curtis, and M. Brenner. 1995. "Possible Role of Climate in the Collapse of Classic Maya Civilization." *Nature* 375:391–394.

Hodges, Glenn. 2000. "The New Cold War: Stalking Arctic Climate Change by Submarine." *National Geographic*, March, 30–41.

Hoffert, M.I., and C. Covey. 1992. "Deriving Global Climate Sensitivity from Paleoclimate Reconstructions." *Nature* 360:573–576.

Hoffert, Martin, Ken Caldeira, Curt Covey, Philip P. Duffy, and Benjamin D. Santer. 1999. "Solar Variability and the Earth's Climate." *Nature* 401 (21 October): 764–765.

Hoffert, Martin I., Ken Caldeira, Atul K. Jain, Erik F. Haites, L.D. Danny Harvey, Seth D. Potter, Michael E. Schlesinger, Stephen H. Schneider, Robert G. Watts, Tom M.L. Wigley, and Donald J. Wuebbles. 1998. "Energy Implications of Future Stabilization of Atmospheric CO2 Content." *Nature* 395 (29 October): 881–884.

Hoffman, J.S., J.J. Wells, D. Keyes, and J.G. Titus. 1983. *Projecting Future Sea Level Rise.* Washington, D.C.: U.S. Environmental Protection Agency.

Hogarth, Murray. 1998. "Sea-warming Threatens Coral Reefs." *Sydney Morning Herald*, 26 November. [http://www.smh.com.au/news/9811/26/text/national13.html]

Hollin, J.T. 1970. "Interglacial Climates and Antarctic Ice Surges." *Quarternary Research* 2:401–408.

Holm-Hansen, O. 1982. *Effects of Increasing CO2 on Ocean Biota.* U.S. Department of Energy 013, vol. 13, part 5. Washington, D.C.: U.S. Department of Energy.

"A Hot '98, Relatively Speaking, in Nation's Northernmost City." 1999. ABC News, 17 January. [http://geography.rutgers.edu/courses/99spring/370sp99/news01_20_99. html#anchor33828]

Houghton, John T. 1997. *Global Warming: The Complete Briefing.* Cambridge, U.K.: Cambridge University Press.

———, ed. 1990. *Climate Change: The IPCC Scientific Assessment.* Report prepared for IPCC [Intergovernmental Panel on Climate Change] by Working Group I. Cambridge, U.K.: Cambridge University Press.

Houghton, J.T., G.J. Jenkins, and J.J. Ephraums. 1990. *Climate Change: The IPCC Scientific Assessment.* New York: Cambridge University Press.

Houghton, J.T., B.A. Callander, and S.K. Varney, eds. 1992. *Climate Change 1992: The Supplementary Report to the IPCC Scientific Assessment.* Cambridge, U.K.: Cambridge University Press.

Houghton, J.T., L.G. Meira Filho, J. Bruce, Hoesung Lee, B.A. Callendar, E. Haites, N. Harris, and K. Maskell, eds. 1995. *Climate Change 1994: Radiative Forcing of Climate Change and an Evaluation of the IPCC IS92 Emission Scenarios.* Cambridge, U.K.: Cambridge University Press.

Houghton, J.T., L.G. Meira Filho, B.A. Callander, N. Harris, A Kattenberg, and K. Maskell, eds. 1996. *Climate Change 1995: The Science of Climate Change.* Cambridge, U.K.: Cambridge University Press.

Houghton, R.A. 1990. "The Future Role of Tropical Forests in Affecting the Carbon-dioxide Concentration in the Atmosphere." *Ambio* 19, no. 4: 204–209.

———. 1991. "Tropical Deforestation and Atmospheric Carbon Dioxide." *Climatic Change* 19:99–118.

———. 1994. "The Worldwide Extent of Land-use Change." *Bioscience* 44, no. 5: 305–314.

———. 1999. "The Annual Net Flux of Carbon to the Atmosphere From Changes in land Use 1850–1990." *Tellus* 51B:298–313.

———. 1987. "Biotic Changes Consistent With the Increased Seasonal Amplitude of Atmospheric CO2 Concentrations." *Journal of Geophysical Research* 92:4223–4230.

———. 1987. "Terrestrial Metabolism and Atmospheric CO2 Concentrations." *Bioscience* 37:672–678.

Houghton, R.A., E.A. Davidson, and G.M. Woodwell. 1998. "Missing Sinks, Feedbacks, and Understanding the Role of Terrestrial Ecosystems in the Global Carbon Balance." *Global Biogeochemical Cycles* 12, no. 1: 25–34.

Houghton, R.A., J.L. Hackler, and K.T. Lawrence. 1999. "The U.S. Carbon Budget: Contributions from Land-use Change." *Science* 285 (23 July): 574–578.

Houghton, R.A., D.L. Skole, Carlos A. Nobre, J.L. Hackler, K.T. Lawrence, and W.H. Chomentowski. 2000. "Annual Fluxes of Carbon From Deforestation and Regrowth in the Brazilian Amazon." *Nature* 403 (20 January): 301–304.

Houghton, R.A., and George M. Woodwell. 1989. "Global Climatic Change." *Scientific American*, April, 36–44.

Howard, J.D., O.H. Pilkey, and A. Kaufman. 1985. "Strategy for Beach Preservation Proposed." *Geotimes* 30:15.

"The Howler Epilogue: A Global Warming." 1999. *The Daily Howler*, 3 May. [http://www.dailyhowler.com/h050399_1.html]

Huang, Shaopeng, Henry N. Pollack, and Po-Yu Shen. 2000 "Temperature Trends Over the Past Five Centuries Reconstructed from Borehole Temperatures." *Nature* 403 (17 February): 756–758.

Huber, Brian T., Kenneth G. MacLeod, and Scott L. Wing. 2000. *Warm Climates in Earth History*. Cambridge, U.K.: Cambridge University Press.

Hughes, T. 1975. "The West Antarctic Ice Sheet: Instability, Disintegration and Initiation of Ice Ages." *Review of Geophysics* 13:502–526.

Hughes, T.J., J.L. Fastook, and G.H. Denton. 1979. *Climatic Warming and the Collapse of the West Antarctic Ice Sheet*. Orono: University of Maine Press.

Hull, C.H.J., and J.G. Titus, eds. 1986. *Greenhouse Effect, Sea-level Rise, and Salinity in the Delaware Estuary*. Washington, D.C.: U.S. Environmental Protection Agency.

Hulme, Mike, Elaine M. Barrow, Nigel W. Arnell, Paula A. Harrison, Timothy C. Johns, and Thomas E. Downing. 1999. "Relative Impacts of Human-induced Climate Change and Natural Climate Variability." *Nature* 397 (25 February): 688–691.

Hulme, Mike, and Mick Kelly. 1993. "Exploring the Links between Desertification and Climate Change." *Environment* 35, no. 6: 5–45.

Hunt, G.E., V. Ramanathan, and R.M. Chervin. 1980. "On the Role of Clouds in the General Circulation of the Atmosphere." *Quarterly Journal of the Royal Meteorological Society* 106:213–215.

Huntley, Brian. 1990. "Lessons from the Climates of the Past." In Jeremy Leggett, ed., *Global Warming: The Greenpeace Report*. New York: Oxford University Press, 133–148.

Hurrell, J.W. 1995. "Decadal Trends in the North Atlantic Oscillation: Regional Temperatures and Precipitation." *Science* 269:676–679.

Hurrell, James W., and Harry Van Loon. 1997. "Decadal Variations in Climate Associated with the North Atlantic Oscillation." *Climatic Change* 36:301–306.

Huybrechts, P., and J. Oerlemans. 1990. "Response of the Antarctic Ice Sheet to Future Greenhouse Warming." *Climate Dynamics* 5:93–102.

Hvidberg, Christine Schott. 2000. "When Greenland Ice Melts." *Nature* 404 (6 April): 551–552.

Hyde, William T., Thomas J. Crowley, Steven K. Baum, and W. Richard Peltier. 2000. "Neoproterozoic 'Snowball Earth' Simulations with a Coupled Climate/Ice-sheet Model." *Nature* 405 (25 May): 425–429.

Idso, Sherwood B. 1982. *Carbon Dioxide: Friend or Foe?* Tempe, Ariz.: IBR Press.

———. 1989. *Carbon Dioxide and Global Change: Earth in Transition*. Tempe, Ariz.: IBR Press.

———. 1980. "The Climatological Significance of a Doubling of the Earth's Atmospheric Carbon Dioxide Concentration." *Science* 207:1462–1463.

Idso, S.B., and R.C. Balling, Jr. 1991. "Evaluating the Climatic Effect of Doubling Atmospheric CO2 via an Analysis of Earth's Historical Temperature Record." *Science of the Total Environment* 106:239–242.

Idso, S.B., R.C. Balling, Jr., and R.S. Cerveny. 1990. "Carbon Dioxide and Hurricanes: Implications of Northern Hemispheric Warming for Atlantic/Caribbean Storms." *Meteorology and Atmospheric Physics* 42:259–63.

Inamdar, A. K., and V. Ramanathan. 1994. "Physics of Greenhouse Effect and Convection in Warm Oceans." *Journal of Climate* 7, no. 5: 715–731.

Indermuhle, A., T.F. Stocker, F. Joos, H. Fischer, H.J. Smith, M. Wahlen, B. Deck, D. Mastroianni, J. Tschumi, T. Blunier, R. Meyer, and B. Stauffer. 1999. "Holocene Carbon-cycle Dynamics Based on CO2 Trapped in Ice at Taylor Dome, Alaska." *Nature* 398 (11 March): 121–126.

Ingram, W.J., C.A. Wilson, and J.F.B. Mitchell. 1989. "Modelling Climate Changes: An Assessment of Sea Ice and Surface Albedo Feedbacks." *Journal of Geophysical Research* 94, no. D6: 8609–8622.

Innes, J.L., 1991. "High-altitude and High-latitude Tree Growth in Relation to Past, Present and Future Global Climate Change." *The Holocene* 1:168–173.

Inouye, David W., Billy Barr, Kenneth B. Armitage, and Brian D. Inouye. 2000. "Climate Change is Affecting Attitudinal Migrants and Hibernating Species." *States of America* 97, no. 4 (15 February): 1630–1634.

Intergovernmental Panel on Climate Change (IPCC). 1990. *Climate Change: The IPCC Scientific Assessment.* Report prepared for IPCC by Working Group I, John T. Houghton et al., eds. Cambridge, U.K.: Cambridge University Press.

International Energy Agency. 1991. *Greenhouse Gas Emissions: The Energy Dimension.* Paris, France: International Energy Agency.

————. 1993. Cars and Climate Change. Paris, France: International Energy Agency.

Ivins, Molly. 2000. "Environment is Ticking, but Will We Hear It?" *Chicago Sun-Times,* 13 July, 39.

Izrael, Yu A. 1991. "Climate Change Impact Studies: The IPCC Working Group II Report." In J. Jager and H.L. Ferguson, eds., *Climate Change: Science, Impacts, and Policy.* Proceedings of the Second World Climate Conference. Cambridge, U.K.: Cambridge University Press, 83–86.

Jackson, Michael B., and Colin R. Black, eds. 1993. *Interacting Stresses on Plants in a Changing Climate.* Berlin: Springer-Verlag.

Jacob, Daniel J. 1999. *Introduction to Atmospheric Chemistry.* Princeton: Princeton University Press.

Jacobs, S.S., H.H. Hellmer, C.S.M. Doake, A. Jenkins, and R.M. Frolich. 1992. "Melting of Ice Shelves and the Mass Balance of Antarctica." *Journal of Glaciology* 38, no. 130: 375–387.

Jager, Jill. 1989. "Developing Policies for Responding to Climate Change." In Dean Edwin Abrahamson, ed., *The Challenge of Global Warming.* Washington, D.C.: Island Press, 96–109.

Jager, Jill, and H.L. Ferguson. 1991. *Climate Change: Science, Impacts, and Policy: Proceedings of the Second World Climate Conference.* Cambridge, U.K.: Cambridge University Press.

James, P.D. 1994. *The Children of Men.* New York: Warner Books.

"Japan Agency Warns CO2 Emissions May Soar." 1997. United Press International, 6 November. [http://benetton.dkrz.de:3688/homepages/georg/kimo/0254.html]

"Japan's Prefectures Taking on Emissions." 1999. *Mainichi Shimbun* (Tokyo), 17 January. [http://geography.rutgers.edu/courses/99spring/370sp99/news01_20_99.html#anchor33828]

Jardine, Kevin. 1994. *The Carbon Bomb: Climate Change and the Fate of the Northern Boreal Forests.* Ontario, Canada: Greenpeace International. [http://dieoff.org/page129.htm]

Jastrow, Robert, William Aaron Nierenberg, and Frederick Seitz. 1990. *Scientific Perspectives on the Greenhouse Problem.* Ottawa: Marshall Press.

Jaworowski, Z. 1997. "Ice Core Data Show No Carbon Dioxide Increase in 21st Century." *Science and Technology* 10, no. 1: 42–52.

Jazcilevich, Aron, Vicente Fuentes, Ernesto Jauregui, and Estaben Luna. 2000. "Simulated Urban Climate Response to Historical Land-use Modification in the Basin of Mexico." *Climatic Change* 44:515–536.

Jenkins, Lyle M. 1993. "Space-based Geoengineering Options for Dealing with Global Change." In Richard A. Geyer, ed., *A Global Warming Forum: Scientific, Economic, and Legal Overview*. Boca Raton, Fla.: CRC Press, 101–109.

Jenkinson, D.S., D.E. Adams, and A. Wild. 1991. "Model Estimates of CO_2 Emissions from Soil in Response to Global Warming." *Nature* 351:304–306.

Jenney, L.L. 1991. "Reducing Greenhouse-gas Emissions From the Transportation Sector." In G.I. Pearman., ed., *Limiting Greenhouse Effects: Controlling Carbondioxide Emissions*. Report of the Dahlem Workshop on Limiting the Greenhouse Effect, Berlin, December 9–14. New York: John Wiley & Sons, 283–302.

Jepma, Catrinus and Mohan Munasinghe. 1998. *Climate Change Policy: Facts, Issues and Analyses*. Cambridge, U.K.: Cambridge University Press.

Johannessen, Ola M., Einar Bjorgo, and Martin W. Miles. 1995. "The Antarctic's Shrinking Sea Ice." *Nature* 376 (13 July): 126–127.

———. 1996. "Global Warming and the Arctic." *Science* 271 (12 January): 129.

Johansen, Bruce E. 2000. "Great Plains Destined for Dust?" *Omaha World-Herald*, 18 April, 23.

———. 1999. Review of John Houghton, *Global Warming: The Complete Briefing*, 1997. In *Native Americas* 16, no. 3/4(Fall/Winter): 96–97.

Johnson, Tim. 1999. "World Out of Balance: In a Prescient Time Native Prophecy Meets Scientific Prediction." *Native Americas* 16, no. 3/4(Fall/Winter): 8–25. [http://nativeamericas.aip.cornell.edu]

Johnston, David Cay. 2000. "Some Need Hours to Start Another Day at the Office." *New York Times*. In *Omaha World-Herald*, 6 February, 1G.

Jones, Elizabeth A., David D. Reed, and Paul V. Desanker. 1994. "Ecological Implications of Projected Climate Change Scenarios in Forest Ecosystems of Central North America." *Agricultural and Forest Meteorology* 72:31–46.

Jones, E.A., D.D. Reed, G.D. Mroz, H.O. Liechty, and P.J. Cattelino. 1993. "Climate Stress as a Precursor to Forest Decline: Paper Birch in Northern Michigan 1985–1990." *Canadian Journal of Forest Research* 23:229–233.

Jones, M.D.H., and A. Henderson-Sellers. 1990. "History of the Greenhouse Effect." *Progress in Physical Geography* 14:1–18.

Jones, Philip D., and Tom M.L. Wigley. 1990. "Global Warming Trends." *Scientific American* 263, no. 2: 84.

Joos, F. 1994. "Imbalance in the Budget." *Nature* 370:181–182.

Joos, Fortunat, Gian-Kasper Plattner, Thomas F. Stocker, Olivier Marchal, and Andreas Schmittner. 1999. "Global Warming and Marine Carbon Cycle Feedbacks on Future Atmospheric CO_2." *Science* 284 (16 April): 464–467.

Kaiser, Jocelyn. 1998. "Possibly Vast Greenhouse Gas Sponge Ignites Controversy." *Science* 282 (16 October): 386–387.

———. 2000. "Panel Estimates Possible Carbon 'Sinks.' " *Science* 288 (12 May): 942–943.

Kalkstein, Laurence S. 1993. "The Impacts of Global Climate Change on Human Health."
    Experientia 49:1–11.
——. 1993. "Direct Impacts in Cities." *Lancet* 342 (4 December): 1397–1400.
Kana, T.W., B.J. Baca, and M.L. Williams. 1986. *Potential Impacts of Sea-level Rise
    Around Charleston, South Carolina.* Washington, D.C.: U.S. Environmental Pro-
    tection Agency.
Kana, T.W., J. Michel, M.O. Hayes, and J.R. Jenson. 1984. "The Physical Impact of Sea
    Level Rise in the Area of Charleston, South Carolina." In M.C. Barth and J.G.
    Titus, eds., *Greenhouse Effect and Sea Level Rise: A Challenge for This Gener-
    ation.* New York: Van Nostrand Reinhold Company, 105–150.
Kane, R.L., and D.W. South. 1991. "The Likely Roles of Fossil Fuels in the Next 15,
    50, and 100 Years, With or Without Active Controls on Greenhouse-gas Emis-
    sions." In G.I. Pearman, ed., *Limiting Greenhouse Effects: Controlling Carbon-
    dioxide Emissions.* Report of the Dahlem Workshop on Limiting the Greenhouse
    Effect, Berlin, September 9–14, 1990. New York: John Wiley & Sons, 189–227.
Kanfoush, Sharon L., David A. Hodell, Christopher D. Charles, Thomas P. Guilderson,
    P. Graham Mortyn, and Ulysses S. Ninnemann. 2000. "Millennial-Scale Insta-
    bility of the Antarctic Ice Sheet During the Last Glaciation." *Science* 288 (9 June):
    1815–1819.
Kaplan, Robert. "The Coming Anarchy." 1994. *Atlantic Monthly*, February, 45–76.
Kareiva, Peter M., Joel G. Kingsolver, and Raymond B. Huey, eds. 1993. *Biotic Inter-
    actions and Global Change.* Sunderland, Mass.: Sinauer Associates.
Karl, Thomas R., Richard R. Heim, Jr., and Robert G. Quayle. 1991. "The Greenhouse
    Effect in Central North America: If Not Now, When?" *Science* 251:1058–1061.
Karl, T.R., P.D. Jones, R.W. Knight, G. Kukla, N. Plummer, V. Razuvayev, K.P. Gallo,
    J. Lindsay, R.J. Charlson, and T.C. Peterson. 1993. "A New Perspective on Re-
    cent Global Warming: Asymmetric Trends of Daily Maximum and Minimum
    Temperature." *Bulletin of the American Meteorological Society* 74:1007–1023.
Karl, T.R., R.W. Knight, R.G. Quayle, and D.R. Easterling. 1995. "Trends in the U.S.
    Climate During the Twentieth Century." *Consequences: The Nature and Impli-
    cations of Environmental Change* 1:2–10.
Karl, T.R., V. Derr, D. Hofmann, D.R. Easterling, C. Folland, S. Levitus, N. Nicholls,
    D. Parker, and G.W. Withee. 1995. "Critical Issues for Long-term Climate Mon-
    itoring." *Climatic Change* 31:185–221.
Karl, T.R., F. Bretherton, W. Easterling, and K. Trenberth. 1995. "Long-term Climate
    Monitoring of the Global Climate Observing Systems (GCOS)." *Climatic Change*
    31:135–147.
Karl, T.R., R.W. Knight, and N. Plummer. 1995. "Trends in High-frequency Climate
    Variability in the Twentieth Century." *Nature* 377:217–220.
Karl, Thomas R., Neville Nicholls, and Jonathan Gregory. 1997. "The Coming Climate:
    Meteorological Records and Computer Models Permit Insights into Some of the
    Broad Weather Patterns of a Warmer World." *Scientific American* 276:79–83.
    [http://www.scientificamerican.com/0597issue/0597karl.htm]
Karl, T.R., and R.W. Knight. 1997. "The 1995 Chicago Heat Wave: How Likely Is a
    Recurrence?" *Bulletin of the American Meteorology Society* 78, no. 6: 1107–1119.
Karl, Thomas R., Richard W. Knight, and Bruce Baker. 2000. "The Record-breaking

Global Temperatures of 1997 and 1998: Evidence for an Increase in the Rate of Global Warming." *Geophysical Research Letters* 27 (1 March): 719–722.

Karner, Daniel B., and Richard A. Muller. 2000. "A Causality Problem for Milankovitch." *Science* 288 (23 June): 2143–2144.

Kasting, James F., and Thomas P. Ackerman. 1986. "Climatic Consequences of Very High Carbon Dioxide Levels in the Earth's Early Atmosphere." *Science* 234: 1383–1385.

Kates, Robert W., and William C. Clark. 1996. "Environmental Surprise: Expecting the Unexpected?" *Environment* 38, no. 2: 6–34.

Katz, Miriam E., Dorothy K. Pak, Gerald R. Dickens, and Kenneth G. Miller. 1999. "The Source and Fate of Massive Carbon Input During the Latest Paleocene Thermal Maximum." *Science* 286 (19 November): 1531–1533.

Kawasaki, Tsuyoshi. 1991. "Effects of Global Climatic Change on Marine Ecosystems and Fisheries." In J. Jager and H.L. Ferguson, eds., *Climate Change: Science, Impacts, and Policy: Proceedings of the Second World Climate Conference.* Cambridge, U.K.: Cambridge University Press, 291–299.

Keeling, Charles D. 1970. "A Chemist Thinks About the Future." *Archives of Environmental Health* 20:764–777.

———. 1973. "Industrial Production of Carbon Dioxide from Fossil Fuels and Limestone." *Tellus* 25:174–198.

———. 1973. "The Carbon Dioxide Cycle." In S.I. Rasool, ed., *Chemistry of the Lower Atmosphere.* New York: Plenum Press, 251–331.

Keeling, C.D., J.A. Adams, C.A. Ekdahl, and P.R. Guenther. 1976. "Atmospheric Carbon Dioxide Variations at South Pole." *Tellus* 28:552–564.

Keeling, C.D., R.B. Bacastow, A.E. Bainbridge, C.A. Ekdahl, P.R. Guenther, L.S. Waterman, and J.F.S. Chin. 1976. "Atmospheric Carbon Dioxide Variations at Mauna Loa Observatory, Hawaii." *Tellus* 28:538–551.

Keeling, C.D., R.B. Bacastow, and T.P. Whorf. 1976. "Atmospheric Carbon Dioxide Variations at Mauna Loa Observatory, Hawaii." *Tellus* 28:28.

———. 1982. "Measurements of the Concentration of Carbon Dioxide at Mauna Loa Observatory, Hawaii." *Carbon Dioxide Review*, 377–384.

Keeling, R.F., S.C. Piper, and M. Heimann. 1996. "Global and Hemispheric CO2 Sinks Deduced From Changes in Atmospheric CO2 Concentration." *Nature* 381:218–221.

Keeling, R.F. and S.R. Shertz. 1992. "Seasonal and Interannual Variations in Atmospheric Oxygen and Implications for the Global Carbon Cycle." *Nature* 358:723–727.

Keeling, Charles D., and Timothy P. Whorf. 2000. "The 1,800-year Oceanic Tidal Cycle: A Possible Cause of Rapid Climate Change." *Proceedings of the National Academy of Sciences of the United States of America* 97, no. 8 (11 April): 3814–3819.

Keigwin, L., W.B. Curry, S.J. Lehman, and S. Johnsen. 1994. "The Role of the Deep Ocean in North Atlantic Climate Change Between 70 and 130 Kyr [Thousand Years] Ago." *Nature* 371:323–329.

Keigwin, L.D., and E.A. Boyle. 2000. "Detecting Holocene Changes in Thermohaline Circulation." *Proceedings of the National Academy of Sciences of the United States of America* 97, no. 4 (15 February): 1343–1346.

Kellogg, William W. 1990. "Theory of Climate Transition from Academic Challenge to Global Imperative." In Terrell J. Minger, ed., *Greenhouse Glasnost: The Crisis of Global Warming.* New York: Ecco Press, 99.

———. 1987. "Mankind's Impact on Climate: The Evolution of an Awareness." *Climate Change* 10:113–136.

———. 1983. "Feedback Mechanisms in the Climate Systems Affecting Future Levels of Carbon Dioxide." *Journal of Geophysical Research* 88:1263–1269.

Kellogg, W.W., and R. Schware. 1981. *Climate Change and Society.* Boulder, Colo.: Westview Press.

Kelly, Mick. 1990. "Halting Global Warming." In Jeremy Leggett, ed., *Global Warming: The Greenpeace Report.* New York: Oxford University Press, 83–112.

Kempton, Willet. 1991. "Public Understanding of Global Warming." *Society and Natural Resources* 4:331–345.

———. 1993. "Will Public Environmental Concern Lead to Action on Global Warming?" *Annual Review Energy Environment* 18:217–245.

Kennett, J.P. 1982. *Marine Geology.* Englewood Cliffs, N.J.: Prentice-Hall.

Kennett, James P., Kevin G. Cannariato, Ingrid L. Hendy, and Richard J. Behl. 2000. "Carbon Isotopic Evidence for Methane Hydrate Instability During Quaternary Interstadials." *Science* 288 (7 April): 128–133.

Kerr, Richard A. 1988. "Is the Greenhouse Here?" *Science* 239:559.

———. 1991. "U.S. Bites Greenhouse Bullet and Gags." *Science* 251 (22 February): 868.

———. 1995. "Study Unveils Climate Cooling Caused by Pollutant Haze." *Science* 268: 802.

———. 1995. "Studies Say—Tentatively—That Greenhouse Warming is Here." *Science* 268:1567–1568.

———. 1995. "It's Official: First Glimmer of Greenhouse Warming Seen." *Science* 270: 1565–1567.

———. 1998. "West Antarctica's Weak Underbelly Giving Way?" *Science* 281 (24 July): 499–500.

———. 1998. "Among Global Thermometers, Warming Still Wins Out." *Science* 281 (25 September): 1948–1949.

———. 1998. "Deep Chill Triggers Record Ozone Hole." *Science* 282 (16 October): 391.

———. 1999. "A Smoking Gun for an Ancient Methane Discharge." *Science* 286 (19 November): 1465.

———. 1999. "Big El Niños Ride the Back of Slower Climate Change." *Science* 283 (19 February): 1108–1109.

———. 1999. "In North American Climate, a More Local Control." *Science* 283 (19 February): 1109.

———. 1999. "Has a Great River in the Sea Slowed Down?" *Science* 286 (5 November): 1061–1062.

———. 1999. "From Eastern Quakes to a Warming's Icy Clues." *Science* 283 (1 January): 28–29.

———. 2000. "Globe's 'Missing Warming' Found in the Ocean." *Science* 287:2126–2127.

———. 2000. "Global Warming: Draft Report Affirms Human Influence." *Science* 288 (28 April): 589–590.

————. 2000. "Viable But Variable Ancient El Niño Spied." *Science* 288 (12 May): 945.

————. 2000. "A North Atlantic Climate Pacemaker for the Centuries." *Science* 288 (16 June): 1984–1985.

————. 2000. "The Sun Again Intrudes on Earth's Decadal Climate Change." *Science* 288 (16 June): 1986.

————. 2000. "Dueling Models: Future U.S. Climate Uncertain." *Science* 288 (23 June): 2113.

————. 2000. "Ice, Mud Point to CO2 Role in Glacial Cycle." *Science* 289 (15 September): 1868.

Kete, Nancy. 1999. "Creating a Climate for Change: Rumours of the Demise of the Kyoto Protocol Are Premature." Review of Grubb et al., *Kyoto Protocol: A Guide and Assessment* (1999). In *Nature* 402 (18 November): 233–234.

Khalil, Mohammad Aslam Khan, ed. 2000. *Atmospheric Methane: Its Role in the Global Environment*. New York: Springer.

Khalil, M.A.K., and Rasmussen, R.A. 1989. "Climate-induced Feedbacks for the Global Cycles of Methane and Nitrous Oxide." *Tellus* 41B:554–559.

Kheshgi, Haroon S., Atul K. Jain, and Donald J. Wuebbles. 1996. "Accounting for the Missing Carbon-sink with the CO2-Fertlization Effect." *Climatic Change* 33:31–62.

Kiehl, Jeffery T. 1999. "Solving the Aerosol Puzzle." *Science* 283 (26 February): 1273–1275.

Kiehl, J.T., and B.P. Brieglab. 1993. "The Radiative Roles of Sulfate Aerosols and Greenhouse Gases in Climate Forcing." *Science* 260:311–314.

Kienast, F. 1991. "Simulated Effects of Increasing Atmospheric CO2 and Changing Climate on the Successional Characteristics of Alpine Forest Ecosystems." *Landscape Ecology* 5, no. 4: 225–238.

Kimball, B.A. 1983. "Carbon Dioxide and Agricultural Yield: An Assemblage and Analysis of 430 Prior Observations." *Agronomy Journal* 75:779–788.

Kimmins, J.P., and D.P. Lavender. 1992. "Ecosystem-level Changes that May be Expected in a Changing Global Climate: A British Columbia Perspective." *Environmental Toxicology and Chemistry* 11:1061–1068.

Kirby, Alex. 1998. "Soar Away U.S. Greenhouse Forecast." British Broadcasting Corp. News, 25 November. [http://news.bbc.co.uk/hi/english/sci/tech/newsid_221000/221734.stm]

————. 1998. "Climate Change: It's the Sun and Us." British Broadcasting Corp. News, 26 November. [http://news.bbc.co.uk/hi/english/sci/tech/newsid_222000/222437.stm]

Kirschbaum, M.U.F., D.A. King, H.N. Comins, R.E. McMurtrie, B.E. Medlyn, S. Pongracic, D. Murty, H. Keith, R.J. Raison, P.K. Khanna, and D.W. Sheriff. 1994. "Modelling Forest Response to Increasing CO2 Concentration Under Nutrient-limited Conditions." *Plant, Cell and Environment* 17:1081–1099.

Kizzia, Tom. 1998. "Seal Hunters Await Late Ice." *Anchorage Daily News*, 28 November. [http://www.adn.com/stories/T98112872.html]

Kleypas, J.A., R.W. Buddemeier, D. Archer, J.P. Gattuso, C. Langdon, and B.N. Opdyke. 1999. Geochemical Consequences of Increased Atmospheric Carbon Dioxide on Coral Reefs." *Science* 284:118–120.

Kling, G.W., and M.C. Grant. 1984. "Acid Precipitation in the Colorado Front Range: An Overview with Time Predictions for Significant Effects." *Arctic and Alpine Research* 16:321–329.

Kling, G.W., G.W. Kipphut, and M.C. Miller. 1992. "The Flux of Carbon Dioxide and Methane from Lakes and Rivers in Arctic Alaska." *Hydrobiologia* 240:23–36.

Klug, Edward C. 1997. "Global Warming: Melting Down the Facts about this Overheated Myth." CFACT Briefing Paper #105, November. [http://www.cfact.org/IssueArchive/greenhouse.bp.n97.txt]

Knox, Joseph P. 1991. "Global Climate Change: Impacts on California: An Introduction and Overview." In Joseph P. Knox and Ann Foley Scheuring, *Global Climate Change and California*. Berkeley: University of California Press, 1–25.

Knox, Joseph P., and Ann Foley Scheuring. 1991. *Global Climate Change and California*. Berkeley: University of California Press.

Knutson, T.R., R.E. Tuleya, and Y. Kurihara. 1999. "Simulated Increase in Hurricane Intensities in a CO2-Warmed Climate." *Science* 279 (13 February): 1018–1020.

Kok, Wim, Prime Minister, Netherlands. 1997. "On Behalf of the European Union." Statements on Climate Change by Foreign Leaders at Earth Summit, 23 June. [http://uneco.org/Global_Warming.html].

Kondratyer, K. Ya., and A.P. Cracknell. 1998. *Observing Global Climate Change*. London: Taylor & Francis.

Koster, E.A., and M.E. Nieuwenhuyzen. 1992. "Permafrost Response to Climatic Change." In M. Boer and E. Koster, eds., *Greenhouse-Impact on Cold-Climate Ecosystems and Landscapes*. Selected Papers of A European Conference on Landscape Ecological Impact: Impact of Climatic Change, Lunteren, the Netherlands, December 3–7, 1989. CARTENA Supplement 22. Cremlingen, Germany: Cartena Verlag, 23–35.

Kowalok, Michael E. 1993. "Common Threads: Research Lessons from Acid Rain, Ozone Depletion and Global Warming."*Environment* 35, no. 6: 12–38.

Krabill, William. 1999. "Rapid Thinning of Parts of the Southern Greenland Ice Sheet." Abstract of an article in *Science* 283 (5 March): 1152–1154. [http://earth.rice.edu/MTPE/cryo/cryosphere/topics/greenthin.html]

Krabill, W., W. Abdalati, E. Frederick, S. Manizade, C. Martin, J. Sonntag, R. Swift, R. Thomas, W. Wright, and J. Yungel. 2000. "Greenland Ice Sheet: High-elevation Balance and Peripheral Thinning." *Science* 289 (21 July): 428–430.

Krabill, W., E. Frederick, S. Manizade, C. Martin, J. Sonntag, R. Swift, R. Thomas, W. Wright, and J. Yungel. 1999. "Rapid Thinning of Parts of the Southern Greenland Ice Sheet." *Science* 283:1522–1524.

Krabill, W., R. Thomas, K. Jezek, K. Kuivinen, S. Manizade. 1995. "Greenland Ice Thickness Changes Measured by Laser Altimetry." *Geophysical Research Letters* 22:2341–2344.

Krause, Florentin. 1992. *Energy Policy in the Greenhouse*. New York: Wiley.

Kristof, Nicholas. 1997. "For Pacific Islanders, Global Warming Is No Idle Threat." *New York Times*, 2 March. [http://sierraactivist.org/library/990629/islanders.html]

Kruse, Gordon H. 1998. "Salmon-run Failures in 1997–1998: A Link to Anomalous Oceanic Conditions?" *Alaska Fishery Research Bulletin* 5, no. 1: 55–63.

Kukla, G., J. Gavin, M. Schlesinger, and T.R. Karl. 1995. "Comparison of Observed Seasonal Temperature Maxima and Diurnal Range in North America with Simulations from Three Global Climate Models." *Atmospheric Research* 37:267–275.

Kurz, Werner A., Michael J. Apps, Brian J. Stocks, and Jan A. Volney. 1995. "Global Climate Change: Disturbance Regimes and Biospheric Feedbacks of Temperate and Boreal Forests." In George M. Woodwell and Fred T. MacKenzie, eds., *Biotic Feedbacks in the Global Climate System: Will the Warming Feed the Warming?* New York: Oxford University Press, 119–133.

Kuypers, Marcel M.M., Richard D. Pancost, and Jaap S. Sinninghe Damste. 1999. "A Large and Abrupt Fall in Atmospheric CO2 Concentration During Cretaceous Times." *Nature* 399 (27 May): 342–345.

Lachenbruch, A.H., and B.V. Marshall. 1986. "Changing Climate: Geothermal Evidence from Permafrost in the Alaskan Arctic." *Science* 234:689–696.

Lacis, Andrew A., and Barbara E. Carlson. 1992. "Keeping the Sun in Proportion." *Nature* 360 (26 November): 297.

Lamb, H. 1985. *Climatic History and the Future.* Princeton, N.J.: Princeton University Press.

Landsea, Christopher W. 1999. NOAA: Report on Intensity of Tropical Cyclones, 12 August, Miami, Fla. [http://www.aoml.noaa.gov/hrd/tcfaq/tcfaqG.html#G3]

———. 1993. "A Climatology of Intense (or Major) Atlantic Hurricanes." *Monthly Weather Review* 121:1703–1713.

Landsea, C.W., N. Nicholls, W.M. Gray, and L.A. Avila. 1996. "Downward Trends in the Frequency of Intense Atlantic Hurricanes During the Past Five Decades." *Geophysical Research Letters* 23:1697–1700.

Lashof, D. 1989. "The Dynamic Greenhouse: Feedback Processes That May Influence Future Concentrations of Atmospheric Trace Gases and Climatic Change." *Climatic Change* 14:213–242.

Lashof, Daniel A., and Eric L. Washburn. 1990. *The Statehouse Effect: State Policies to Cool the Greenhouse.* Washington, D.C.: Natural Resources Defense Council.

Lave, Lester B. 1988. "The Greenhouse Effect: What Government Actions are Needed?" *Journal of Policy Analysis and Management* 7:460–470.

Lawler, Andrew. 1995. "NASA Mission Gets Down to Earth." *Science* 269:1208–1210.

Lawrence, Mark G., and Paul J. Crutzen. 1999. "Influence of Nitrous Oxide Emissions from Ships on Tropospheric Photochemistry and Climate." *Nature* 402 (11 November): 167–168.

Lazaroff, Cat. 2000. "Greenland Ice Sheet Melting Away." Environmental News Service (Washington, D.C.), 21 July.

Lean, J., J. Beer, and R. Bradley. 1995. "Reconstruction of Solar Irradiance Since 1610: Implications for Climate Change." *Geophysical Research Letters* 22:3195–3198.

Lean, J., and D.A. Warrilow. 1989. "Simulation of the Regional Climate Impact of Amazon Deforestation." *Nature* 342 (23 November): 411–413.

Lear, A. 1989. "Potential Health Effects of Global Climate and Environmental Changes." *New England Journal of Medicine* 321:1577–1583.

Leatherman, Stephen P. 1992. "Coastal Land Loss in the Chesapeake Bay Region: An Historical Analog Approach to Global Change Analysis." In Jurgan Schmandt and Judith Clarkson, eds., *The Regions and Global Warming: Impacts and Response Strategies.* New York: Oxford University Press, 17–27.

LeBlanc, D.C., and J.K. Foster. 1992. "Predicting Effects of Global Warming on Growth and Mortality of Upland Oak Species in the Midwestern United States: A Physiologically Based Dendroecological Approach." *Canadian Journal of Forest Research* 22:1739–1752.

Lefor, M.W., W.C. Kennard, and D.L. Civco. 1990. "Relationships of Salt-marsh Plant Distributions to Tidal Levels." In Jeremy Leggett, ed., *Global Warming: The Greenpeace Report*. New York: Oxford University Press, 83–112.

Leggett, Jeremy, ed. 1990. *Global Warming: The Greenpeace Report*. New York: Oxford University Press.

———. 1993. "Who Will Underwrite the Hurricane?" *New Scientist* 139:28.

———. 2000. *The Carbon War: Global Warming and the End of the Oil Era*. London: Penguin.

Levins, R.T. Awerbuch, and U. Brinkman. 1994. "The Emergence of New Diseases." *American Scientist* 82:52–60.

Levitus, Sydney, John I. Antonov, Timothy P. Boyer, and Cathy Stephens. 2000. "Warming of the World Ocean." *Science* 287:2225–2229.

Levy, D.A. 1994. *Potential Impacts of Global Warming on Salmon Production in the Fraser River Watershed*. Downsview, Ontario: Climate Products and Publications Division.

Lewis, Lowell, William Rains, and Lynne Kennedy. 1991. "Global Climate Change and California Agriculture." In Joseph P. Knox and Ann Foley Scheuring. *Global Climate Change and California*. Berkeley: University of California Press, 97–122.

Liesman, Steve. 2000. "Texaco Appears to Moderate Stance on Global Warming: New Hires and Investments Move the Oil Giant in a Greener Direction." *Wall Street Journal*, 15 May, B-4.

Liesman, Steve, and Jacob M. Schlesinger. 1999. "The Price of Oil Has Doubled This Year; So, Where's the Recession?" *Wall Street Journal*, 15 December, A-1, A-10.

Lighthill, J., G. Holland, W. Gray, C. Landsea, G. Craig, J. Evans, Y. Kurihara, and C. Guard. 1994. "Global Climate Change and Tropical Cyclones." *Bulletin of the American Meteorological Society* 75:2147–2157.

Limbaugh, Rush. 1992. *The Way Things Ought to Be*. New York: Pocket Books.

Lindsay, S., and M. Birley. 1996. "Climatic Change and Malaria Transmission." *Annals of Tropical Medicine and Parasitology* 90, no. 6: 714–717.

Lindzen, Richard S. 1990. "Some Coolness Concerning Global Warming." *Bulletin of the American Meteorological Society* 71:288–299.

———. 1993. "Absence of Scientific Basis." *Research and Exploration* 9, no. 2: 191–200.

———. 1997. "Can Increasing Carbon Dioxide Cause Climate Change?" *Proceedings of the National Academy of Sciences of the United States of America* 94:8335–8342.

Liss, P.S. 1983. *Man-made Carbon Dioxide and Climatic Change: A Review of Scientific Problems*. Norwich, U.K.: Geo Books.

Lobitz, Brad, Louisa Beck, Anwar Huq, Byron Wood, George Fuchs, A.S.G. Faruque, and Rita Colwell. 2000. "Climate and Infectious Disease: Use of Remote Sensing for Detection of *Vibrio Cholerae* by Indirect Measurement." *Proceedings of the National Academy of Sciences of the United States of America* 97, no. 4 (15 February): 1438–1441.

Loe, Rob C. de, and Reid D. Kreutzwiser. 2000. "Climate Variability, Climate Change and Water Resource Management in the Great Lakes." *Climatic Change* 45:163–179.

Loevinsohn, M.E. 1994. "Climatic Warming and Increased Malaria Incidence in Rwanda." *Lancet* 343:714–718.

Logan, Leslie. 2000. "Intense Destruction: Storms of the Warming Waters." *Native Americas* 17, no. 1 (Spring): 28–29.

Lonergan, Stephen. 1991. "Climate Warming, Water Resources and Geopolitical Conflict: A Study of Nations Dependent on the Nile, Litani, and Jordan River Systems." Extra-Mural Paper No. 55. Ottawa: Operational Research and Analysis Establishment.

Lorius, C., J. Jouzel, D. Raynaud, J. Hansen, and H. Le Treut. 1990. "The Ice-core Record: Climate Sensitivity and Future Greenhouse Warming." *Nature* 347:139–145.

Lovejoy, Thomas E., and Robert L. Peters. 1992. *Global Warming and Biological Diversity*. New Haven, Conn.: Yale University Press.

Lovins, A.B., and L.H. Lovins. 1991. "Least-cost Climatic Stabilization." In G.I. Pearman, ed., *Limiting Greenhouse Effects: Controlling Carbon-dioxide Emissions*. Report of the Dahlem Workshop on Limiting the Greenhouse Effect, Berlin, December 9–14, 1990. New York: John Wiley & Sons, 351–442.

Lowell, Thomas V. 2000. "As Climate Changes, So Do Glaciers." *Proceedings of the National Academy of Sciences of the United States of America* 97, no. 4 (15 February): 1351–1354.

Lunde, Leiv. 1991. "North/South and Global Warming—Conflict or Cooperation?" *Bulletin of Peace Proposals* 22, no. 2: 199–210.

Luxmoore, R.J., S.D. Wullschleger, and P.J. Hanson. 1993. "Forest Responses to CO2 Enrichment and Climate Warming." *Water, Air, & Soil Pollution* 70:309–323.

Lyman, Francesca, Irving Mintzer, Kathleen Courrier, and James MacKenzie. 1990. *The Greenhouse Trap: What We're Doing to the Atmosphere and How We Can Slow Global Warming*. Boston: Beacon Press.

Mabey, N., S. Hall, C. Smith, and S. Gupta. 1997. *Argument in the Greenhouse*. New York: Routledge.

MacAyeal, D.R. 1992. "Irregular Oscillations of the West Antarctic Ice Sheet." *Nature* 359:29–32.

MacCracken, Michael. 1991. "Greenhouse Gases: Changing the Global Climate." In Joseph P. Knox and Ann Foley Scheuring, eds., *Global Climate Change and California*. Berkeley: University of California Press, 26–39.

MacCracken, Michael C., Alan D. Hecht, Mikhail I. Budyko, and Yuri A. Izrael. 1990. *Prospects for Future Climate Change: A Special US/USSR Report on Climate and Climate Change*. Chelsea, Mich.: Lewis Publishers.

MacDonald, Gordon J. 1990. "The Role of Methane Clathrates in Past and Future Climates." *Climatic Change* 16:247–281.

Machado, S., and R. Piltz. 1988. *Reducing the Rate of Global Warming: The States' Role*. Washington, D.C.: Renew America.

MacKenzie, Debora, and Fred Pearce. 1989. "A Sudden Thaw in the Arctic." *New Scientist* 122:25.

MacKenzie, James J., and Michael P. Walsh. 1990. *Driving Forces: Motor Vehicle Trends and Their Implications for Global Warming, Energy Strategies, and Transportation Planning*. Washington, D.C.: World Resources Institute.

MacLean, Jayne T. 1992. *Global Warming and the Greenhouse Effect: January 1986– January 1992*. Beltsville, Md.: National Agricultural Library.

Madden, R.A., and V. Ramanathan. 1980. "Detecting Climate Change Due to Increasing CO2 in the Atmosphere." *Science* 209:763–768.

Magnuson, John J., Dale M. Robertson, Barbara J. Benson, Randolph H. Wynne, David M. Livingstone, Tadashi Arai, Raymond A. Assel, Roger G. Barry, Virginia Card, Esko Kuusisto, Nick G. Granin, Terry D. Prowse, Kenton M. Stewart, and Valery S. Vuglinski. 2000. "Historical Trends in Lake and River Ice Cover in the Northern Hemisphere." *Science* 289 (8 September): 1743–1746.

Magoon, O.T., H. Converse, D. Miner, D. Clark, and L.T. Tobin, eds. 1985. *Coastal Zone '85*. New York: American Society of Civil Engineers.

Mahtab, Fasih Uddin. 1992. "The Delta Regions and Global Warming: Impact and Response Strategies for Bangladesh." In Jurgan Schmandt and Judith Clarkson, *The Regions and Global Warming: Impacts and Response Strategies*. New York: Oxford University Press, 28–43.

Malcolm, Jay R., and Adam Markham. 2000. *Global Warming and Terrestrial Biodiversity Decline*. Washington, D.C.: World Wildlife Fund. [http://www.library. adelaide.edu.au/cgi-bin/director?id=1379565]

Malmer, Nils. 1992. "Peat Accumulation and the Global Carbon Cycle." In M. Boer and E. Koster, eds., *Greenhouse-Impact on Cold-Climate Ecosystems and Landscapes*. Selected Papers of a European Conference on Landscape Ecological Impact: Impact of Climatic Change, Lunteren, the Netherlands, December 3–7, 1989. CARTENA Supplement 22. Cremlingen, Germany: Cartena Verlag, 97–110.

Malone, Thomas F., Edward D. Goldberg, and Walter H. Munk N.d. "Roger Randall Dougan Revelle, 1909–1991." [http://www.nap.edu/readingroom/books/biomems/ rrevelle.html]

Manabe, Syukuro. 1989. "Changes in Soil Moisture." In Dean Edwin Abrahamson, ed., *The Challenge of Global Warming*. Washington, D.C.: Island Press, 146–150.

Manabe, S., and F. Mollner. 1961. "On the Radiative Equilibrium and Heat Balance of the Atmosphere." *Monthly Weather Review* 89:503–532.

Manabe, Syukuro, and Ronald J. Stouffer. 1993. "Century-scale Effects of Increased Atmospheric CO2 on the Ocean-atmosphere System." *Nature* 364:215–218.

———. 1995. "Simulation of Abrupt Climate Change Induced by Freshwater Input to the North Atlantic Ocean." *Nature* 378 (9 November): 165–167.

Manabe, S., and R.T. Wetherald. 1986. "Reduction in Summer Soil Wetness Induced by an Increase in Atmospheric Carbon Dioxide." *Science* 232:626–628.

Manabe, S., R.T. Wetherald, and R.J. Stauffer. 1981. "Summer Dryness Due to an Increase in Atmospheric CO2 Concentration." *Climatic Change* 3:347–386.

Mann, Barbara. 1999. Personal communication. 3 August.

Mann, Michael E. 2000. "Climate Change: Lessons for a New Millennium." *Science* 289 (14 July): 253–254.

Mann, Michael E., Raymond S. Bradley, and Michael K. Hughes. 1998. "Global-scale Temperature Patterns and Climate Forcing over the Past Six Centuries." *Nature* 392 (23 April): 779–787.

Mann, Michael E., and Richard J. Richels. 1990. "CO2 Emission Limits: An Economic Cost Analysis for the USA." *The Energy Journal* 11, no. 2.

Manning, Anita. 2000. "Ragweed Warms to Climate Change; Increasing CO2 has Doubled Pollen, Misery." *USA Today*, 15 August, 8D.

Mantua, N.J., S.R. Hare, Y. Zhang, J.M. Wallace, and R.C. Francis. 1997. "A Pacific

Interdecadal Climate Oscillation With Impacts on Salmon Production." *Bulletin of the American Meteorological Society* 78, no. 6: 1069–1079.

"Many More Hot, Sticky Days to Come, Forecasters Predict." 1999. Knight-Ridder Washington Bureau. In *Omaha World-Herald*, 7 November, 23A.

Marland, Gregg. 2000. "The Future Role of Reforestation in Reducing Buildup of Atmospheric CO_2." In T.M.L. Wigley and D.S. Schimel, eds., *The Carbon Cycle*. Cambridge U.K: Cambridge University Press, 190–198.

Marotzke, Jochem. 2000. "Abrupt Climate Change and Thermohaline Circulation: Mechanisms and Predictability." *Proceedings of the National Academy of Sciences of the United States of America* 97, no. 4 (15 February): 1347–1350.

Marston, J.B., M. Oppenheimer, R.M. Fujita, and S.R. Gaffin. 1991. "Carbon Dioxide and Temperature." *Nature* 349:573–574.

Martens, Pim. 1999. "How Will Climate Change Affect Human Health?" *American Scientist* 87, no. 6 (November/December): 534–541.

———. 1998. *Health and Climate Change: Modelling the Impacts of Global Warming and Ozone Depletion*. London: Earthscan Publications.

Martens, Willem J.M., Theo H. Jetten, and Dana A. Focks. 1997. "Sensitivity of Malaria, Schistosomiasis and Dengue to Global Warming." *Climatic Change* 35:145–156.

Martin, J.H., and R.M. Gordon. 1988. "Northeast Pacific Iron Distributions in Relation to Phytoplankton Productivity." *Deep-sea Resources* 35:177–196.

Maskell, Kathy, and Irving M. Mintzer. 1993. "Basic Science of Climate Change." *Lancet* 342:1027–1032.

Maslin, Mark. 1993. "Waiting for the Polar Meltdown." *New Scientist* 139:36.

Mass, C., and S.H. Schneider. 1977. "Influence of Sunspots and Volcanic Dust on Long-term Temperature Records Inferred by Statistical Investigations." *Journal of Atmospheric Science* 34:1995–2004.

Mastio, David. 1999. "Global Warming Propaganda Trumps Science." *USA Today*, 16 December, 19A.

Mathews-Amos, Amy, and Ewann A. Berntson. 1999. "Turning up the Heat: How Global Warming Threatens Life in the Sea." World Wildlife Fund and Marine Conservation Biology Institute. [http://www.worldwildlife.org/news/pubs/wwf_ocean.htm]

Matsuoka, Y., and K. Kai. 1994. "An Estimation of Climatic Change Effects on Malaria." *Journal of Global Environment Engineering* 1:1–15.

Maugh, Thomas H. 2000. "Great Lakes Water Levels Show Spring is Arriving Earlier." *Los Angeles Times*, 25 May, B-2.

Maunder, W.J. 1992. *Dictionary of Global Climate Change*. New York: Chapman & Hall.

Maxwell, Barrie. 1992. "Arctic Climate: Potential for Change Under Global Warming." In F. Stuart Chapin III, Robert L. Jefferies, James F. Reynolds, Gaius R. Shaver, and Josef Svoboda, eds., *Arctic Ecosystems in a Changing Climate: An Ecophysiological Perspective*. San Diego: Academic Press, 11–34.

Mayewski, Paul A., Barry Keim, Gregory A. Zielinski, Cameron Wake, Kevan Carpenter, Justin Cox, Joe Souney, Iola Hubbard, Paul Sanborn, and Mark Rodgers. 1998. *New England's Changing Climate, Weather, and Air Quality*. Climate Change Research Center. [http://heed.unh.edu/heed/coral/tom22.html]

McBeath, J.H., ed. 1984. *The Potential Effects of Carbon Dioxide-induced Climatic*

Changes in Alaska. Conference Proceedings, University of Alaska, Fairbanks, April 7–8, 1982. Fairbanks: University of Alaska, Fairbanks.

McConnell, J.R., R.J. Arthern, E. Mosley-Thompson, C.H. Davis, R.C. Bales, R. Thomas, J.F. Burkhart, and J.D. Kyne. 2000. "Changes in Greenland Ice Sheet Elevation Attributed Primarily to Snow Accumulation Variability." *Nature* 406 (24 August): 877–879.

McCrea, Steve. 1996. "Air travel: Eco-tourism's Hidden Pollution." *San Diego Earth Times,* August. [http://www.sdearthtimes.com/et0896/et0896s13.html]

McFarling, Usha Lee. 2000. "Scientists Warn of Losses in Ozone Layer over Arctic." *Los Angeles Times,* 27 May, A-20.

McKibben, Bill. 1989. *The End of Nature.* New York: Random House.

———. 1993. "James Hansen: Getting Warmer." *Outside,* May, 116–189.

McMichael, A.J. 1993. *Planetary Overload: Global Environmental Change and the Health of the Human Species.* Cambridge: Cambridge University Press.

———. 1996. "Human Population Health." In R.T. Watson, M.C. Zinyowera, R.H. Moss, and D.J. Dokken, eds., *Climate Change 1995: Impacts, Adapations, and Migration of Climate Change, Scientific and Technical Analysis.* New York: Cambridge University Press.

———. 1997. "Integrated Assessment of Potential Health Impact of Global Environmental Change: Prospects and Limitations." *Environmental Modeling and Assessment* 2:129–137.

McMichael, A.J., A. Haines, R. Sloff, and S. Kovats. 1996. *Climate Change and Human Health: An Assessment Prepared by a Task Force on Behalf of the World Health Organization, the World Meterological Association, and the United Nations Environmental Programme.* Geneva: World Health Organization.

McMichael, A.J., and W.J.M. Martens. 1995. "The Health Impacts of Global Climate Change: Grappling with Scenarios, Predictive Models, and Multiple Uncertainties." *Ecosystem Health* 1:23–33.

McNeill, J.R. 2000. *Something New Under the Sun: An Environmental History of the Twentieth-century World.* New York: W.W. Norton.

Meacher, Michael. 2000. "This is the World's Chance to Tackle Global Warming." *London Times,* 3 September, in LEXIS.

Mead, Margaret, and William Kellogg, eds. 1980. *The Atmosphere: Endangered and Endangering.* Tunbridge Wells, England: Castle House.

Meehl, G.A., and W.M. Washington. 1996. "El Niño-like Climate Change in a Model with Increased Atmospheric CO2 Concentrations." *Nature* 382:56–60.

Meier, M.F. 1984. "Contribution of Small Glaciers to Global Sea Level." *Science* 226: 1418–1421.

Melillo, Jerry M. 1999. "Warm, Warm on the Range." *Science* 283 (8 January): 183.

"Melting Antarctic Glacier Could Raise Sea Level." 1998. Reuters, 24 July. [http://bonanza.lter.uaf.edu/~davev/nrm304/glbnews.htm]

Mendelsohn, Robert. 1999. *The Greening of Global Warming.* Washington, D.C.: American Enterprise Institute.

———. 1999. *The Impact of Climate Change on the United States Economy.* Cambridge, U.K.: Cambridge University Press.

Mendelsohn, R., W. Nordhaus, and D. Shaw. 1994. "The Impact of Global Warming on Agriculture: A Ricardian Analysis." *American Economic Review* 84:753–771.

Menocal, Peter de, Joseph Ortiz, Tom Guilderson, and Michael Sarnthein. 2000. "Coherent High- and Low-Latitude Climate Variability During the Holocene Warm Period." *Science* 288 (23 June): 2198–2202.

Mercer, J.H. 1970. "Antarctic Ice and Interglacial High Sea Levels." *Science* 168:1605–1606.

————. 1978. "West Antarctic Ice Sheet and CO2 Greenhouse Effect: A Threat of Disaster." *Nature* 271:321–325.

Michaels, Patrick J. 1993. "Benign Greenhouse." *Research and Exploration* 9, no. 2: 222–233

————. 1996. "Solar Energy." In "What's Hot," *World Climate Report Archives* 1, no. 15. [http://www.nhes.com/back_issues/Vol1and2/WH/hot.html]

Michaels, P.J., and P.C. Knappenberger. 1996. "Human Effect on Global Climate?" *Nature* 384:522–523.

Michaelson, G.J., C.L. Ping, G.W. Kling, and J.E. Hobbie. 1998. "The Character and Bioactivity of Dissolved Organic Matter at Thaw and in the Spring Runoff Waters of the Arctic Tundra North Slope, Alaska." *Journal of Geophysical Research* 103: 28,939–28,946.

"Mideast Snow Disrupts Life, Prayers." 2000. *Omaha World-Herald*, 29 January, 4.

Miller, Kathleen A. 2000. "Pacific Salmon Fisheries: Climate, Information, and Adaptation in a Conflict-ridden Context." *Climatic Change* 45:37–61.

Miller, T., J.C. Walker, G.T. Kingsley, and W.A. Hyman. 1989. "Impact of Global Climate Change on Urban Infrastructure." In J.B. Smith and D.A. Tirpak, eds., *Potential Effects of Global Climate Change on the United States: Appendix H, Infrastructure.* Washington, D.C.: U.S. Environmental Protection Agency.

Mintzer, Irving W. 1992. *Confronting Climate Change: Risks, Implications, and Responses.* New York: Cambridge University Press.

Mitchell, George J. 1991. *World on Fire: Saving an Endangered Earth.* Toronto: Collier MacMillan Canada.

Mitchell, J.F.B. 1989. "The Greenhouse Effect and Climate Change." *Reviews of Geophysics* 27, no. 1: 115–139.

Mitchell, J.F.B., R.A. Davis, W.J. Ingram, and C.A. Senior. 1995. "On Surface Temperature, Greenhouse Gases, and Aerosols: Models and Observations." *Journal of Climate* 8:2364–2385.

Mitchell, J.F.B., and T.C. Johns. 1997. "On Modification of Global Warming by Sulfate Aerosols." *Journal of Climate* 10, no. 2: 245–266.

Mitchell J.F.B., and D.A. Warrilow. 1987. "Summer Dryness in Northern Mid-latitudes Due to Increased CO2." *Nature* 330:238–240.

Monastersky, Richard. 1998. "Greenhouse Warming Hurts Arctic Ozone." *Science News*, 11 April, 228.

Montague, Peter. N.d. "A New Disinfomation Campaign." Environmental Research Foundation. [http://www.eieio.org/airquality/news/rachelsglobalwarming.html]

————, ed. 1992. "Global Warming Part I: How Global Warming is Sneaking Up on Us." *Rachel's Environment and Health Weekly* No. 300 (26 August). Annapolis, Md.: Environmental Research Foundation. [http://www.monitor.net/rachel/r300.html]

Mooney, Harold A., and George W. Koch. 1994. "The Impact of Rising CO2 Concentrations on the Terrestrial Biosphere." *Ambio* 23, no. 1: 74–76.

Moore, Thomas Gale. 1996. "Health and Amenity Effects of Global Warming." Working Papers in Economics, E-96–1. The Hoover Institution.

———. 1995. "Why Global Warming Could be Good for You." *Public Interest* 118 (Winter): 83–99. [http://www.cycad.com/cgi-bin/Upstream/Issues/science/WARMIN.html]

———. 1997. "Happiness Is A Warm Planet." *Wall Street Journal*, 7 October. [http://www.freerepublic.com/forum/a182.htm]

———. 1998. *A Politically Incorrect View of Global Warming: Foreign Aid Masquerading as Climate Policy.* Washington, D.C.: Cato Institute.

Moreno, Fidel. 1999. "In the Arctic, Ice is Life, and It's Disappearing." *Native Americas* 16, no. 3/4(Fall/Winter): 42–45. [http://nativeamericas.aip.cornell.edu]

Moritz, Richard E., Cecilia M. Bitz, Konstantin Y. Vinnikov, Alan Robock, Ronald J. Stouffer, John E. Walsh, Claire L. Parkinson, Donald J. Cavalieri, John F.B. Mitchell, Donald Garrett, and Victor F. Zakharov. 2000. "Northern Hemisphere Sea Ice Extent." *Science* 288 (12 May): 927.

Motavalli, Jim. 2000. *Forward Drive: The Race to Build "Clean" Cars for the Future.* San Francisco: Sierra Club Books.

"Much Deforestation Driven By Population, Poverty." 1999. *Global Futures Bulletin* 84 (15 May). [http://www.gsreport.com/articles/art000149.html]

Muller, Frank. 1996. "Mitigating Climate Change: The Case for Energy Taxes." *Environment* 38, no. 2: 13–42.

Murgia, Joe. 1999. "NASA: Greenland's Glaciers Are Shrinking: A New Study Suggests that Rapid Thinning and Excess Run-off from Greenland's Southeastern Glaciers May be Partly Caused by Climate Changes." GS Report, 10 March. [http://www.gsreport.com/articles/art000078.html]

Murray, Alan. 1999. "The American Century: Is it Going or Coming?" *Wall Street Journal*, 27 December, 1.

Murray, William. 1979. *The Greenhouse Theory and Climate Change.* Ottawa: Library of Parliament Research Branch.

Musil, Robert. 2000. "Death By Degrees: The Health Threats of Climate Change in Washington [State]." Physicians for Social Responsibility, July. [www.psr.org/washington]

Myers, Norman. 1993. "Environmental Refugees in a Globally Warmed World." *Bioscience* 43:752–761.

Nakicenovic, Nebojsa, Arnulf Grubler, Atsushi Inaba, Sabine Messner, Sten Nilsson, Yoichi Nishimura, Hans-Holger Rogner, Andreas Schafer, Leo Schrattenholzer, Manfred Stubegger, Joel Swisher, David Victor, and Deborah Wilson. 1993. "Long-Term Strategies for Mitigating Global Warming." *Energy* 18, no. 5: 401–609.

Nance, John J. 1991. *What Goes Up: The Global Assault on Our Atmosphere.* New York: William Morrow and Co.

National Academy of Sciences of the United States of America. 1991. *Policy Implications of Greenhouse Warming.* Washington, D.C.: National Academy Press.

———. 1979. *CO2 and Climate: A Scientific Assessment.* Washington, D.C.: National Academy Press.

National Energy Board of Canada. 1994. *Canadian Energy Supply and Demand: 1993–2010, Appendix to Technical Report.* Ottawa: Public Works Canada.

National Research Council. 1987. *Responding to Changes in Sea Level: Engineering Implications.* Washington, D.C.: National Academy Press.

National Research Council, Polar Research Board. 1985. *Glaciers, Ice Sheets, and Sea Level: Effects of a CO2-induced Climatic Change.* Report of a Workshop Held in Seattle, Washington, September 13–14, 1984. Washington, D.C.: National Academy Press.

Native Americas. 1999. Vol. 16, no. 3/4(Fall/Winter).

Natural Resources Canada. 1992. *Canada's Energy Outlook: 1990 to 2020.* Ottawa: NRCan.

———. 1994. *Canada's Energy Outlook 1990 to 2020: Update 1994.* Ottawa: NRCan.

Neilson, R.P., G.A. King, and J. Lenihan. 1994. "Modeling Forest Response to Climatic Change: The Potential for Large Emissions of Carbon from Dying Forests." In M. Kanninen, ed., *Carbon Balance of World's Forested Ecosystems: Towards a Global Assessment.* Helsinki, Finland: Academy of Finland, 150–162.

Nemani, Ramakrishna R., and Steven W. Running. 1995. "Satellite Monitoring of Global Land-cover Changes and Their Impact on Climate." *Climatic Change* 31 (December): 395–413.

Nepstad, Daniel C. 1998. "Report from the Amazon: May, 1998." Woods Hole Research Center. [http://terra.whrc.org/science/tropfor/fire/report2.htm]

Nepstad, D.C., A. Veríssimo, A. Alencar, C. Nobre, E. Lima, P. Lefebvre, P. Schlesinger, C.S. Potter, P. Moutinho, and E. Mendoza. 1999. "Large-Scale Impoverishment of Amazonian Forests by Logging and Fire." *Nature* 398:505–507.

"New England May Lose Sugar Maples to Global Warming." 1998. Reuters. [http://bonanza.lter.uaf.edu/~davev/nrm304/glbxnews.htm]

Newell, R.E., and T.G. Dopplick. 1979. "Questions Concerning the Possible Influence of Anthropogenic CO2 on Atmospheric Temperature." *Journal of Applied Meteorology* 18:822–825.

Newman, Cathy. 2000. "Prescott Warns U.S. Over Climate." *Financial Times* (London), 27 April, 7.

Newman, P.W.G., and J.A. Kenworthy. 1989. *Cities and Automobile Dependence: A Sourcebook.* Aldershot, Hants, United Kingdom: Gower Technical.

Newton, David E. 1993. *Global Warming: A Reference Handbook.* Santa Barbara, Calif.: ABC-CLIO.

Nicholls, N., G. Gruza, J. Jouzel, T.R. Karl, L. Ogallo, and D. Parker. 1995. *Observed Variability and Change, Intergovernmental Panel on Climate Change (IPCC) Second Scientific Assessment of Climate Change.* Cambridge, U.K.: Cambridge University Press.

Nisbit, E.G. 1990. Quoted in *Canadian Journal of Earth Sciences* 27:148.

Nisbit, E.G., and B. Ingham. 1995. "Methane Output from Natural and Quasinatural Sources: A Review of the Potential for Change and for Biotic and Abiotic Feedbacks." In George M. Woodwell and Fred T. MacKenzie, eds., *Biotic Feedbacks in the Global Climate System: Will the Warming Feed the Warming?* New York: Oxford University Press, 188–218.

Nnoli, O. 1990. "Desertification, Refugees and Regional Conflict in West Africa." *Disasters* 14:132–139.

Nordhaus, William D. 1991. "Economic Approaches to Greenhouse Warming." In Rudiger Dornbusch and James M. Poterba, eds., *Global Warming: Economic Policy Responses.* Cambridge, Mass.: MIT Press, 33–66.

———. 1991. "To Slow or Not to Slow: The Economics of the Greenhouse Effect."

Coulves Foundation for Research in Economics, Paper No. 791. New Haven, Conn.: Yale University. (Also published in *Economic Journal* 101, no. 6 [1991]: 920–937.)

———. 1991. "A Sketch of the Economics of the Greenhouse Effect." *American Economic Review* 81, no. 2 (May): 146–150.

———. 1992. "An Optimal Transition Path for Controlling Greenhouse Gases." *Science* 258:1315–1319.

———. 1994. *Managing the Global Commons: The Economics of Climate Change.* Cambridge, Mass.: MIT Press.

———. 1994. "Expert Opinion on Climate Change." *American Scientist* 82:45–51.

Normile, Dennis. 2000. "Some Coral Bouncing Back From El Niño." *Science* 288 (12 May): 941–942.

Norris, Richard D., and Ursula Rohl. 1999. "Carbon Cycling and Chronology of Climate Warming During the Palaeocene/Eocene Transition." *Nature* 401 (21 October): 775–778.

North, Gerald R., Jurgen Schmandt, and Judith Clarkson. 1995. *The Impact of Global Warming on Texas.* Austin: University of Texas Press.

Norwegian Government Falls on Global Warming Issue. 2000. Environment News Service, 9 March. [http://ens.lycos.com/ens/mar2000/2000L-03-09-05.html]

Nuttall, Nick. 1998. "Global Warming 'Will Turn Rainforests into Deserts.' " *The Times of London,* 3 November. [http://bonanza.lter.uaf.edu/~davev/nrm304/glbxnews. htm]

———. 2000. "Climate Change Lures Butterflies Here Early." *The Times of London,* 24 May.

———. 2000. "Experts are Poles Apart Over Ice Cap." *The Times of London,* 21 August, in LEXIS.

Oberthur, Sebastian, and Hermann E. Ott. 1999. *The Kyoto Protocol: International Climate Policy for the 21st Century.* Berlin: Springer.

O'Brien, S.T., B.P. Hayden, and H.H. Shugart. 1992. "Global Climatic Change, Hurricanes, and a Tropical Forest." *Climatic Change* 22:175–190.

Oechel, W.C., S.J. Hastings, G. Vourlitis, M. Jenkins, G. Riechers, and N. Grulke. 1993. "Recent Change of Arctic Tundra Ecosystems from a Net Carbon Dioxide Sink to a Source." *Nature* 361:520–523.

Oechel, Walter C., George L. Vourlitis, Steven J. Hastings, Rommel C. Zulueta, Larry Hinzman, and Douglas Kane. 2000. "Acclimation of Ecosystem CO2 Exchange in the Alaskan Arctic in Response to Decadal Climate Warming." *Nature* 406 (31 August): 978–981.

Oerlemans, Johannes. 1994. "Quantifying Global Warming from the Retreat of Glaciers." *Science* 264:243–245.

———. 1986. "Glaciers as Indicators of a Carbon-dioxide Warming." *Nature* 320:607–609.

Oerlemans, J., and J.P.F. Fortuin. 1992. "Sensitivity of Glaciers and Small Ice Caps to Greenhouse Warming." *Science* 258:115–117.

Ohmura, A., and Reeh Ohmura. 1991. "New Precipitation and Accumulation Maps for Greenland." *Journal of Glaciology* 37:140–148.

Okken, P.A., R.J. Swart, and S. Zwerver. 1989. *Climate and Energy: The Feasibility of Controlling CO2 Emissions.* Dordrecht, Germany: Kluwer Academic Pubishers.

Oldale, R. 1985. "Late Quaternary Sea Level History of New England: A Review of Published Sea Level Data." *Northeastern Geology* 7:192–200.

Oppenheimer, Michael, and Robert H. Boyle. 1990. *Dead Heat: The Race Against the Greenhouse Effect.* New York: Basic Books.

Oppo, D.W., J.F. McManus, and J.L. Cullen. 1998. "Abrupt Climate Events 500,000 to 340,000 Years Ago: Evidence from Subpolar North Atlantic Sediments." *Science* 279:1335–1338.

Ostrander, Gary K., Kelley Meyer Armstrong, Edward T. Knobbe, David Gerace, and Erik P. Scully. 2000. "Rapid Transition in the Structure of a Coral Reef Community: The Effects of Coral Bleaching and Physical Disturbance." *Proceedings of the National Academy of Sciences of the United States of America* 97, no. 10 (9 May): 5297–5302.

Overpeck, Jonathan T. 1996. "Warm Climate Surprises." *Science* 271:1820–1821.

———. 2000. "Climate Change: The Hole Record." *Nature* 403 (17 February): 714–715.

Overpeck, J., D. Rind, A. Lacis, and R. Healy. 1996. "Possible Role of Dust-induced Regional Warming in Abrupt Climate Change During the Last Glacial Period." *Nature* 384:447–449.

Overpeck, J., K. Hughen, D. Hardy, R. Bradely, R. Case, M. Douglas, B. Finney, K. Gajewski, G. Jacoby, A. Jennings, S. Lamoureux, A. Lasca, G. MacDonald, J. Moore, M. Retelle, S. Smith, A. Wolfe, and G. Zielinski. 1997. "Arctic Environmental Change of the Last Four Centuries." *Science* 278:1251–1256.

Overpeck, Jonathan, and Robert Webb. 2000. "Non-glacial Rapid Climate Events: Past and Future." *Proceedings of the National Academy of Sciences of the United States of America* 97, no. 4 (15 February): 1335–1338.

Owen, T., R.D. Cess, and V. Ramanathan. 1979. "Early Earth: An Enhanced Carbon-dioxide Greenhouse to Compensate for Reduced Solar Luminosity." *Nature* 277: 640–642.

Pacific Salmon Commission. 1999. *Report of the Fraser River Panel to the Pacific Salmon Commission on the 1997 Fraser River Sockeye and Pink Salmon Fishing Season.* Prepared in March by the Pacific Salmon Commission, Vancouver, British Columbia.

Packard, Kimberly O'Neill, and Forest Reinhardt. 2000. "What Every Executive Needs to Know About Global Warming." *Harvard Business Review* 78, no. 4 (July/August): 128–135.

Pagani, M., M.A. Arthur, and K.H. Freeman. 1999. *Paleoceanography* 14:273–292.

Palmer, Frederick. 1998. "Not So Hot." *Washington Post*, 12 December, A-23.

Parker, David E. 2000. "Perspectives on Climate Change: Temperatures High and Low." *Science* 287:1216–1217.

Parmesan, C. 1996. "Climate and Species' Range." *Nature* 382:765–766.

Parmesan, Camille, Nils Ryrholm, Constanti Stefanescu, Jane K. Hill, Chris D. Thomas, Henri Descimon, Brian Huntley, Lauri Kaila, Jaakko Kulberg, Toomas Tammaru, W. John Tennent, Jeremy A. Thomas, and Martin Warren. 1999. "Poleward Shifts in Geographical Ranges of Butterfly Species Associated with Regional Warming." *Nature* 399 (10 June): 579–583.

Parry, Martin Lewis. 1990. *Climate Change and World Agriculture.* London: Earthscan.

Parry, Martin and Zhang Jiachen. 1991. "The Potential Effect of Climate Changes on Agriculture." In J. Jager and H.L. Ferguson, *Climate Change: Science, Impacts, and Policy*. Proceedings of the Second World Climate Conference. Cambridge, U.K.: Cambridge University Press, 279–289.

Parry, Martin, Timothy Carter, and Nicholas Konijn, eds. 1988. *The Impact of Climatic Variations on Agriculture*. 2 vols. Dordrecht, Germany: Kluwer.

Parry, M.L., J.E. Hossell, P.J. Jones, T. Rehman, R.B. Tranter, J.S. Marsh, C. Rosen-zweig, G. Fischer, I.G. Carson, and R.G.H. Bunce. 1996. "Integrating Global and Regional Analyses of the Effects of Climate Change: A Case Study of Land Use in England and Wales." *Climatic Change* 32, nos. 1–2: 185–198.

Parsons, Michael L. 1995. *Global Warming: The Truth Behind the Myth*. New York: Plenum Press/Insight.

Pascual, Mercedes, Xavier Rodó, Stephen P. Ellner, Rita Colwell, and Menno J. Bouma. 2000. "Cholera Dynamics and El Niño-Southern Oscillation." *Science* 289 (8 September): 1766–1769

Passell, Peter. 1999. "Global Warming: China Perplex." *New York Times*, 7 March, Economic Scene.

Pastor, J., and W.M. Post. 1988. "Response of Northern Forests to CO_2-induced Climate Change." *Nature* 334:55–58.

Patz, Jonathan A., Paul R. Epstein, Thomas A. Burke, and John M. Balbus. 1996. "Global Climate Change and Emerging Infectious Diseases." *Journal of the American Medical Association* 275 (17 January): 217–223.

Patz, J.A., W.J.M. Martens, D.A. Focks, and T.H. Jetten. 1998. "Dengue Fever Epidemic Potential as Projected by General Circulation Models of Global Climate Change." *Environmental Health Perspectives* 106, no. 3: 147–153.

Pauli, H., M. Gottfried, and M. Grabherr. 1996. "Effects of Climate Change on Mountain Ecosystems—Upward Shifting of Alpine Plants." *World Resources Review* 8: 382–390.

Paulson, Tom. 1999. "Global Warming Could Bring Us Wetter Winters." *Seattle Post-Intelligencer*, 9 November. [http://www.post-intelligencer.com/local/clim09.shtml]

Pearce, D. 1991. "The Role of Carbon Taxes in Adjusting to Global Warming." *Economic Journal* 101:935–948.

Pearce, Fred. 1998. "Nature Plants Doomsday Devices." *The Guardian* (England), 25 November. [http://go2.guardian.co.uk/science/912000568-disast.html]

———. 1997. "Greenhouse Wars." *New Scientist* 139:38–44.

———. 1994. "Will Global Warming Plunge Europe into an Ice Age?" *New Scientist* 144:20–21.

Pearman, G.I., ed. 1991. *Limiting Greenhouse Effects: Controlling Carbon-dioxide Emissions*. Report of the Dahlem Workshop on Limiting the Greenhouse Effect, Berlin, December 9–14. New York: John Wiley & Sons.

Pearman, G.I., and P. Hyson. 1981. "The Annual Variation of Atmospheric CO_2 Concentration Observed in the Northern Hemisphere." *Journal of Geophysical Research* 86:9839–9843.

Pearman, Graeme, ed. 1988. *Greenhouse: Planning for Climate Change*. Leiden, Germany: E.J. Brill.

Pearson, Paul N., and Martin B. Palmer. 2000. "Atmospheric Carbon Dioxide Concentrations over the Past 60 Million Years." *Nature* 406 (August 17): 695–699.

Peltier, W.R., and A.M. Tushingham. 1989. "Global Sea Level Rise and the Greenhouse Effect: Might They Be Connected?" *Science* 244:806–810.

Peng, Tsung-Hung. 1993. "Possible Reduction of Atmopsheric CO2 by Iron Fertilization in the Antarctic Ocean." In Richard A. Geyer, ed., *A Global Warming Forum: Scientific, Economic, and Legal Overview.* Boca Raton, Fla.: CRC Press, 263–285.

Perry, Michael. 1998. "Global Warming Devastates World's Coral Reefs." Reuters, 26 November. [http://www.gsreport.com/articles/art000023.html]

Peteet, Dorothy. 2000. "Sensitivity and Rapidity of Vegetational Response to Abrupt Climate Change." *Proceedings of the National Academy of Sciences of the United States of America* 97, no. 4 (15 February): 1359–1362.

Peters, Robert L. 1989. "Effects of Global Warming on Biological Diversity." In Dean Edwin Abrahamson, ed., *The Challenge of Global Warming.* Washington, D.C.: Island Press, 82–95.

Peters, R.L., and J.D.S. Darling. 1985. "The Greenhouse Effect and Nature Reserves: Global Warming Would Diminish Biological Diversity by Causing Extinctions Among Reserve Species." *Bioscience* 35, no. 11: 707–717.

Peters, R.L., and T. Lovejoy, eds. 1992. *Global Warming and Biological Diversity.* New Haven, Conn.: Yale University Press.

Peterson, David L. N.d. "Response of Earth's Ecosystem to Global Change." NASA-Ames Research Center. [http://astrobiology.arc.nasa.gov/workshop/speakers/peterson/peterson_abstract.html]

Peterson, David L., and Darryll R. Johnson. 1995. *Human Ecology and Climate Change: People and Resources in the Far North.* Washington, D.C.: Taylor & Francis.

Petit, Charles W. 2000. "Polar Meltdown: Is the Heat Wave on the Antarctic Peninsula a Harbinger of Global Climate Change?" *U.S. News and World Report,* 28 February, 64–74.

Petit, J.R., J. Jouzel, D. Raynaud, N.I. Barkov, J.-M. Barnola, I. Basile, M. Benders, I. Chappellaz, M. Davis, G. Delaygue, M. Delmotte, V.M. Kotlyakov, M. Legrand, V.Y. Lipenkov, C. Lorius, L. Pepin, C. Ritz, E. Saltzman, and M. Stievenard. 1999. "Climate and Atmospheric History of the Past 420,000 Years From the Vostok Ice Core, Antarctica." *Nature* 399 (3 June): 429–436.

Pew Center. 1999. "Experts Say Global Warming More Than Predicted." A New Study Released July 10 by the Pew Center on Global Climate Change Foresees Greater Global Warming than Previously Predicted, Along with Greater Extremes of Weather and Faster Sea-level Rise. [http://www.gsreport.com/articles/art000175.html]

Philander, S. George. 1998. *Is the Temperature Rising? The Uncertain Science of Global Warming.* Princeton, N.J.: Princeton University Press.

Phillips, John. 2000. "Tropical Fish Bask in Med's Hot Spots." *London Times,* 15 July.

Phillips, Oliver L., Yadvinder Malhi, Niro Higuchi, William F. Laurance, Percy V. Nunez, Rodolfo M. Vasquez, Susan G. Laurance, Leandro V. Ferreira, Margaret Stern, Sandra Brown, and John Grace. 1998. "Changes in the Carbon Balance of Tropical Forests: Evidence from Long-term Plots." *Science* 282 (16 October): 439–442.

Pienitz, Reinhard, and Warwick F. Vincent. 2000. "Effect of Climatic Change Relative to Ozone Depletion on UV Exposure in Subarctic Lakes." *Nature* 404 (30 March): 484–487.

Pierrehumbert, R.T. 2000. "Climate Changes and the Tropical Pacific: A Sleeping Dragon Wakes." *Proceedings of the National Academy of Sciences of the United States of America* 97, no. 4 (15 February): 1355–1358.

Plantico, M.S., T.R. Karl, G. Kukla, and J. Gavin. 1990. "Is Recent Climate Change Across the United States Related to Rising Levels of Anthropogenic Greenhouse Gases?" *Journal of Geophysical Research* 95, no. 16: 617–637.

Plass, G.N. 1956. "The Carbon Dioxide Theory of Climatic Change." *Tellus* 8:141–154.

———. 1956. "Effect of Carbon Dioxide Variations on Climate." *American Journal of Physics* 24:376–387.

Pomerance, Rafe. 1989. "The Dangers From Climate Warming: A Public Awakening." In Dean Edwin Abrahamson, ed., *The Challenge of Global Warming*. Washington, D.C.: Island Press, 259–269.

Porter, Larry. 2000. "Birds May Face Rude Welcome." *Omaha World-Herald*, 12 February, 1.

Poterba, James. 1991. "Tax Policy to Combat Global Warming." In Rutiger Dornbusch and James M. Poterba, eds., *Global Warming: Economic Policy Responses*. Cambridge, Mass.: MIT Press, 71–98.

Predictions of Warming Continue to Drop. N.d. [http://www.marshall.org/globalfax.html1#fax2]

"Prehistoric Extinction Linked to Methane." 2000. Associated Press. In *Omaha World-Herald*, 27 July, 9.

Prentice, I.C., and M. Sarnthein. 1993. "Self-regulatory Processes in the Biosphere in the Face of Climate Change." In J. Eddy and H. Oeschger, eds., *Global Changes in the Perspective of the Past*. Chichester, U.K.: Wiley, 29–38.

President's Science Advisory Committee. 1965. *Restoring the Quality of Our Environment*. Washington, D.C.: President's Science Advisory Committee, The White House.

Pringle, Laurence P. 1990. *Global Warming*. New York: Arcade Publishing.

Quammen, David. 1998. "Planet of Weeds: Tallying the Losses of Earth's Animals and Plants." *Harpers*, October, 57–69.

Quattrochi, D.A., and M.K. Ridd. 1994. "Measurement and Analysis of Thermal Energy Responses from Discrete Urban Surfaces Using Remote Sensing Data." *International Journal of Remote Sensing*, 15:1991–2022.

Rabb, Theodore K. 1983. "Bibliography to Accompany Climate and Society in History: A Research Agenda." In Robert S. Chen, Elise Boulding, and Stephen H. Schneider, eds., *Social Science Research and Climate Change: An Interdisciplinary Appraisal*. Boston: D. Reidel Publishing Co., 77–114.

"Rachel's #466: Warming and Infectious Diseases." 1995. *Rachel's Environment and Health Weekly*, 2 November. Annapolis, Md.: Environmental Research Foundation. [http://www.igc.apc.org/awea/wew/othersources/rache1466.html]

Radford, Tim. 2000. "Greenhouse Buildup Worst for 20m [million] Years." *The Guardian* (London), 17 August, 9.

Rahmstorf, S. 1994. "Rapid Climate Transitions in a Coupled Ocean-atmosphere Model." *Nature* 372:82–85.

Rahmstorf, Stefan. 1999. "Shifting Seas in the Greenhouse?" *Nature* 399 (10 June): 523–524.

Rahmstorf, S., and A. Ganopolski. 1999. "Long-term Global Warming Scenarios Computed with an Efficient Coupled Climate." *Climatic Change* 43:353–67.

Ramanathan, Veerabhaadran. 1975. "Greenhouse Effect Due to Chlorofluorocarbons: Climatic Implications." *Science* 190:50–52.

———. 1976. "Atmospheric Fluorocarbons: Possible Effects of a Large Increase on the Global Climate." *Environmental Conservation* 3:90.

———. 1981. "The Role of Ocean-Atmospheric Interactions in the CO2 Climate Problem." *Journal of Atmospheric Science* 38:918–930.

———. 1988. "The Greenhouse Theory of Climate Change: A Test by Inadvertent Global Experiment." *Science* 240:293–299.

———. 1989. "Observed Increases in Greenhouse Gases and Predicted Climatic Changes." In Dean Edwin Abrahamson, ed., *The Challenge of Global Warming*. Washington, D.C.: Island Press, 239–247.

———. 1991. "Trace Gas Trends and Change." Testimony before the Senate Subcommittee on Environmental Protection, 23 January 1987. Cited in Lynne T. Edgerton and the Natural Resources Defense Council, *The Rising Tide: Global Warming and World Sea Levels*. Washington, D.C.: Island Press, 11, 108.

———. 1997. Selected publications. [http://www.cirrus.ucsd.edu/personnel/vramanathan/ram_select_pubs.html]

———. 1998. "Trace Gas Greenhouse Effect and Global Warming: Underlying Principles and Outstanding Issues. Volvo Environmental Prize Lecture, 1997." *Ambio* 27, no. 3: 187–197.

Ramanathan, V., and R.D. Cess. 1974. "Radiative Transfer Within the Mesospheres of Mars and Venus." *Astrophysics Journal* 188:407–416.

Ramanathan, V., and W. Collins. 1993. "A Thermostat in the Tropics?" *Nature* 361: 410–411.

Ramanathan, V., H.B. Singh, R.J. Cicerone, and J.T. Kiehl. 1985. "Trace Gas Trends and Their Potential Role in Climate Change." *Journal of Geophysical Research* 90:5547–5566.

Ramanathan, V., and A.M. Vogelmann. 1997. "Greenhouse Effect, Atmospheric Solar Absorption and the Earth's Radiation Budget: From the Arrhenius-Langley Era to the 1990s." In H. Rodhe and R. Charlson, eds., *The Legacy of Svante Arrhenius: Understanding the Greenhouse Effect*. Stockholm, Sweden: Royal Swedish Academy of Sciences, Stockholm University, 85–103.

Range, Stacey. 1999. "Year of the Drought (and the Floods): Extremes Mark Weather of '99." *Omaha World-Herald*, 30 December, 11, 15.

———. 2000. "Climatologists Say Midlands in Dust Bowl-like Drought." *Omaha World-Herald*, 12 January, 18.

Rao, P.K., ed. 2000. *The Economics of Global Warming*. Armonk, N.Y.: M.E. Sharp.

Raval, A., and V. Ramanathan. 1989. "Observational Determination of the Greenhouse Effect." *Nature* 342:758.

Raynor, Steve, and Elizabeth L. Malone. 1998. *Human Choice and Climate Change*. 4 vols. Columbus, Ohio: Battelle Press.

Read, Peter. 1994. *Responding to Global Warming: The Technology, Economics and Politics of Sustainable Energy*. London: Atlantic Highlands.

Reaka-Kudla, M.L. 1996. "The Global Biodiversity of Coral Reefs: A Comparison With Rainforests." In M.L. Reaka-Kudla, D.E. Wilson, and E.O. Wilson, eds., *Biodiversity II: Understanding and Protecting our Natural Resources*. Washington, D.C.: Joseph Henry/National Academy Press, 83–108.

Reckman, Alexander. 1991. *Global Warming*. New York: Gloucester Press.

Reid, Stephen J. 2000. *Ozone and Climate Change: A Beginner's Guide*. Amsterdam: Gordon and Breach.

Reid, W.V., and M.C. Trexler. 1991. *Drowning the National Heritage: Climate Change and U.S. Coastal Biodiversity*. Washington, D.C.: World Resources Institute.

"Report: Last Ice Age Had Quick End." 1999. Associated Press. In *Omaha World-Herald*, 29 October, 15.

The Republic of the Marshall Islands Climate Change Website. N.d. [http://www.unfccc.de/resource/ccsites/marshall/activity/seaframe.htm]

Reukin, Andrew. 1992. *Global Warming: Understanding the Forecast*. New York: Abbeville Press.

Revelle, Roger. 1982. "Carbon Dioxide and World Climate." *Scientific American*, August, 35–43.

Revelle, R., and A.E. Maxwell. 1953. "Heat Flow through the Floor of the Eastern North Pacific Ocean." *Nature* 170:199–200.

Revelle, Roger, and Hans S. Suess. 1957. "Carbon Dioxide Exchanges Between Atmosphere and Ocean and the Question of an Increase of Atmospheric CO_2 During the Past Decades." *Tellus* 9:18–27.

Revkin, Andrew C. 1988. "Endless Summer: Living with the Greenhouse Effect." *Discover* 9, no. 10: 50.

———. 2000. "Study Faults Humans for Large Share of Global Warming." *New York Times*, 14 July, A-12.

———. 2000. "Study Proposes New Strategy to Stem Global Warming." *New York Times*, 19 August, A-13.

Rich Nations, Multinationals Profit From Fossil Fuel Funding. 1998. Global Exchange, 15 May. [http://www.globalexchange.org/economy/rulemakers/ClimateChange.html]

Ridenour, David. 1998. "Hyprocrisy in Buenos Aires: Millions of Gallons of Fuel to be Burned By Those Seeking Curbs on Fuel Use." National Policy Analysis: A Publication of the National Center for Public Policy Research. No. 217 (October). [http://nationalcenter.org/NPA217.html]

Ridley, Michael A. 1998. *Lowering the Cost of Emission Reduction: Joint Implementation in the Framework Convention on Climate Change*. Dordrecht, Germany: Kluwer Academic Publishers.

Riebsame, William E. 1988. "Adjusting Water Resources Management to Climate Change." *Climatic Change* 13:69–97.

Rignot, E.J. 1998. "Fast Recession of a West Antarctic Glacier." *Science* 281:549–551.

Rind, D., E.W. Chiou, W. Chu, J. Larsen, S. Oltmans, J. Lerner, M.P. McCormick, and L. McMaster. 1991. "Positive Water Vapour Feedback in Climate Models Confirmed by Satellite Data." *Nature* 349:500–503.

Rind, D., and J. Overpeck. 1993. "Hypothesized Causes of Decade-to-century-scale Climate Variability: Climate Model Results." *Quaternary Science Reviews* 12:357–374.

Rinfret, Alex. 1996. "Disastrous Chinook Season Predicted for the Charlottes." *Queen Charlotte Islands Observer*, 28 March, 1.

Rittenour, Tammy M., Julie Brigham-Grette, and Michael E. Mann. 2000. "El Niño-Like Climate Teleconnections in New England During the Late Pleistocene." *Science* 288 (12 May): 1039–1042.

Rizzo, B., and E. Wiken. 1992. "Assessing the Sensitivity of Canada's Ecosystems to Climatic Change." *Climatic Change* 21:37–55.

Roberts, Leslie. 1989. "Global Warming: Blaming the Sun." *Science* 246:992–993.

———. 1989. "How Fast Can Trees Migrate?" *Science* 243:735–737.

Robertson, G. Philip., Eldor A. Paul, and Richard R. Harwood. 2000. "Greenhouse Gases in Intensive Agriculture: Contributions of Individual Gases to the Radiative Forcing of the Atmosphere." *Science* (15 September): 1922–1925.

Robock, Alan, K.Y. Vinnikov, R. Wetherald, S. Manabe, J. Entin, R. Stouffer, V. Zabelin, and A. Namkhai. 1999. Summer Desiccation as a Global Warming Fingerprint? The Tenth Symposium on Global Change Studies. [http://www.confex2.com/ams/99annual/abstracts/1023.htm]

Rodhe, Henning. 1999. "Clouds and Climate." *Nature* 401:223–225.

Rodwell, M.J., D.R. Rowell, and C.K. Folland. 1999. "Oceanic Forcing of the Wintertime North Atlantic Oscillation and European Climate." *Nature* 398 (25 March): 320–323.

Roemmich, D., and J. McGowan. 1995. "Climatic Warming and the Decline of Zooplankton in the California Current." *Science* 267:1324–1326.

Rogers, Caroline S. 2000. "Confounding Factors in Coral-reef Recovery." *Science* 289 (21 July): 391.

Rogers, David J., and Sarah E. Randolph. 2000. "The Global Spread of Malaria in a Future, Warmer World." *Science* 289 (8 September): 1763–1766.

Rolfe, Christopher. 1996. "Comments on the British Columbia Greenhouse Gas Action Plan." West Coast Environmental Law Association. A Presentation to the Air and Water Management Association. 17 April. [http://www.wcel.org/wcelpub/11026.html]

Rooney, C.A., J. McMichael, R.S. Kovats, and M. Coleman. 1998. "Excess Mortality in England and Wales, and in Greater London, During the 1995 Heatwave." *Journal of Epidemiology and Community Health* 52:482.

Root, T.L., and S.H. Schneider. 1995. "Ecology and Climate: Research Strategies and Implications." *Science* 269:331–341.

Rosenburg, Norman J., Pierre R. Crossman, William E. Easterling III, Mary S. McKenney, Kenneth D. Frederick, and Michael Bowes. 1992. "Methodology for Assessing Regional Economic Impacts of and Responses to Climate Change: the MINK Study." In Jurgan Schmandt and Judith Clarkson, *The Regions and Global Warming: Impacts and Response Strategies*. New York: Oxford University Press, 132–153.

Rosenburg, Norman J., Pierre R. Crossman, Kenneth D. Frederick, William E. Easterling III, Mary S. McKenney, Michael D. Bowes, Roger A. Sedjo, Joel Darmstadter, Laura A. Katz, and Kathleen M. Lemon. 1993. "The MINK [Missouri-Iowa-Nebraska-Kansas] Methodology: Background and Baseline." *Climatic Change* 24: 7–22.

Rosenburg, Norman J., Daniel J. Epstein, David Wang, Raghavan Srinivasan, and Jeffrey G. Arnold. 1999. "Possible Impacts of Global Warming on the Hydrology of the Ogallala Aquifer Region." *Climatic Change* 42:677–692.

Rosenzweig, Cynthia, and Daniel Hillel. 1993. "Agriculture in a Greenhouse World." *Research and Exploration* 9, no. 2: 208–221.

Rosenzweig, Cynthia, and Martin L. Parry. 1994. "Potential Impact of Climate Change on World Food Supply." *Nature* 367:133–138.

Rosman, Veronica. 1999. "November Among Warmest Ever." *Omaha World-Herald*, 2 December, 11.

Rouse, W.R., M. Douglas, R.E. Hecky, A. Hershey, G.W. Kling, L. Lesack, P. Marsh, M. McDonald, B. Nicholson, N. Roulet, and J. Smol. 1997. "Effects of Climate Change on the Fresh Waters of Arctic and Subarctic North America." *Hydrological Processes* 11:873–902.

Rowlands, Ian H. 1995. *The Politics of Global Atmospheric Change*. Manchester, U.K.: Manchester University Press.

Rudel, Thomas K. 1989. "Population, Development, and Tropical Deforestation: A Cross-national Study." *Rural Sociology* 54:327–338.

Ruhlemann, Carsten, Stefan Mulitza, Peter J. Muller, Gerold Wefer, and Rainer Zahn. 1999. "Warming of the Tropical Atlantic Ocean and Slowdown of Thermohaline Circulation During the Last Deglaciation." *Nature* 402 (2 December): 511–514.

Runnegar, Bruce. 2000. "Loophole for Snowball Earth." *Nature* 405 (25 May): 403–404.

Sacherst, R.J. 1990. "Impact of Climate Change on Pests and Diseases in Australasia." *Search* 21:230–232.

Sailor, William C., David Bodansky, Chaim Braun, Steve Fetter, and Bob van der Zwaan. 2000. "A Nuclear Solution to Climate Change?" *Science* 288 (19 May): 1177–1178.

Salawitch, Ross J. 1998. "A Greenhouse Warming Connection." *Nature* 392 (9 April): 551–552.

Santer, Benjamin D., Karl E. Taylor, Tom M.L. Wigley, Joyce E. Penner, Philip D. Jones, and Ulrich Cubasch. 1995. "Towards the Detection and Attribution of an Anthropogenic Effect on Climate." *Climate Dynamics* 12:79–100.

Santer, B.D., K.E. Taylor, T.M.L. Wigley, T.C. Johns, P.D. Jones, D.J. Karoly, J.F.B. Mitchell, A.H. Oort, J.E. Penner, V. Ramaswamy, M.D. Schwarzkopf, R.J. Stouffer, and S. Tett. 1996. "A Search for Human Influences on the Thermal Structure of the Atmosphere." *Nature* 382:39–46.

Sarmiento, J.L., T.M.C. Hughes, R.J. Stouffer, and S. Manabe. 1998. "Simulated Response of the Ocean Carbon Cycle to Anthropogenic Climate Warming." *Nature* 393:245–249.

Satellite Images Show Chunk of Broken Antarctic Ice Shelf. 1998. Eurekalert.org, 16 April. [http://www.eurekalert.org/releases/brkantartice.html]

Satheesh, S.K., and V. Ramanathan. 2000. "Large Differences in Aerosol Forcing at the Top of the Atmosphere and Earth's Surface." *Nature* 405 (4 May): 60–63.

Schelling, T. 1992. "Some Economics of Global Warming." *American Economic Review* 82, no. 1: 1–14.

Schemo, Diana Jean. 1998. "Amazon Fires Threaten Intact Forest, Indigenous People." *New York Times*. In *The Age* (Melbourne, Australia), 14 September.

Schimel, David, Jerry Melillo, Hanqin Tian, A. David McGuire, David Kicklighter, Timothy Kittel, Nan Rosenbloom, Steven Running, Peter Thornton, Dennis Ojima, William Parton, Robin Kelly, Martin Sykes, Ron Neilson, and Brian Rizzo. 2000. "Contribution of Increasing CO2 and Climate to Carbon Storage by Ecosystems in the United States." *Science* 287 (17 March): 2004–2006.

Schindler, David W. 1999. "Carbon Cycling: The Mysterious Missing Sink." *Nature* 398 (11 March): 105–106.

Schlesinger, M.E., and X. Jiang. 1991. "Revised Projection of Future Greenhouse Warming." *Nature* 350:219–221.

Schlesinger, Michael E., Navin Ramankutty, Natalia Andronova, Michael Margolis, and Richard A. Kerr. 2000. "Temperature Oscillations in the North Atlantic." *Science* 289 (28 July): 547–548.

Schmandt, J., S. Hadden, and D. Ward. 1992. *Texas and Global Warming: Emissions, Surface Water Supplies, and Sea Level Rise.* Austin: University of Texas Press.

Schmandt, Jurgen, and Judith Clarkson. 1992. *The Regions and Global Warming: Impacts and Response Strategies.* New York: Oxford University Press.

Schneider, David. 1997. "The Rising Seas." *Scientific American*, March, 112–117.

Schneider, Stephen H. 1975. "On the Carbon Dioxide-Climate Confusion." *Journal of the Atmospheric Sciences* 32:2060–2066.

———. 1976. *The Genesis Strategy: Climate and Global Survival.* New York: Plenum Press.

———. 1983. "CO2, Climate, and Society: A Brief Overview." In Robert S. Chen, Elise Boulding, and Stephen H. Schneider, eds., *Social Science Research and Climate Change: An Interdisciplinary Appraisal.* Boston: D. Reidel Publishing Co., 9–15.

———. 1989. *Global Warming: Are We Entering the Greenhouse Century?* San Francisco: Sierra Club Books.

———. 1990. "The Global Warming Debate: Science or Politics?" *Environmental Science and Technology* 24, no. 4: 432–435.

———. 1990. "Prudent Planning for a Warmer Planet." *New Scientist* 128:49.

———. 1993. "Degrees of Certainty." *Research and Exploration* 9, no. 2: 173–190.

———. 1994. "Detecting Climatic Change Signals: Are There Any Fingerprints?" *Science* 263:341–347.

———. 1997. *Laboratory Earth: The Planetary Gamble We Can't Afford to Lose.* New York: Harper Collins.

———. 1997. "Integrated Assessment Modeling of Global Climate Change: Transparent Rational Tool for Policy Making or Opaque Screen Hiding Value-laden Assumptions?" *Environmental Modeling and Assessment* 2:229–249.

———. 1999. Modeling Climate Change Impacts and Their Related Uncertainties. Paper for Conference on Social Science and the Future, Somerville College, Oxford, July 7–8. Ms. copy provided by the author.

———. 2000. Personal communication, 18 March.

———. 2001. "No Therapy for the Earth: When Personal Denial Goes Global." In Michael Aleksiuk and Thomas Nelson, eds., *Nature, Environment & Me: Explorations of Self In A Deteriorating World.* Montreal: McGill-Queens University Press.

———. N.d. Talk Abstract: Surprises and Scaling Connections between Climatology and Ecology. Institute for Mathematics and Its Applications. [http://www.ima.umn. edu/biology/wkshp_abstracts/schneider1.html]

———. N.d. Kyoto Protocol: The Unfinished Agenda: An Editorial Essay. Unpublished mss. supplied by the author.

———, ed. 1996. *Encyclopedia of Climate and Weather.* 2 vols. New York: Oxford University Press.

Schneider, Stephen H., and R.S. Chen. 1980. "Carbon Dioxide Warming and Coastline Flooding: Physical Factors and Climatic Impact." *American Review of Energy* 5: 107–140.

Schneider, S.H., and R. Londer. 1984. *The Coevolution of Climate and Life.* San Francisco: Sierra Club Books.

Schneider, Stephen H., Paul Schroeder, and Lew Ladd. 1991. "Slowing the Increase of Atmospheric Carbon Dioxide: A Biological Approach." *Climatic Change* 19:283–290.

Schneider, S.H., and L. Goulder. 1997. "Achieving Carbon Dioxide Concentration Targets: What Needs to be Done Now?" *Nature* 389:13–14.

Schrader, Ann. 1999. "Ice Dam's Collapse in Canada Cooled Europe for Centuries; Torrent was 15 Times Greater than Amazon, CU Scientist Says." *Denver Post*, 24 July, A-28.

Schwartzman, Stephen. 1999. "Reigniting the Rainforest: Fires, Development and Deforestation." *Native Americas* 16, no. 3/4(Fall/Winter): 60–63.

"Scientific Consensus: Villach (Austria) Conference." 1989. In Dean Edwin Abrahamson, ed., *The Challenge of Global Warming*. Washington, D.C.: Island Press, 63–67.

Scientists at the Smithsonian's National Museum of Natural History Find Global Warming to be Major Factor in Early Blossoming [of] Flowers in Washington. 2000. Press Release, Smithsonian Institution, March. [http://www.mnh.si.edu/feature.html]

"Scientists Report Large Ozone Loss." 2000. *USA Today*, 6 April, 3A.

Scientists' Statement on Global Climatic Disruption. 1997. Statements on Climate Change by Foreign Leaders at Earth Summit, 23 June. [http://uneco.org/Global_Warming.html]

Scripps Study Discovers Possible New Role of Clouds in Global Climate. 1995. Press Release, Scripps Center for Clouds, Chemistry, and Climate, 23 January. [http://www.sio.ucsd.edu/supp_groups/siocomm/pressreleases/ramanthan.html]

Seabacher, D.J., R.C. Harriss, K.B. Bartlett, S.M. Sebacher, and S.S. Grice. 1986. "Atmospheric Methane Sources: Alaska Tundra Bogs, an Alpine Fen, and a Subarctic Boreal Marsh." *Tellus* 38B:1–10.

Seitz, Frederick, Karl Bendetsen, Robert Jastrow, and William A. Nierenberg. 1989. *Scientific Perspectives on the Greenhouse Problem*. Washington, D.C.: George C. Marshall Institute.

Sellers, P.J., L. Bounoua, G.J. Collatz, D.A. Randall, D.A. Dazlich, S.O. Los, J.A. Berry, I. Fung, C.J. Tucker, C.B. Field, and T.G. Jensen. 1996. "Comparison of Radiative and Physiological Effects of Doubled Atmospheric CO2 on Climate." *Science* 271:1402–1406.

Shabtaie, S., and C.R. Bentley. 1987. "West Antarctic Ice Streams Draining to the Ross Ice Shelf: Configuration and Mass Balance. *Journal of Geophysical Research* 92: 1311–1336.

Shackleton, Nicholas J. 2000. "The 100,000-Year Ice-Age Cycle Identified and Found to Lag Temperature, Carbon Dioxide, and Orbital Eccentricity." *Science* 289 (15 September): 1897–1902.

Shaver, G.R., W.D. Billings, F.S. Chapin III, A.E. Giblin, K.J. Nadelhoffer, W.C. Oechel, and E.B. Rastetter. 1992. "Global Change and the Carbon Balance of Arctic Ecosystems." *Bioscience* 42:433–441.

Sherry, Mike. 2000. "Western High Presure Keeps Snow, Cold Out." *Omaha World-Herald*, 10 January, 9.

Shindell, Drew T., Ron L. Miller, Gavin A. Schmidt, and Lionell Pandolfo. 1999. "Simulation of Recent Northern Climate Trends by Greenhouse-gas Forcing." *Nature* 399 (3 June): 452–455.

Shindell, Drew T., David Rind, and Patrick Lonergan. 1998. "Increased Polar Stratospheric Ozone Losses and Delayed Eventual Recovery Owing to Increasing Greenhouse-gas Concentrations." *Nature* 392 (9 April): 589–592.

Shope, R. 1991. "Global Climate Change and Infectious Disease." *Environmental Health Perspectives* 96:171–174.

Siegenthaler, U. and H. Oeschger. 1978. "Predicting Future Atmospheric Carbon Dioxide Levels." *Science* 199:388–395.

Siegenthaler, U., and J.L. Sarmiento. 1993. "Atmospheric Carbon Dioxide and the Ocean." *Nature* 365:119–125.

Sierra Club. 1999. "Global Warming: The High Costs of Inaction." [http://www.sierraclub.org/global-warming/resources/innactio.htm]

Silver, Cheryl Simon, and Ruth S. DeFries. 1990. *One Earth, One Future: Our Changing Global Environment*. Washington, D.C.: National Academy Press.

Simons, Paul. 1992. "Why Global Warming Could Take Britain by Storm." *New Scientist* 136:35.

Singer, S. Fred. 1997. *Hot Talk, Cold Science: Global Warming's Unfinished Debate*. Oakland, Calif.: Independent Institute.

———. 1999. "Global Warming Whining." *Washington Times*, 16 April. [http://www.cop5.org/apr99/singer.htm]

———. 1992. "Global Climate Change: Facts and Fiction." In Jay H. Lehr, ed., *Rational Readings on Environmental Concerns*. New York: Van Nostrand Reinhold.

———. 1992. "Foreword." In Robert C. Balling, Jr., *The Heated Debate: Greenhouse Predictions Versus Climate Reality*. San Francisco: Public Research Institute for Public Policy.

———. 2000. "Sure, the North Pole is Melting. So What?" *Wall Street Journal*, 28 August, A-18.

Smith, Amy Tetlow. 1995. "Environmental Factors Affecting Global Atmospheric Methane Concentrations." *Progress in Physical Geography* 19:322–335.

Smith, Joel B., and D.A. Tirpak. 1990. *The Potential Effects of Global Climate Change in the United States*. New York: Hemisphere. (Also published 1989, Washington, D.C.: U.S. Environmental Protection Agency.)

Socci, Anthony D. 1993. "The Climate Continuum: An Overview of Global Warming and Cooling Throughout the History of Planet Earth." In Richard A. Geyer, ed., *A Global Warming Forum: Scientific, Economic, and Legal Overview*. Boca Raton, Fla.: CRC Press, 161–207.

Soden, Brian J. 1997. "Variations in the Tropical Greenhouse Effect during El Niño." *Journal of Climate* 10:1050–1055.

Soden, B.J., and R. Fu. 1995. "A Satellite Analysis of Deep Convection, Upper Tropospheric Humidity, and the Greenhouse Effect." *Journal of Climate* 8:2333–2351.

Sokolik, I.N., and O.B. Toon. 1996. "Direct Radiative Forcing by Anthropogenic Airborne Mineral Aerosols." *Nature* 380:419–422.

Solomon, A.M. 1986. "Transient Response of Forests to CO2-induced Climate Change: Simulation Modeling Experiments in Eastern North America." *Oecologia* 68:567–579.

Solomon, A.M., and D.C. West. 1987. "Simulating Forest Ecosystem Responses to Expected Climate Change in Eastern North America: Applications to Decision Making in the Forest Industry." In W.E. Shands and J.S. Hoffman, eds., *The Greenhouse Effect, Climate Change, and the U.S. Forests*. Washington, D.C.: The Conservation Foundation, 189–217.

Somerville, Richard C.J. 1996. *The Forgiving Air: Understanding Environmental Change*. Berkeley: University of California Press.

South, Eileen L. 1990. *The Changing Atmosphere: A Global Challenge*. New Haven: Yale University Press.

Spencer, R.W., and W.D. Braswell. 1997. "How Dry is the Tropical Free Tropical? Implications for Global Warming Theory." *Bulletin of the American Meteorological Society* 78:1097–1106.

Spotts, Peter N. 2000. "As Arctic Warms, Scientists Rethink Culprits." *Christian Science Monitor*, 22 August, 4.

"Spring 2000 is Warmest on Record." 2000. Associated Press. In *Omaha World-Herald*, 17 June, 1.

Stauffer, Bernhard. 1999. "Cornucopia of Ice Core Results." *Nature* 399 (3 June): 412–413.

Steitz, David E. 1999. "NASA Researchers Document Shrinking of Greenland's Glaciers." NASA Press Release 99–33, 4 March. [http://www.earthobservatory.nasa.gov/Newsroom/NasaNews/19990304207.html]

Stevens, William K. 1996. "Ice Shelves Melting as Forecast, but Disaster Script is in Doubt." *New York Times*, 30 January, C-4.

———. 1997. "Computers Model World's Climate, But How Well?" *New York Times*, 11 April. [http://benetton.dkrz.de:3688/homepages/georg/kimo/0254.html]

———. 1998. "New Evidence Finds this is the Warmest Century in 600 years." *New York Times*, 28 April, C-3.

———. 1999. "1998 and 1999 Warmest Years Ever Recorded." *New York Times*, 19 December, 1, 38.

———. 1999. *The Change in the Weather: People, Weather, and the Science of Climate*. New York: Delacorte Press.

———. 2000. "The Oceans Absorb Much of Global Warming, Study Confirms." *New York Times*, 24 March, A-16.

———. 2000. "New Survey Shows Growing Loss of Arctic Atmosphere's Ozone." *New York Times*, 6 April, A-19.

Stevenson, Ian R., and David M. Bryant. 2000. "Avian Phenology: Climate Change and Constraints on Breeding." *Nature* 406 (27 July): 366–367.

Stirling, I., and A.E. Derocher. 1993. "Possible Impacts of Climate Warming on Polar Bears." *Arctic* 46, no. 3: 240–245.

Stirling, I., and N.J. Lunn. 1997. "Environmental Fluctuation in Arctic Marine Ecosystems as Reflected by Variability in Reproduction of Polar Bears and Ringed Seals." In S.J. Woodlin and M. Marquiss, eds., *Ecology of Arctic Environments*. Oxford: Blackwell Science Ltd., 167–181.

Stocker, T.F., and A. Schmittner. 1997. "Influence of CO2 Emission Rates on the Stability of the Thermohaline Circulation." *Nature* 388:862–864.

Stocker, Thomas F., and Olivier Marchal. 2000. "Abrupt Climate Change in the Computer: Is it Real?" *Proceedings of the National Academy of Sciences of the United States of America* 97, no. 4 (15 February): 1362–1364.

Stocks, B.J. 1993. "Global Warming and Forest Fires in Canada." *The Forestry Chronicle* 69:290.

Stone, Richard. 1995. "If the Mercury Soars, So May Health Hazards." *Science* 267: 957–958.

Street-Perrott, F.A., and R.A. Perrott. 1990. "Abrupt Climate Fluctuations in the Tropics: The Influence of Atlantic Ocean Circulation." *Nature* 343:607–611.

Sturges, W.T., T.J. Wallington, M.D. Hurley, K.P. Shine, K. Sihra, A. Engel, D.E. Oram, S.A. Penkett, R. Mulvaney, and C.A.M. Brenninkmeijer. 2000. "A Potent Greenhouse Gas Identified in the Atmosphere: SF5CF3." *Science* 289 (28 July): 611–613.

Sudetic, Chuck. 1999. "As the World Burns." *Rolling Stone*, 2 September, 97–106, 129.

Suffling, R. 1992. "Climate Change and Boreal Forest Fires in Fennoscandia and Central Canada." In M. Boer and E. Koster, eds., *Greenhouse Impact on Cold-climate Ecosystems and Landscapes*. Selected papers of A European Conference on Landscape Ecological Impact: Impact of Climatic Change, Lunteren, the Netherlands, December 3–7, 1989. CARTENA Supplement 22. Cremlingen, Germany: Cartena Verlag, 111–132.

Sun, D.Z., and I.M. Held. 1996. "A Comparison of Modelled and Observed Relationships Between Interannual Variations of Water Vapor and Temperature." *Journal of Climate* 9:665–675.

Sundquist, Eric T. 1990. "Long-term Aspects of Future Atmospheric CO2 and Sea-level Changes." In Roger R. Revelle, ed., *Sea-level Change*. Washington, D.C.: National Resource Council and National Academy Press.

Suplee, Curt. 2000. "For 500 Million, a Sleeper on Greenland's Ice Sheet." *Washington Post*, 10 July, A-9.

———. 2000. "Historical Records Provide a Growing Sense of Global Warmth." *Washington Post*, 8 September, A-2.

Sveinbjornsson, Bjartmar. 1992. "Arctic Tree Line in a Changing Climate." In F. Stuart Chapin III, Robert L. Jefferies, James F. Reynolds, Gaius R. Shaver, and Josef Svoboda, eds., *Arctic Ecosystems in a Changing Climate: An Ecophysiological Perspective*. San Diego: Academic Press, 239–256.

Sykes, M.T., and I.C. Prentice. 1995. "Boreal Forest Futures: Modelling the Controls on Tree Species Range Limits and Transient Responses to Climate Change." *Water, Air and Soil Pollution* 82:415–428.

Tabazadeh, A., M.L. Santee, M.Y. Danilin, H.C. Pumphrey, P.A. Newman, P.J. Hamill, and J.L. Mergenthaler. 2000. "Quantifying Denitrification and its Effect on Ozone Recovery." *Science* 288 (26 May): 1407–1411.

Taliman, Valerie. 1999. "Reading the Clouds: Native Perspectives on Southwestern Environments." *Native Americas* 16, no. 3/4(Fall/Winter): 34–41. [http://nativeamericas.aip.cornell.edu]

Taubes, Gary. 1997. "Apocalypse Not." [http://www.junkscience.com/news/taubes2.html]

Taylor, K.C., R.B. Alley, G.A. Doyle, P.M. Grootes, P.A. Mayewski, G.W. Lamorey, J.W.C. White, and L.K. Barlow. 1993. "The 'Flickering Switch' of Late Pleistocene Climate Change." *Nature* 361:432–436.

Taylor, K.E., and J.E. Penner. 1994. "Response of the Climate System to Atmospheric Aerosols and Greenhouse Gases." *Nature* 369:734–737.

Tegen, I., A.A. Lacis, and I. Fung. 1996. "The Influence of Mineral Aerosols from Disturbed Soils on the Global Radiation Budget." *Nature* 380:419–422.

Tesar, Jenny E. 1991. *Global Warming*. New York: Facts on File.

Tett, Simon F.B., Peter A. Stott, Myles R. Allen, William J. Ingram, and John F.B.

Mitchell. 1999. "Causes of Twentieth-century Temperature Change Near the Earth's Surface." *Nature* 399 (10 June): 569–572.

Thomas, Chris D., and Jack J. Lennon. 1999. "Birds Extend Their Ranges Northwards." *Nature* 399 (20 May): 213.

Thomas, R. 1986. "Future Sea-level Rise and its Early Detection by Satellite Remote Sensing." In *Effects of Changes in Atmospheric Ozone and Global Climate*. Vol. 4. New York: United Nations Environment Programme/United States Environmental Protection Agency.

Thomas, R., T. Akins, B. Csatho, M. Fahnestock, P. Gogineni, C. Kim, and J. Sonntag. 2000. "Mass Balance of the Greenland Ice Sheet at High Elevations." *Science* 289 (July 21): 426–428.

Thomas, R., T.J.O. Sanderson, and K.E. Rose. 1979. "Effect of Climatic Warming on the West Antarctic Ice Sheet." *Nature* 277:355–362.

Thomas, William. 1998. "Salmon Dying in Hot Waters." Environmental News Service, 22 September. [http://www.econet.apc.org/igc/en/hl/9809244985/hl11.html]

Thompson, L.G., T. Yao, E. Mosley-Thompson, M.E. Davis, K.A. Henderson, and P.-N. Lin. 2000. "A High-Resolution Millennial Record of the South Asian Monsoon from Himalayan Ice Cores." *Science* 289 (15 September): 1916–1919.

Thompson, Russell D. 1995. "The Impact of Atmospheric Aerosols on Global Climate: A Review." *Progress in Physical Geography* 19:336–350.

Thornes, John. 1995. "Global Environmental Change and Regional Response: The European Mediterranean." *Transactions of the Institute of British Geographers* 20: 357–367.

Thornton, Joe. 2000. *Pandora's Poison: On Chlorine, Health, and a New Environmental Strategy*. Cambridge, Mass.: MIT Press.

Thurow, Charles. 1983. *Improving Street Climate Through Urban Design*. Planning Advisory Service #376. Chicago, Ill.: American Planning Association.

Tian, H., J.M. Mellilo, D.W. Kicklighter, A.D. McGuire, J.V.K. Helfrich III, B. Moore III, and C.J. Vorosmarty. 1998. "Effect of Interannual Climate Variability on Carbon Storage in Amazonian Ecosystems." *Nature* 396 (17 December): 664–667.

Tichy, Milos. 1996. "Greenhouse Gas Emissions Projections and Mitigation Options for the Czech Republic, 1990–2010." *Environmental Management* 20:S47–S55.

Timmermann, A., J. Oberhuber, A. Bacher, M. Esch, M. Latif, and E. Roeckner. 1999. "Increased El Niño Frequency in a Climate Model Forced by Future Greenhouse Warming." *Nature* 398 (22 April): 694–696.

Tisdale, Robert, Dave Glandorf, Margaret Tolbert, and Owen B. Toon. 1998. "Infrared Optical Constants of Low Temperature H2SO4 Solutions Representative of Stratospheric Sulfate Aerosols." *Journal of Geophysical Research* 103:25,353–25,370.

Titus, J.G. 1984. "Planning for Sea Level Rise Before and After a Coastal Disaster." In M.C. Barth and J.G. Titus, eds., *Greenhouse Effect and Sea Level Rise: A Challenge for this Generation*. New York: Van Nostrand Reinhold Company, 253–270.

———. 1986. "Greenhouse Effect, Sea Level Rise, and Coastal Zone Management." *Coastal Zone Management* 14, no. 3: 147–171.

———. 1990. "Greenhouse Effect, Sea Level Rise, and Barrier Islands: Case Study of Long Beach Island, New Jersey." *Coastal Management* 18:165–190.

————, ed. 1985. *Potential Impacts of Sea Level Rise on the Beach at Ocean City, Maryland.* Washington, D.C.: U.S. Environmental Protection Agency.

————, ed. 1988. *Greenhouse Effect, Sea Level Rise, and Coastal Wetlands.* Washington, D.C.: U.S. Environmental Protection Agency.

Titus, J.G., and M.S. Greene. 1989. "An Overview of Studies Estimating the Nationwide Cost of Holding Back the Sea." In J.B. Smith and D.A. Tirpak, eds., *Potential Effects of Climate Change on the United States: Appendix B, Sea Level Rise.* Washington, D.C.: U.S. Environmental Protection Agency.

Titus, J.G., R.A. Park, S.P. Leatherman, J.R. Weggel, M.S. Greene, P.W. Mausel, S. Brown, C. Gaunt. M. Trehan, and G. Yohe. 1991. "Greenhouse Effect and Sea Level Rise: The Cost of Holding Back the Sea." *Coastal Management* 19:171–210.

Titus, James G., and Vijay Narayanan. 1996. "The Risk of Sea Level Rise: A Delphic Monte Carlo Analysis in which Twenty Researchers Specify Subjective Probability Distributions for Model Coefficients within their Respective Areas of Expertise." *Climatic Change* 33, no. 2: 151–212. (Also published N.d., U.S. Environmental Protection Agency. [http://users.erols.com/jtitus/Risk/CC.html])

Today's Science: Global Warming and Ozone Hole Linked. 1998. Facts on File. Today's Science on File, June. [http://facts.com/cd/s70026.htm]

Toon, Owen B., and Richard C. Miake-Lye. 1998. "Subsonic Aircraft: Contrail and Cloud Effects Special Study (SUCCESS)." *Geophysical Research Letters* 25:1109–1112.

Travis, J. 1996. "The Loitering El Niño: Greenhouse Guest?" *Science News* 149 (27 January): 54.

Tregguer, Paul, and Philippe Pondaven. 2000. "Global Change: Silica Control of Carbon Dioxide." *Nature* 406 (27 July): 358–359.

Trenberth, Kevin E. 1999. "Conceptual Framework for Changes of Extremes of the Hydrological Cycle with Climate Change." *Climatic Change* 42:327–339.

Trexler, M.C., and K.M. McFall. "Building a CO2 Mitigation Portfolio: The Future is Now for Carbon Sinks." *Electricity Journal* 6, no. 8: 60–69.

Turco, Richard P. 1997. *Earth Under Siege: From Air Pollution to Global Change.* Oxford, England: Oxford University Press.

Turkenberg, W.C., E.A. Alsema, and K. Blok. 1989. "The Prospects of Photovoltaic Solar Energy Conversion." In P.A. Okken, R.J. Swart, and S. Zwerver, *Climate and Energy: The Feasibility of Controlling CO2 Emissions.* Dordrecht, Germany: Kluwer Academic Pubishers, 95–106.

Turner, B.L. II, William C. Clark, Robert W. Kates, John F. Richards, Jessica T. Mathews, and William B. Meyer, eds. 1990. *The Earth as Transformed by Human Action: Global and Regional Changes in the Biosphere Over the Past 300 Years.* New York: Cambridge University Press with Clark University.

Tyler, Patrick. 1995. "China's Inevitable Dilemma: Coal Equals Growth." *New York Times,* 29 November.

Tyndall, John. 1861. "On the Aborption and Radiation of Heat by Gases and Vapours, and on the Physical Connexion of Radiation, Absorption, and Conduction." *The London, Edinburgh, and Dublin Philosophical Magazine and Journal of Science,* 4th Ser. (September): 169–194.

————. 1863. "On Radiation Through the Earth's Atmosphere." *The London, Edinburgh, and Dublin Philosophical Magazine and Journal of Science* 4:200–207.

Union of Concerned Scientists. N.d. "Has the Climate Already Changed?" [http://www.ucsusa.org/warming/gw.science.html#already]

———. N.d. "Focus on Mosquito-borne Diseases." [http://www.ucsusa.org/warming/gw.newfind.html#mosquito]

———. N.d. "Global Warming." [http://www.ucsusa.org/warming/index.html]

United States Congress Office of Technology Assessment. 1991. *Changing by Degrees: Steps to Reduce Greenhouse Gases.* Washington, D.C.: Government Printing Office.

U.S. Department of Energy. 1993. *Trends '93: A Compendium of Data on Global Climate Change.* Washington, D.C.: U.S. Department of Energy.

———. 1983. Proceedings to Carbon Dioxide Research Conference: Carbon Dioxide, Science, and Consensus. Washington, D.C.: U.S. Department of Energy.

U.S. Department of State. 1999. *Coral Bleaching, Coral Mortality, and Global Climate Change: A Report Presented by Rafe Pomerance, Deputy Assistant Secretary of State for the Environment and Development to the U.S. Coral Reef Task Force.* Washington D.C.: Bureau of Oceans and International Environmental and Scientific Affairs, U.S. Department of State.

United States Environmental Protection Agency (EPA). *Ecological Impacts from Climate Change: An Economic Analysis of Freshwater Recreational Fishing.* Washington, D.C.: EPA.

———. 1997. *The Cost of Holding Back the Sea.* Washington, D.C.: EPA. [http://users.erols.com/jtitus/Holding/NRJ.html#causes]

U.S. Environmental Protection Agency (EPA) and Louisiana Geological Survey. 1987. *Saving Louisiana's Coastal Wetlands: The Need for a Long-term Plan of Action.* Washington, D.C.: EPA.

Valentini, R., G. Matteucci, A.J. Dolman, E.D. Schulze, C. Rebmann, E.J. Moors, A. Granier, P. Gross, N.O. Jensen, K. Pilegaard, A. Lindroth, A. Grelle, C. Bernhofer, T. Grunwald, M. Aubinet, R. Ceulmans, A.S. Kowalski, T. Vesala, U. Rannik, P. Berbigler, D. Loustau, J. Gudmundsson, H. Thorgiersson, A. Ibrom, K. Morgenstern, R. Clement, J. Moncrieff, L. Montagnani, S. Minerbi, and P.G. Jarvis. 2000. "Respiration as the Main Determinant of Carbon Balance in European Forests." *Nature* 404 (20 April): 861–865.

Van der Veen, C.J. 1987. "Ice Sheets and the CO2 Problem." *Surveys in Geophysics* 9:1–42.

Vaughan, D., and T. Lachlan-Cope. 1995. "Recent Retreat of Ice Shelves on the Antarctic Peninsula." *Weather* 50 (November): 374–376, U.K.: Royal Meteorological Society.

Vaughan, D.G., and C.S.M. Doake. 1996. "Recent Atmospheric Warming and Retreat of Ice Shelves on the Antarctic Peninsula." *Nature* 379 (25 January): 328–331.

Victor, David G., and Julian E. Salt. 1994. "From Rio to Berlin: Managing Climate Change." *Environment* 36, no. 10: 7–36.

Vörösmarty, Charles J., Pamela Green, Joseph Salisbury, and Richard B. Lammers. 2000. "Global Water Resources: Vulnerability from Climate Change and Population Growth." *Science* 289 (14 July): 284–288.

Vrolijk, Christiaan. 1999. "The Buenos Aries Climate Conference: Outcome and Implications." The Royal Institute of International Affairs. 2 April. [http://www.riia.org/briefingpapers/bp53.html]

Wadman, Meredith. 2000. "Car Maker Joins Exodus from Anti-Kyoto Coalition." *Nature* 404:322.

Walsh, J.F., D.H. Molyneux, and M.H. Birley. 1993. "Deforestation: Effects on Vector-borne Disease." *Parasitology* 106:S55–S75.

Wang, W.C., Y.L. Yung, A.A. Lacis, T. Mo, and J.E. Hansen. 1976. "Greenhouse Effects Due to Man-made Perturbations of Trace Gases." *Science* 194:685–690.

"Warming Affects Ocean Algae." 1999. ABC News, 14 January. [http://geography. rutgers.edu/courses/99spring/370sp99/news01_20_99.html#anchor33828]

Warrick, Jody. 1998. "Earth at Its Warmest in Past 12 Centuries; Scientist Says Data Suggest Human Causes." *Washington Post*, 8 December, [http://www.asoc.org/ currentpress/1208post.htm]

Watson, R.T., M.C. Zinyowera, R.H. Moss, and D.J. Dokken, eds. 1996. *The Regional Impacts of Climate Change—An Assessment of Vulnerability.* A Special Report of IPCC Working Group II. New York: Cambridge University Press.

Weaver, Andrew J. 1993. "The Oceans and Global Warming." *Nature* 364:192–193.

————. 1993. "Looking Far Ahead into the Greenhouse." *Science News* 14 (August): 111.

Webb, Jason. 1998. "World Forests Said Vulnerable to Global Warming." Reuters, 4 November. [http://bonanza.lter.uaf.edu/~davev/nrm304/glbxnews.htm]

————. 1998. "Small Islands Say Global Warming Hurting Them Now." Reuters. [http: //bonanza.lter.uaf.edu/~davev/nrm304/glbxnews.htm]

————. 1998. "World Temperatures Could Jump Suddenly." Reuters, 4 November. [http: //bonanza.lter.uaf.edu/~davev/nrm304/glbxnews.htm]

————. 1998. "Mosquito Invasion as Argentina Warms." Reuters. [http://bonanza. lter.uaf.edu/~davev/nrm304/glbxnews.htm]

Weber, Rudolf O., Peter Talkner, Ingeborg Auer, Reinhard Bohm, Marjana Gajic-Capka, Ksenija Zaninovic, Rudolf Brazdil, and Pavel Fasko. 1997. "20th-century Changes of Temperature in the Mountain Regions of Central Europe." *Climatic Change* 36:327–344.

Weggel, J.R., S. Brown, J.C. Escajadillo, P. Breen, and E.L. Doheny. 1989. "The Cost of Defending Developed Shorelines Along Sheltered Waters of the United States from a Two-meter Rise in Mean Sea Level." In J.B. Smith and D.A. Tirpak, eds., *Potential Effects of Global Climate Change on the United States: Appendix B, Sea Level Rise.* Washington, D.C.: U.S. Environmental Protection Agency.

Weinburg, Bill. 1999. "Hurricane Mitch, Indigenous Peoples and Mesoamerica's Climate Disaster." *Native Americas* 16, no. 3/4(Fall/Winter): 50–59. [http://nativeamericas. aip.cornell.edu/fall99/fall99weinberg.html]

Weiner, Jonathan. 1990. *The Next One Hundred Years: Shaping the Fate of Our Living Earth.* New York: Bantam Books.

Weissert, Helmut. 2000. "Global Change: Deciphering Methane's Fingerprint." *Nature* 406 (27 July): 356–357.

Welsh, Jonathan. 2000. "Drive Buys: Honda Insight. A Car With an Extra Charge." *Wall Street Journal*, 10 March, W-15C.

Wenzel, George. 1992. "Global 'Warming,' the Arctic, and Inuit: Some Sociocultural Implications." *The Canadian Geographer* 36:78–79.

West Antarctic Ice Sheet Not in Jeopardy. 1998. [http://www.enn.com/news/ennstories/ 1998/12/120198/antarc.asp]

Westing, Arthur H., ed. 1998. *Cultural Norms, War and the Environment.* Oxford: Oxford University Press.

————, ed. 1990. *The Environmental Hazards of War: Releasing Dangerous Forces in an Industrialized World.* Newbury Park, Calif.: Sage.

————, ed. 1980. *Warfare in a Fragile World.* London: Taylor & Francis.

Westing, Arthur H., and Malvern Lumsden. 1978. *Threat of Modern Warfare to Man and His Environment: An Annotated Bibliography.* No. 40. Paris: UNESCO.

Wetherald, R.J., and S. Manabe. 1986. "An Investigation of Cloud Cover Change in Response to Thermal Forcing." *Climatic Change* 8:5–23.

Whalley, John, and Randall Wigle. 1991. "The International Incidence of Carbon Taxes." In Rudiger Dornbusch and James M. Poterba, eds., *Global Warming: Economic Policy Responses.* Cambridge, Mass.: MIT Press, 233–263. (Also presented September 1990 as a paper at a Conference on Economic Policy Responses to Global Warming, Rome, Italy.)

Whelan, S.C., and W.S. Reeburgh. 1990. "Consumption of Atmospheric Methane by Tundra Soils." *Nature* 346:160–162.

White, James C., ed., William R. Wagner and Carole N. Beal, assoc. eds. 1992. *Global Climate Change: Linking Energy, Environment, Economy, and Equity.* New York: Plenum Press.

Wigley, T.M.L. 1989. "Possible Climate Change Due to SO2-derived Cloud Condensation Nuclei." *Nature* 339:365–367.

Wigley, T.M.L., and S.C.B. Raper. 1987. "Thermal Expansion of Seawater Associated with Global Warming." *Nature* 330:324–331.

————. 1990. "Natural Variability of the Climate System and Detection of the Greenhouse Effect." *Nature* 344:324–327.

Wigley, T.M., P.D. Jones, and S.C.B. Raper. 1997. "The Observed Global Warming Record: What Does it Tell Us?" *Proceedings of the National Academy of Sciences of the United States of America* 94:8314–8320.

Wigley, T.M.L., R. Richels, and J.A. Edmonds. 1996. "Economic and Environmental Choices in the Stabilization of Atmospheric CO_2 Concentrations." *Nature* 379: 240–243.

Wigley, T.M.L., and D.S. Schimel. 2000. *The Carbon Cycle.* Cambridge, U.K.: Cambridge University Press.

Wilfred, John Noble. 2000. "Ages-Old Polar Icecap Is Melting, Scientists Find." *New York Times,* 19 August, 1.

Wilkinson, C., O. Linden, H. Cesar, G. Hodgson, J. Rubens, and A.E. Strong. 1999. "Ecological and Socioeconomic Impacts of 1998 Coral Mortality in the Indian Ocean: An ENSO Impact and a Warning of Future Change?" *Ambio* 28:188–196.

Williams, Jack. 2000. "Rising Ocean Temperatures Aren't Breaking the Ice." *USA Today,* 25 April, 10D.

Williams, Jill, ed. 1978. *Carbon Dioxide, Climate, and Society: Proceedings of a IIASA Workshop Co-sponsored by WMO, UNEP, and SCOPE, February 21–24, 1978.* Oxford, U.K.: Pergamon Press.

Willis, Ian, and Jean-Michel Bonvin. 1995. "Climate Change in Mountain Environments: Hydrological and Water Resource Implications." *Geography* 80:247–261.

Willougby, H.E. 1999. "Hurricane Heat Engines." *Nature* 401 (14 October): 649–650.

"Wind Facility Provides Clean, Renewable Energy to Thousands in Midwest." 1998.

EarthVision Reports, 25 September. [http://climatechangedebate.com/archive/ 09–18_10–27_1998.txt]

Wirth, D. 1989. "Climate Chaos." *Foreign Policy* 7:3–22.

Witze, Alexandra. 2000. "Evidence Supports Warming Theory." *Dallas Morning News* In *Omaha World-Herald*, 12 January, Metro extra, 4.

Wojick, David E. 1998. "Hansen: Experiment Shows Chaos Trumps Global Warming." *Electricity Daily* 10, no. 29 (13 February). [http://climatechangedebate.com/ archive/09–18_10–27_1998.txt]

Wood, Richard A., Anne B. Keen, John F.B. Mitchell, and Jonathan M. Gregory. 1999. "Changing Spatial Structure of the Thermohaline Circulation in Response to Atmospheric CO2 Forcing in a Climate Model." *Nature* 399 (10 June): 572–575.

Woodard, Colin. 1998. "Glacial Ice is Slip-sliding Away." *Christian Science Monitor*, 10 December. [http://benetton.dkrz.de:3688/homepages/georg/kimo/0254.html]

———. 2000. "Slowly, but Surely, Iceland is Losing its Ice; Global Warming is Prime Suspect in Meltdown." *San Francisco Chronicle*, 21 August, A-1.

Woodhouse, C.A., and J.T. Overpeck. 1998. "2000 Years of Drought Variability in the Central United States." *Bulletin of the American Meteorological Society* 79, no. 12 (December): 2693–2714.

Woodwell, George M. 1978. "The Carbon Dioxide Question." *Scientific American* 238: 234–243.

———. 1983. "Global Deforestation: Contribution to Atmospheric Carbon Dioxide: A Review and a Projection." *Science* 222:1081–1086.

———. 1986/1987. "Forests and Climate: Surprises in Store." *Oceanus* 29 (Winter): 71– 75.

———. 1989. "The Warming of the Industrialized Middle Latitudes 1985–2050: Causes and Consequences." *Climate Change* 15:31–50.

———. 1990. "The Effects of Global Warming." In Jeremy Leggett, ed., *Global Warming: The Greenpeace Report*. New York: Oxford University Press, 116–132.

———. 1991. "Forests in a Warming World: A Time for New Policies." *Climatic Change* 19:245–251.

———. 1995. "Biotic Feedbacks from the Warming of the Earth." In George M. Woodwell and Fred T. MacKenzie, eds., *Biotic Feedbacks in the Global Climate System: Will the Warming Feed the Warming?* New York: Oxford University Press, 3–21.

———. 1999. "The Global Warming Issue." [http://www.gibbons.freeonline.co.uk/ Articles/The_Global_warming_issue.htm]

Woodwell, G.M., J.E. Hobbie, R.A. Houghton, J.M. Melillo, B. Moore, B.J. Peterson, and G.R. Shaver. 1983. "Global Deforestation: Contribution to Atmospheric Carbon Dioxide." *Science* 222:1081–1086.

Woodwell, G.M., and R.A. Houghton. 1997. "Biotic Influences on the World Carbon Budget." In W. Stumm, ed., *Global Chemical Cycles and Their Alterations by Man*. Berlin: Dahlem Konferenzen, 61–72.

Woodwell, George M., and Fred T. MacKenzie, eds. 1995 *Biotic Feedbacks in the Global Climate System: Will the Warming Feed the Warming?* New York: Oxford University Press.

Woodwell, George M., Fred T. MacKenzie, R.A. Houghton, Michael J. Apps, Eville Gorham, and Eric A. Davidson. 1995. "Will the Warming Speed the Warming?" in George M. Woodwell and Fred T. MacKenzie, eds., *Biotic Feedbacks in the*

Global Climate System: Will the Warming Feed the Warming? New York: Oxford University Press, 393–411.

———. 1998. "Biotic Feedbacks in the Warming of the Earth." *Climatic Change* 40: 495–518.

World Meteorological Organization. 1988. *The Changing Atmosphere: Implications for Global Security.* Conference Proceedings, Toronto, Canada, June 27–30, 1998. WMO, No. 710. Geneva, Switzerland: World Meteorological Organization.

Worldwatch. 1998. Pre-Buenos Aires Briefing Press Release: Kyoto Protocol Faces Crucial Test. Washington, D.C.: Worldwatch Institute. [http://bonanza.lter.uaf.edu/~davev/nrm304/glbnews.htm]

Worster, Donald. 2000. "Climate and History: Lessons from the Great Plains." In Jill Ker Conway, Kenneth Keniston, and Leo Marx, eds., *Earth, Air, Fire, Water: Humanistic Studies of the Environment.* Amherst: University of Massachusetts Press, 51–77.

Wuebbles, Donald J. 1991. *Primer on Greenhouse Gases.* Chelsea, U.K.: Lewis Publishers.

Wuethrich, Bernice. 1999. "Lack of Icebergs Another Sign of Global Warming?" *Science* 285 (2 July): 37.

———. 2000. "How Climate Change Alters Rhythms of the Wild." *Science* 287 (4 February): 793–795.

Wyman, Richard L., ed. 1991. *Global Climate Change and Life on Earth.* New York: Routledge, Chapman and Hall.

Wynne, Brian. 1993. "Implementation of Greenhouse Gas Reductions in the European Community: Institutional and Cultural Factors." *Global Environmental Change: Human and Policy Dimensions* 3, no. 1: 101–128.

Ye, H., and J.R. Mather. 1997. "Polar Snow Cover Changes and Global Warming." *International Journal of Climatology* 17:155–162.

Yellen, Janet, Chair, White House Council of Economic Advisers. 1998. Statement on the Economics of the Kyoto Protocol before the Committee on Agriculture, Nutrition, and Forestry, U.S. Senate, 5 March, Washington, D.C. [http://www.fetc.doe.gov/products/gcc/research/economic.html]

Yohe, Gary. 1996. "Exercises in Hedging Against Extreme Consequences of Global Change and the Expected Value of Information." *Global Environmental Change* 6, no. 2: 87–101.

Yohe, G., J. Neumann, P. Marshall, and H. Ameden. 1996. "The Economic Cost of Greenhouse-induced Sea-level Rise for Developed Property in the United States." *Climatic Change* 32:387–410.

Yoon, C.K. 1994. "Warming Moves Plants up Peaks, Threatening Extinction." *New York Times,* 21 June, C-4.

Young, Stephen. 1994. "Insects that Carry a Global Warming." *New Scientist* 142:32.

Yozwiak, Steve. 1998. " 'Island' Sizzle, Growth May Make Valley an Increasingly Hot Spot." *Arizona Republic* (Phoenix), 25 September. [http://www.sepp.org/reality/arizrepub.html]

Zimmermann, M., and W. Haeberli. 1992. "Climatic Change and Debris-flow Activity in High Mountain Areas: A Case Study in the Swiss Alps." In M. Boer and E. Koster, eds., *Greenhouse-Impact on Cold-Climate Ecosystems and Landscapes.* Selected papers of A European Conference on Landscape Ecological Impact: Im-

pact of Climatic Change, Lunteren, the Netherlands, December 3–7, 1989. CAR-
TENA Supplement 22. Cremlingen, Germany: Cartena Verlag, 59–72.
Zhang, Y., and W.C. Wang. 1997. "Model-simulated Northern Winter Cyclone and An-
ticyclone Activity Under a Greenhouse Warming Scenario." *Journal of Climate*
10:1616–1634.
Zwally, H.J. 1991. "Breakup of Antarctic Ice." *Nature* 350 (28 March): 274.

Index

Abbey, Edward, 21
Abrahamson, Dean Edwin, 1, 69, 237
Abu-Asab, Mones, 243
Adams, Jonathan, 139, 140
Aerosols, as climate forcing, 90
Agarwal, Anil, 16
Agriculture and warming, 199–203; crop
  yields, 200, 201, 202, 202; growing
  seasons, 202, 203; heat stress, 200–202;
  industrial scale, 200
Air travel: commuting by air, 26–27; de-
  pendence on fossil fuels, 25; fossil-fuel
  emissions of, 25–27; increases in traf-
  fic, 25; jet fuel, chemical composition,
  25; jet fuel, pollution by, 25
Albedo, 125
Albuquerque Declaration, 227–228
Alley, Richard B., 68, 127, 128
Alliance of Small Island States (AOSIS),
  156, 157
Almendares, J., 189
Alternative energy, 252
American Petroleum Institute, 109
Andolaro, Franco, 164
Animal adaptations to warming: birds,
  196, 197; butterflies, 196–197
Antarctic ice, flow dynamics, 130–131
Antarctic Peninsula; glacial retreat on,
  129; temperature rises, 128–129
Apangalook, Leonard, Sr., 224

Arctic: as carbon sink, 136; ice-increase
  areas, 223; icemelt, 132–138; ice pack,
  decreases in coverage, 135–136, 137,
  174, 223, 224; Inuit, and warming,
  222, 223, 224; North Pole, open water
  at, 137–138; Northwest Passage, 138;
  polar bears and warming, 136; seals, in
  food web, 224, 225; temperature rises,
  134, 135; walrus, and warming, 222;
  water temperatures, 138
Aristotle, 18
Armyworms, 195–196
Arrhenius, Svante August, 38
Arthur, M.A., 54
Asthma, 213
Atlanta, as urban heat island, 97
Atmospheric modification, 269–270
Auken, Svend, 254
Automobile: early electric models, 24;
  gas-electric hybrid, 264; gasoline
  engine, inefficiency of, 263; and greenhouse-gas generation, 23–25; hydrogen-
  powered, 263; increases in traffic,
  1990s, 24; mileage standards, 263–264

de Baar, Hein J.W., 269
Bach, Wilford, 164
Baliunas, Sallie, 91
Balling, Robert C., 58, 86, 101, 107–108,
  172

About the Author

BRUCE E. JOHANSEN is Robert T. Reilly Professor of Communication and Native American Studies at the University of Nebraska–Omaha. He has written on modern industrialism's toll on Native Americans in *Ecocide of Native America* (1995) and has written on environmental themes in *The Nation, The Progressive,* and *The Washington Post*.